Mobility and Degradation of Organic Contaminants in Subsurface Environments

Warren J. Lyman
Patrick J. Reidy
Benjamin Levy
Camp Dresser & McKee Inc.

Chi-Yuan Fan
EPA Project Officer

C. K. SMOLEY, INC.

Library of Congress Cataloging-in-Publication Data

Lyman, Warren J.
Mobility and degradation of organic contaminants in subsurface environments / Warren J. Lyman, Patrick J. Reidy, Benjamin Levy
p. cm.
Includes bibliographical references and index.
1. Oil pollution of soils—Environmental aspects. 2. Oil pollution of water—Environmental aspects. 3. Petroleum products—Underground storage—Environmental aspects. 4. Oil storage tanks—Environmental aspects. 5. Transport theory. I. Reidy, Patrick J. II. Levy, Benjamin (Benjamin S.) III. Title.
TD879.P4L964 1992
628.5′5—dc20 92-3838
ISBN 0-87371-800-3

C. K. SMOLEY, INC.
121 South Main Street
Chelsea, Michigan 48118

PRINTED IN THE UNITED STATES OF AMERICA
1 2 3 4 5 6 7 8 9 0
Printed on acid-free paper

Abstract

The problems associated with leakage of motor fuels and organic chemicals from underground storage tanks (USTs) are compounded by a general lack of understanding of the partitioning, retention, transformation, and transport of these contaminants in the subsurface environment. The research material presented herein is the result of an intensive data collection and evaluation effort. The objective of this research was to compile the most up-to-date and comprehensive knowledge of contaminant behavior in the subsurface into a single document. It provides an understanding of micro-scale fate and transport processes as a means to understanding the larger scale movement of contaminants. This, in turn, leads to a more thorough understanding of the application of corrective measures and remediation techniques. The micro-scale analysis focuses on 13 loci, each of which represent a location and conditions in the subsurface environment, where and how contaminants may exist after a UST release.

The extensive research culminating in this report identifies the most important "rules" governing transport, partitioning, retention, and transformation of leaked motor fuels in the underground environment. The report is a data base of comprehensive scientific knowledge that can be drawn upon to define the key parameters, and to increase the level of sophistication in the approach to the problem of motor fuel leaking from USTs.

Environmental scientists, engineers and managers of varying levels of technical expertise may find this technical handbook useful. For the less technical user, it serves to strengthen an understanding of the fate and transport processes vital to effective remedial response. For the more experienced user, it serves as a source book of information, data, and equations to support more quantitative assessments of pollutant fate and transport.

The information in this book is from *Mobility and Degradation of Organic Contaminants in Subsurface Environments*, prepared by Warren J. Lyman, Patrick J. Reidy, and Benjamin Levy, Camp Dresser & McKee, Inc. for the U.S. Environmental Protection Agency.

Acknowledgment

The authors would like to thank Chi-Yuan Fan, the U. S. Environmental Protection Agency Project Officer, for his continued help and support throughout the program.

Notice

The materials in this book were prepared as accounts of work sponsored by the U.S. Environmental Protection Agency. This information has been reviewed in accordance with the U.S. Environmental Protection Agency's review policies and approved for publication. On this basis, the Publisher assumes no responsibility nor liability for errors or any consequences arising from the use of the information contained herein.

Mention of trade names or commercial products does not constitute endorsement or recommendation for use by the Agency or the Publisher. Final determination of the suitability of any information or product for use contemplated by any user, and the manner of that use, is the sole responsibility of the user. The book is intended for informational purposes only. The reader is warned that caution must always be exercised when dealing with hazardous chemicals, hazardous wastes, or hazardous processes, and that expert consultation should be obtained before implementation of countermeasures for hazardous chemical contamination.

Contents

Introduction 1
Section 1: Contaminant Vapors as a Component of Soil Gas in the Unsaturated Zone 11
Section 2: Liquid Contaminants Adhering to "Water-Dry" Soil Particles in the Unsaturated Zone 49
Section 3: Contaminants Dissolved in the Water Film Surrounding Soil Particles in the Unsaturated Zone 73
Section 4: Contaminants Sorbed to "Water-Wet" Soil Particles or Rock Surface (After Migrating Through the Water) in Either the Unsaturated or Saturated Zone 107
Section 5: Liquid Contaminants in the Pore Spaces Between Soil Particles in the Saturated Zone 131
Section 6: Liquid Contaminants in the Pore Spaces Between Soil Particles in the Unsaturated Zone 153
Section 7: Liquid Contaminants Floating Upon the Water Tables 181
Section 8: Contaminants Dissolved in Groundwater 207
Section 9: Contaminants Sorbed onto Colloidal Particles in Water in Either the Unsaturated or Saturated Zone 237
Section 10: Contaminants That Have Diffused into Mineral Grains or Rocks in Either the Unsaturated or Saturated Zone 257
Section 11: Contaminants Sorbed onto or into Soil Microbiota in Either the Saturated or Unsaturated Zone 281
Section 12: Contaminants Dissolved in the Mobile Pore Water of the Unsaturated Zone 309
Section 13: Liquid Contaminants in Fractured Rock or Karstic Limestone in Either the Unsaturated or Saturated Zone 335
Glossary 357
Index 369

Mobility and Degradation of Organic Contaminants in Subsurface Environments

Introduction

CONTENTS

Background and Purpose 3
Objectives of Document 6
Research Approach 6
Organization of Document 8

TABLES

1 Brief Descriptions of the 13 Physiochemical-Phase Loci 4
2 Matrix of Fate and Transport Process Considered for Each Locus 9
3 Generic Outline for Locus Section 10

FIGURES

1 Schematic Representation of the 13 Loci in Terms of Unsaturated and Saturated Zones 5

Introduction

BACKGROUND AND PURPOSE

The U.S. Environmental Protection Agency (EPA) has developed a comprehensive program for regulating certain underground storage tanks (UST) that contain regulated substances. With this program comes the need for development of guidance to assist those parties involved in complying with regulatory requirements. One significant portion of the legislation that has been developed pertains to corrective actions for releases of petroleum products such as gasoline and other motor fuels. EPA estimates that over 95 percent of the estimated 1.4 million UST systems are used to store petroleum products.

Current regulations require that owners and operators take corrective measures to mitigate releases from underground storage tanks. To date, guidance documents developed by the EPA and other organizations present various technologies applicable to remediation of the subsurface environment. However, only minimal amounts of information have been generated with regard to evaluating the movement and disposition of motor fuel contaminants in the subsurface environment, and only recently has significant research been devoted to this effort. Currently, investigators have no guidance available to determine corrective actions at UST sites that is based upon the scientific principles governing the behavior and degradation of motor fuel constituents in the subsurface environment. Recognizing this, EPA sponsored the comprehensive, scientific literature research effort resulting in this document.

As part of its philosophy, EPA developed the concept that a substance leaking from an UST will be present in and transient between one or more locations or settings in the subsurface environment. A total of 13 of these locations, referred to as physiochemical-phase loci, were identified as the focal points for this research report. Each of the 13 loci represents a point in space (or location) and the physical state of the leaked substance that together describe where and how these contaminants may exist in the subsurface environment after an UST release. For example, contaminants may be dispersed as a component of soil gas, or they may be dissolved in the water film surrounding a wet soil particle in the unsaturated zone. These are just two examples of the 13 loci. Collectively, the 13 loci represent all locations/states where and how leaked

Table 1. Brief Descriptions of the 13 Physicochemical-Phase Loci

Locus Number	Description
1	Contaminant vapors as a component of soil gas in the unsaturated zone.
2	Liquid contaminants adhering to "water-dry" soil particles in the unsaturated zone.
3	Contaminants dissolved in the water film surrounding soil particles in the unsaturated zone.
4	Contaminants sorbed to "water-wet" soil particles or rock surface (after migrating through the water) in either the unsaturated or saturated zone.
5	Liquid contaminants in the pore spaces between soil particles in the saturated zone.
6	Liquid contaminants in the pore spaces between soil particles in the unsaturated zone.
7	Liquid contaminants floating on the groundwater table.
8	Contaminants dissolved in groundwater (i.e., water in the saturated zone).
9	Contaminants sorbed onto colloidal particles in water in either the unsaturated or saturated zone.
10	Contaminants that have diffused into mineral grains or rocks in either the unsaturated or saturated zone.
11	Contaminants sorbed onto or into soil microbiota in either the unsaturated or saturated zone.
12	Contaminants dissolved in the mobile pore water of the unsaturated zone.
13	Liquid contaminants in rock fractures in either the unsaturated or saturated zone.

material may be present in the subsurface environment. The distribution of contaminants among these loci is constantly changing, as contaminants tend to move between different loci at varying rates over time. Table 1 presents a brief description of each locus. Figure 1 presents a schematic cross-section of the subsurface environment and identifies where each locus may exist in terms of the unsaturated and saturated zones.

The primary focus of this work assignment was to formulate a data base of comprehensive scientific knowledge that can be drawn upon to define the key parameters and to increase the level of sophistication in the approach to the problem of motor fuel leaking from USTs. The focus on organic contaminants is based on the reality that over 95 percent of materials stored in underground tanks is some type of petroleum product. As a starting point, EPA sponsored a two-day seminar where recognized experts and specialists in eight specific disciplines presented

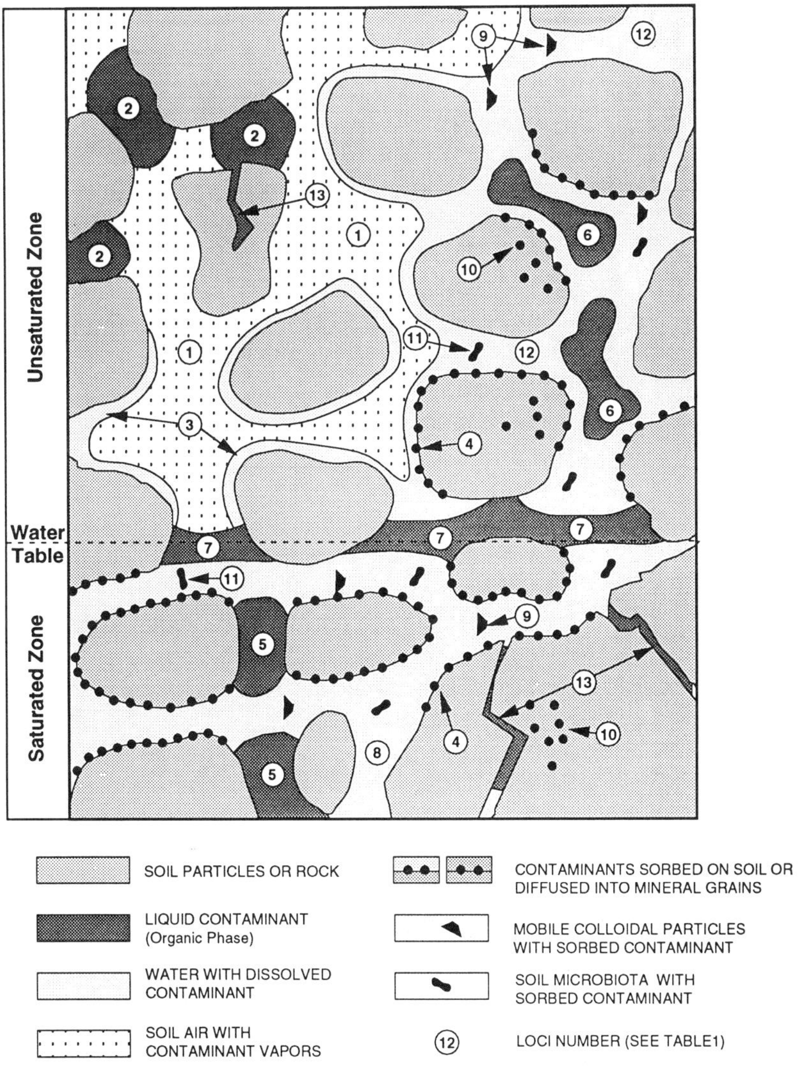

Figure 1. Schematic representation of the 13 loci in terms of unsaturated and saturated zones.

information on the physical, chemical, and biological properties and fundamentals of motor fuel and organic chemical behavior in the subsurface environment. Based on the information presented at this seminar, an intensive evaluation of available scientific data was conducted to provide a comprehensive, up-to-date understanding of the physical and chemical "rules" governing each locus for each of the 13 loci.

OBJECTIVES OF DOCUMENT

The EPA's Office of Underground Storage Tanks (OUST) is responsible for establishing the Agency's regulatory program for managing underground storage tanks. The Risk Reduction Engineering Laboratory (RREL) of EPA's Office of Research and Development is responsible for providing engineering and scientific support to OUST. One means of providing this technical support is through the preparation of guidance materials such as handbooks, manuals, and technical reports. To date, there have been a number of such documents developed by the EPA and other organizations on various aspects of the UST corrective action process. A review of these and other currently available documents indicates that there was no pre-existing guidance which directly relates fate and transport processes to corrective actions, a shortcoming deemed serious by EPA. Therefore, the primary objective of this document was to provide a comprehensive, in-depth research report with detailed coverage of fate and transport mechanisms that also includes timely discussion of these mechanisms as they relate to currently used or innovative approaches to UST releases. This report, in turn, provided the technical basis for the preparation of the desired guidance documents recently released by RREL:

- ASSESSING UST CORRECTIVE ACTION TECHNOLOGIES: Site Assessment and Selection of Unsaturated Zone Treatment Technologies. Report No. EPA/600/2-90/011, Risk Reduction Engineering Laboratory, Cincinnati, OH, March 1990.
- ASSESSING UST CORRECTIVE ACTION TECHNOLOGIES: Early Screening of Clean-up Technologies for the Saturated Zone Report. No. EPA/600/2-90/027, Risk Reduction Engineering Laboratory, Cincinnati, OH, June 1990.

RESEARCH APPROACH

The approach for this research effort is based on the concept that a chemical leaking from an UST will be present in and transient between one or more of the 13 physicochemical-phase loci (locations or settings in the subsurface environment) that were identified earlier. These loci, as described and presented in Table 1 and Figure 1, can be used in reference to any type of contaminant that enters the subsurface environment.

However, for the purpose of this research effort, the contaminants of interest are petroleum product (e.g., gasoline) constituents, i.e., hydrocarbons and other chemicals, including common gasoline additives. As pointed out previously, the reason for focusing on petroleum product constituents is that these products make up over 95 percent of the materials managed in underground storage tanks. All references to contaminants in this document are, therefore, with regard to hydrocarbons and associated organics.

Each locus, after being initially defined, was researched and evaluated in terms of the mobilization, immobilization, transformation, and bulk transport processes that pertain to it and in terms of how these various processes affect the locus. The objective was to evaluate and understand micro-scale partitioning and transformation as a means to understanding the larger scale movement of contaminants which will lead, in turn, to a greater understanding of the application of corrective measures and remediation techniques. Table 2 is a matrix that identifies the different processes that affect each locus. For example, if the reader refers to the column "a" heading it identifies two phases (liquid contaminant water) and a partitioning process in which they may be involved (dissolution). The dissolution process, when it involves air and water, appears as the heading for column "d". Bulk transport processes refer to advection and dispersion. These heading names identify only one direction of each partitioning process, while the reverse process is inferred. The checkmarks that appear in each column identify those loci that may be affected by that particular process. Table 2 also identifies where detailed discussions are presented on each process in terms of the loci sections contained in this report. Because there are many processes that are important to more than one locus, it was necessary to minimize redundancy in the material presented. This was accomplished by identifying, up front, individual loci sections in which the detailed process discussions would be included. The section(s) containing detailed discussions of a given topic would then be referenced in other sections rather than repeat the material. For example, if the reader refers to Table 2, locus no. 7, column "a", a circle is shown around the checkmark. This circled checkmark indicates that the detailed discussion of the process involving partitioning from the liquid contaminant phase to water and vice versa (i.e., dissolution/phase separation) is included in the section on locus no. 7 (Section 7). This process is also discussed in the sections on locus nos. 2, 3, 5, 6, 8, 12, and 13, denoted solely by a checkmark, but in less detail than locus no. 7.

The one locus that is treated differently from others in the body of this report is locus no. 11 (contaminants sorbed into/onto biota). It is treated differently in terms of section organization and contents because locus no. 11 is considered to be a transformation process in itself, i.e., it is considered to be both a locus and a process.

Following discussion of partitioning, transformation, and transport processes, each section provides guidance on calculating maximum and average values for the contaminant storage capacity of the locus, including estimation of parameter values and example calculations. In addition, example calculations are provided where appropriate for mass transport processes (e.g., advection, dispersion, and diffusion) and partitioning equilibrium processes (e.g., calculations for volatilization, dissolution, and adsorption). Lastly, the relative importance of each locus with regard to fate and transport of contaminants is discussed in terms of remediation, interaction with other loci, identification of critical information gaps, and recommendations for future research.

The ultimate objective in conducting research on the fate and transport of hydrocarbons in the manner presented, i.e., locus by locus, and process by process from a micro-scale perspective, is to gain a better understanding of the basis for larger scale contaminant movement. The ways in which the behavior of hydrocarbons may be altered are directly related to remediation techniques. Therefore, a comprehensive understanding of the fundamental "rules" of contaminant behavior engenders a better understanding of how this behavior may be induced or prohibited to optimize a given remedial strategy.

Even though remediation is not the subject of this report, indirectly the ultimate purpose of conducting the loci research is to increase the sophistication of current approaches to site assessment and corrective action selection. Brief discussions of remediation and corrective measures are presented in each locus section in three places; in each introductory subsection, in the subsections discussing effective partitioning processes, and in the overview of each locus' relative importance presented in the last subsection.

ORGANIZATION OF DOCUMENT

The main body of this document is organized into 13 major locus sections. Each locus section, except for Section 11 (locus no. 11), consists of the same five main subsections, which, in turn, contain discussions that are presented similarly. Table 3 presents a generic outline of the locus sections. As previously explained, locus no. 11 is dealt with differently in terms of the structure of its section because of its nature, i.e., it is both a locus (biota) and a process (biodegradation). Because of

Table 2. Matrix of Fate and Transport Processes Considered for Each Locus

LOCUS NO.	a	b	c	d	e	f	g	h	i	j	k	l	m	n	o	p	q
	Mobilization, Immobilization, Transformation, and Bulk Transport Processes																
	Partitioning and Transformation Processes								Bulk Transport Processes								
	Liquid Contaminant <–> Water (Dissolution)	Liquid Contaminant <–> Soil Air (Volatilization)	Liquid Contaminant <–> "Dry" Soil (Sorption)	Air <–> Water (Dissolution)	Water <–> Soil (Sorption)	Air <–> Soil (Condensation)	( ) <–> Biota (Biodegradation)	( ) <–> Chemical Oxidation	Liquid Contaminant "Water-Wet"	Liquid Contaminant "Oil-Wet"	Liquid Water in Unsaturated Zone	Liquid Water in Saturated Zone	Soil Air	Transport of Biota or Colloidal Particles with Water	Diffusion in Air	Diffusion in Liquids	Diffusion in Solids
1		⊘		✓		⊘	✓	✓					⊘		⊘		
2	(✓)	✓	⊘				✓	✓		⊘						✓	
3	✓			⊘	✓		✓	⊘			(✓)					⊘	
4					⊘		✓	✓									(✓)
5	✓		(✓)				✓	✓	⊘	(✓)						✓	
6	✓	✓	(✓)				✓	✓	⊘	(✓)						✓	
7	⊘	✓	(✓)				✓	✓	⊘	⊘						✓	
8	✓			(✓)	✓		✓	✓				⊘				✓	
9			(✓)		⊘		✓	✓						⊘			(✓)
10							✓	✓									⊘
11							⊘	✓						⊘		(✓)	
12	✓			✓	(✓)		✓	✓			⊘					✓	
13	✓	✓	(✓)				✓	✓	⊘	⊘						✓	

✓ = Important Process
(✓) = Relatively Less Important Process
⊘ (circled check) = Detailed Discussion Provided in Corresponding Locus Section
( ) = All Media/Phases

Table 3. Generic Outline for Locus Sections

		SECTION X – LOCUS NO. X	
X.1	Locus Description		
	X.1.1	Short Definition	
	X.1.2	Expanded Definition and Comments (Includes two figures; Figure X-1 is schematic cross-sectional diagram of locus, and Figure X-2 is schematic diagram showing fate and transport processes and interaction with other loci)	
X.2	Evaluation of Criteria for Remediation		
	X.2.1	Introduction	
	X.2.2	Mobilization/Remobilization	
		X.2.2.1	Partitioning onto/into Mobile Phase
		X.2.2.2	Transport of/with Mobile Phase
	X.2.3	Fixation	
		X.2.3.1	Partitioning onto Immobile (Stationary) Phase
		X.2.3.2	Other Fixation Approaches
	X.2.4	Transformation	
		X.2.4.1	Biodegradation
		X.2.4.2	Chemical Oxidation
X.3	Storage Capacity in Locus		
	X.3.1	Introduction and Basic Equations	
	X.3.2	Guidance on Inputs for, and Calculation of, Maximum Value	
	X.3.3	Guidance on Inputs for, and Calculation of, Average Value	
X.4	Example Calculations		
	X.4.1	Storage Capacity Calculations	
	X.4.2	Transport Rate Calculations (Includes calculations of applicable mass transport rates, e.g., advection, diffusion, dispersion, and example calculations of important partitioning equilibrium process equations)	
X.5	Summary of Relative Importance of Locus		
	• Remediation		
	• Loci Interactions		
	• Information Gaps		

its nature, the evaluation and discussion of locus no. 11 was organized in a different way as compared to the other loci. Although many sections deviate slightly from the standard outline, the same types of information and evaluations have been included in all sections. The specific contents of each locus section is presented at the beginning of each section as individual tables of contents. The main table of contents identifies only the main section headings.

SECTION 1

Contaminant Vapors as a Component of Soil Gas in the Unsaturated Zone

CONTENTS

List of Tables 12
List of Figures 12
1.1 Locus Description 15
  1.1.1 Short Definition 15
  1.1.2 Expanded Definition and Comments 15
1.2 Evaluation of Criteria for Remediation 15
  1.2.1 Introduction 15
  1.2.2 Mobilization/Remobilization 16
    1.2.1.1 Partitioning into Mobile Phase 16
      Pure Chemical Vapor Pressure 17
      Partial Pressure 22
      Volatilization Rate 22
      Soil and Moisture Conditions 22
      Changes in Physiocochemical Properties . 23
    1.2.2.2 Transport of the Mobile Phases 23
      Diffusion 24
      Impulse Input 26
      Step Input 27
      Tortuosity 28
      Advection 29
      Gravity-Driven Transport 30
      Dispersion 31
      Enhancement of Mobilization 32
  1.2.3 Fixation 33
    1.2.3.1 Partitioning onto Immobile Phase 33
      Adsorption in Water 34
      Adsorption on Dry Soils 34
    1.2.3.2 Other Fixation Approaches 38
  1.2.4 Transformation 38
    1.2.4.1 Biodegradation 38
    1.2.4.2 Chemical Oxidation 39

1.3 Storage Capacity in Locus 40
1.3.1 Introduction and Basic Equations 40
1.3.2 Guidance on Inputs for, and Calculation of, Maximum Value 40
1.3.3 Guidance on Inputs for, and Calculation of, Average Values 41
1.4 Example Calculations 42
1.4.1 Storage Capacity Calculations 42
1.4.1.1 Maximum Quantity 42
1.4.1.2 Average Quantity 42
1.4.2 Transport Rate Calculations 43
Vapor Sorption of Benzene 43
1.5 Summary of Relative Importance of Locus 43
1.5.1 Remediation 43
1.5.2 Loci Interaction 44
1.5.3 Information Gaps 45
1.6 Literature Cited 45

TABLES

1-1A. Estimated Properties for a Synthetic Gasoline and its Constituents at 10°C 18
1-1B. Estimated Properties for a Synthetic Gasoline and its Constituents at 20°C 20
1-2 BET Monolayer Adsorption Capacities (X_m) and Associated Molar Adsorption Heats (ΔH_m) of Adsorbates on Dry Woodburn Soil at 20°C 39

FIGURES

1-1 Schematic Cross-Sectional Diagram of Locus No. 1 - Contaminant Vapors as a Component of Soil Gas in the Unsaturated Zone 16
1-2 Schematic Representation of Important Transformation and Transport Processes Affecting Other Loci 17
1-3 Illustration of VOC Adsorption with three Moisture Regimes 24
1-4 Boiling Point Distribution Curves for Samples of "Fresh" and "Weathered" Gasolines 26
1-5 Decay through Time of the Impulse Input 27
1-6 Description of the Time-History of Diffusion Away from a Constant Source of Contamination 28
1-7 Relation Between Peclet Number and Ratio of the Longitudinal Dispersion Coefficient and the Coefficient of Molecular Diffusion in a Sand of Uniform-sized Grains 32

1–8 Schematic Diagram Showing the Contribution of Molecular Diffusion and Mechanical Dispersion to the Spread of a Concentration Front in a Column with a Step-Function Input 33
1–9 Effect of Relative Humidity on Soil Sorption of Organic Chemical Vapors 35
1–10 Adsorption of Various Gases as a Function of their Normal Boiling Points 36
1–11 Uptake of Organic Vapors and Moisture by Dry Woodburn Soil at 20°C vs Relative Vapor concentration 37
1–12 Vapor Uptake of m-Dichlorobenzene and 1,2,4-Trichlorobenzene by Dry Woodburn Soil at 20 and 30°C 39

SECTION 1

Contaminant Vapors as a Component of Soil Gas in the Unsaturated Zone

1.1 LOCUS DESCRIPTION

1.1.1 Short Definition

Contaminant vapors as a component of soil gas in the unsaturated zone.

1.1.2 Expanded Definition and Comments

Locus no. 1 includes soil gas with contaminant vapors anywhere in the unsaturated zone, including the fissures of fractured bedrock. The contaminants in this locus are quite mobile, moving via both diffusion through air pores (a concentration gradient-driven process) and advection, which may be driven by density gradients, barometric pressure pumping or sweep flow from in situ gas generation. Figures 1–1 and 1–2 present a schematic cross-sectional diagram of locus no. 1 and a schematic representation of the transformation and transport processes affecting other loci, respectively.

1.2 EVALUATION OF CRITERIA FOR REMEDIATION

1.2.1 Introduction

The remediation of vapor-phase contaminant from locus no. 1, the soil gas in the unsaturated zone, should be based on an understanding of the physical and chemical processes that control the volatilization and movement of vapor-phase contaminants. These physical and chemical parameters (e.g., soil air permeability, water content, vapor pressure) are discussed in Sections 1.2.2 and 1.2.3, which discuss the effects that these parameters have on the mobilization and fixation of volatilized liquid contaminants in locus no. 1.

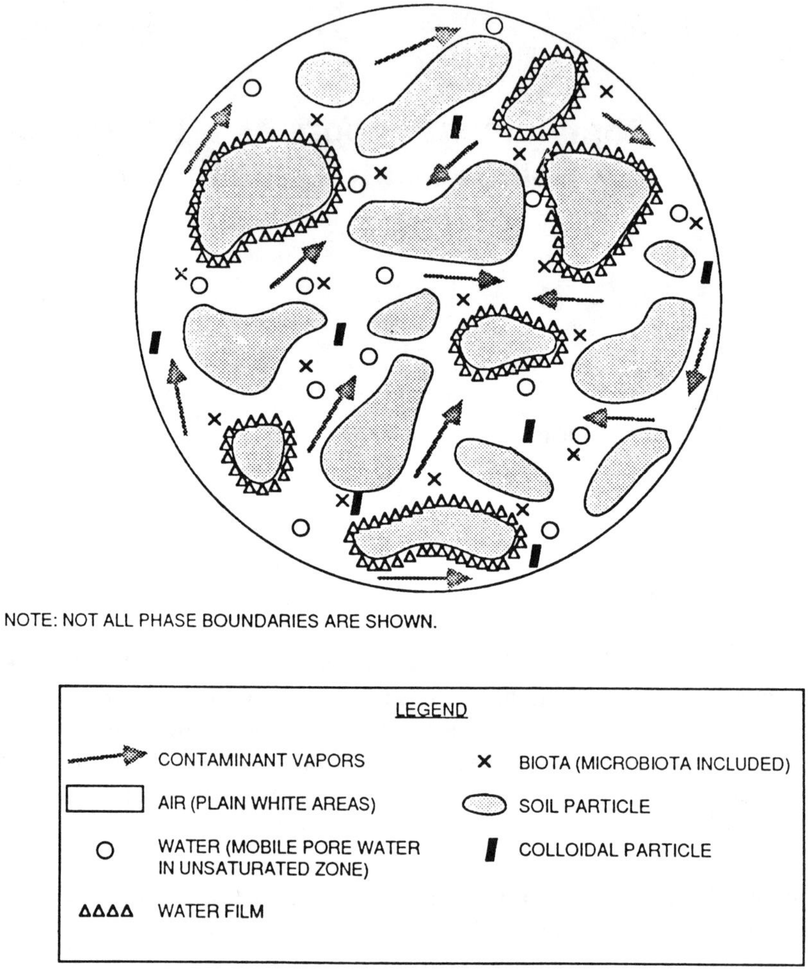

Figure 1–1. Schematic cross-sectional diagram of locus no. 1 - contaminant vapors as a component of soil gas in the unsaturated zone.

1.2.2 Mobilization/Remobilization

1.2.2.1 Partitioning into Mobile Phase

Contaminant chemicals enter the soil gas of locus no. 1 by volatilization of pure contaminant liquid (i.e., from loci 2, 6, 7, and 10). The process of volatilization is discussed in detail in locus no. 7. Contaminants may also enter locus no. 1 by volatilization from contaminated water (locus no. 3). This process is discussed in detail in locus no. 3.

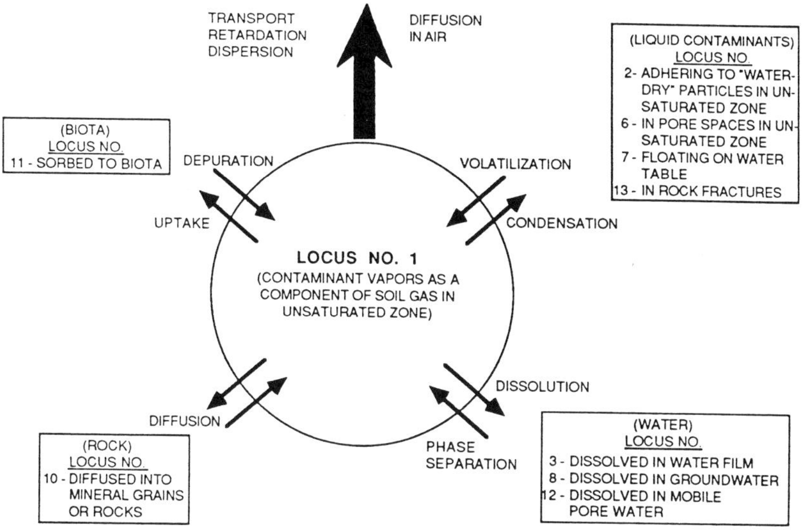

Figure 1–2. Schematic representation of important transformation and transport processes affecting other loci.

Pure Chemical Vapor Pressure. An important factor that determines the degree to which a chemical volatilizes is its pure chemical vapor pressure, P°. Tables 1–1A and 1–1B show values of vapor pressure for several hydrocarbons comprising a hypothetical, 23-component gasoline. The C_4-C_6 alkanes (e.g., isobutane, n-butane, and isopentane) have the highest vapor pressure of any of the constituents of gasoline. The BTEX compounds (benzene, toluene, ethylbenzene, and xylenes), normally regarded as volatile, have quite low vapor pressures relative to other gasoline constituents.

Tables 1–1A and 1–1B show a range of physical and chemical properties for a hypothetical gasoline developed by CDM (1987). This gasoline mixture is based on data from several sources in the literature, and is an attempt to define the constituents in gasoline to allow for estimation of various properties of the gasoline mixture. The chemicals comprising the gasoline blend were chosen to represent classes of chemicals typically found in gasoline. Below each table are listed calculated properties for the bulk liquid. Table 1–1A is calculated at 10°C and Table 1–1B is calculated at 20°C; comparing the two tables shows how temperature influences the properties.

Vapor pressure is strongly temperature-dependent. Tables 1–1A and 1–1B show that the vapor pressure of isobutane, for example, rises from

Table 1–1A. Estimated Properties for a Synthetic Gasoline and its Constituents at 10°C

Representative Chemical	Percent Composition	Gram Mol. Weight (Gm/Mol)	Liq. Phase Mol Fract.	Air Diffusion Coefficient (CM^2/Sec)	Liquid Density (GM/CM^3)	Pure Chemical Vapor Pressure (as Hg)	Partial Pressure Over Gasoline (as Hg)
Isobutane[a]	2	58.12	0.0326	0.0857	0.5728	1647.77	53.7745
n-Butane[a]	1	58.12	0.0163	0.0857	0.5931	1112.75	18.1571
Isopentane	14	72.15	0.1840	0.0769	0.6311	392.47	72.2231
n-Pentane	3	72.15	0.0394	0.0769	0.6364	283.84	11.1925
n-Octane	1	114.23	0.0083	0.0563	0.7096	5.63	0.0468
Benzene	3	78.11	0.0364	0.0851	0.8957	45.53	1.6583
Toluene	5	92.14	0.0515	0.0776	0.8758	12.43	0.6397
Xylene (a)	7	106.17	0.0625	0.0629	0.8701	3.26	0.2039
n-Hexane	9	86.18	0.0990	0.0669	0.6676	75.70	7.4976
2-Methylpentane	8	86.18	0.0880	0.0669	0.6620	109.55	9.6446
Cyclohexane	3	84.16	0.0338	0.0677	0.7884	47.51	1.6060
n-Heptane	1.5	100.20	0.0142	0.0620	0.6914	20.66	0.2933
2-Methylhexane	5	100.20	0.0473	0.0620	0.6867	30.94	1.4642
Methylcyclohexane	1	98.19	0.0097	0.0627	0.7789	21.45	0.2071
2,4-Dimethylhexane	8	114.23	0.0664	0.0581	0.7071	13.30	0.8833
Ethylbenzene	2	106.17	0.0179	0.0690	0.8748	3.77	0.0673
1-Pentene	1.5	70.14	0.0203	0.0780	0.6513	359.48	7.2909
2,2,4-Trimethylhexane	2	128.26	0.0148	0.0572	0.7240	6.18	0.0915
2,2,5,5-Tetramethylhexane	1.5	142.29	0.0100	0.0543	0.7264	3.40	0.0340
1,4-Diethylbenzene	5	134.22	0.0353	0.0559	0.8683	0.32	0.0113
1-Hexane	1.5	84.16	0.0169	0.0677	0.6821	94.87	1,6036
1,3,5-Trimethylbenzene	5	120.20	0.0394	0.0591	0.8718	0.84	0.0333
Cl2-aliphatic	10	170.00	0.0558	0.0493	0.8656	0.03	0.0016
Total	100	102.20	1.0000		0.7457		188.6252

Temperature = 283.15 deg. K
Pressure = 760 as Hg
Gas Constant = 0.06236 as Hg*a^3/sol*K
Molecular Weight of Air = 28.96 ga/sol
Average Molecular Weight of Gasoline Vapors = 68.92 ga/sol

Molecular weight of Gasoline-Air Mixture = 38.88 ga/sol
Vapor Density of Gasoline-Air Mixture = 1673.38 ga/a^3
Weighted average air diffusion coefficient = 0.0684 ca^2/sec
Weighted average liquid density 0.7271 ga/ca^3
Weighted average gas density 736.25 ga/a^3

Table 1–1A. (cont'd.)

Representative Chemical	Pure Chemical Vapor Density (GM/M^3)	Concentration Over Liquid Gasoline (pps)	Boiling Point (deg. K)	Henry's Law Constant (dis.)	Gas Viscosity (μPoise)
Isobutane[a]	5423.76	70755.9	261.25	3259.64	67.62
n-Butane[a]	3662.69	23890.9	272.65	2521.65	68.48
Isopentane	1603.70	95030.4	300.95	3387.63	66.56
n-Pentane	1159.80	14727.0	309.15	2977.29	61.67
n-Octane	36.43	61.5	398.75	5200.06	58.98
Benzene	201.40	2182.0	353.25	11.32	69.76
Toluene	64.86	841.7	383.75	12.60	71.49
Xylene (a)	19.61	268.3	466.25	11.83	51.89
n-Hexane	369.48	9865.2	341.85	3661.34	56.50
2-Methylpentane	534.70	12690.2	333.35	3785.69	57.37
Cyclohexane	226.44	2113.2	353.85	376.99	68.11
n-Heptane	117.23	385.9	371.55	4459.78	52.35
2-Methylhexane	175.57	1926.5	363.15	6962.93	53.12
Methylcyclohexane	119.26	272.6	374.05	797.24	54.09
2,4-Dimethylhexane	86.03	1162.2	382.55	5808.15	50.42
Ethylbenzene	22.66	88.6	409.25	13.65	52.16
1-Pentene	1427.99	9593.3	303.05	977.71	64.23
2,2,4-Trimethylhexane	44.92	120.3	399.65	6214.17	48.08
2,2,5,5-Tetramethylhexane	27.37	44.7	410.55	21405.95	46.4
1,4-Diethylbenzene	2.43	14.9	456.85	19.24	45.98
1-Hexane	452.20	2110.1	336.55	917.75	58.44
1,3,5-Trimethylbenzene	5.74	43.8	437.85	9.02	48.49
Cl2-Aliphatic	0.27	2.1	489.15	7742.54	39.99
		248191.2		3202.673	

a. Liquid properties are for the hypothetical superheated liquid.

Table 1–1B. Estimated Properties for a Synthetic Gasoline and its Constituents at 20°C

Representative Chemical	Percent Composition	Gram Mol. Weight (Gm/Mol)	Liq. Phase Mol Fract.	Air Diffusion Coefficient (CM^2/Sec)	Liquid Density (GM/CM^3)	Pure Chemical Vapor Pressure (as Hg)	Partial Pressure Over Gasoline (as Hg)
Isobutane[a]	2	58.12	0.0326	0.0911	0.5570	2252.75	73.5178
n-Butane[a]	1	58.12	0.0163	0.0911	0.5790	1555.33	25.3788
Isopentane	14	72.15	0.1840	0.0817	0.6200	574.89	105.7925
n-Pentane	3	72.15	0.0394	0.0817	0.6260	424.38	16.7346
n-Octane	1	114.23	0.0083	0.0598	0.7030	10.46	0.0869
Benzene	3	78.11	0.0364	0.0905	0.8850	75.20	2.7390
Toluene	5	92.14	0.0515	0.0824	0.8670	21.84	1.1237
Xylene (a)	7	106.17	0.0625	0.0668	0.8640	6.16	0.3852
n-Hexane	9	86.18	0.0990	0.0711	0.6590	121.24	12.0080
2-Methylpentane	8	86.18	0.0880	0.0711	0.6530	171.50	15.0985
Cyclohexane	3	84.16	0.0338	0.0719	0.7790	77.55	2.6218
n-Heptane	1.5	100.20	0.0142	0.0659	0.6840	35.55	0.5047
2-Methylhexane	5	100.20	0.0473	0.0659	0.6790	51.90	2.4563
Methylcyclohexane	1	98.19	0.0097	0.0666	0.7707	36.21	0.3497
2,4-Dimethylhexane	8	114.23	0.0664	0.0617	0.7000	23.32	1.5492
Ethylbenzene	2	106.17	0.0179	0.0733	0.8670	7.08	0.1264
1-Pentene	1.5	70.14	0.0203	0.0829	0.6400	530.80	10.7654
2,2,4-Trimethylhexane	2	128.26	0.0148	0.0608	0.7173	11.30	0.1671
2,2,5,5-Tetramethylhexane	1.5	142.29	0.0100	0.0577	0.7200	6.47	0.0647
1,4-Diethylbenzene	5	134.22	0.0353	0.0594	0.8620	0.70	0.0246
1-Hexane	1.5	84.16	0.0169	0.0719	0.6730	149.97	2.5348
1,3,5-Trimethylbenzene	5	120.20	0.0394	0.0628	0.8650	1.73	0.0681
Cl2-aliphatic	10	170.00	0.0558	0.0524	0.8600	0.08	0.0042
Total	100	102.20	1.0000		0.7373		

Temperature = 293.15 deg. K
Pressure = 760 as Hg
Gas Constant = 0.06236 as Hg*a^3/sol*K
Molecular Weight of Air = 28.96 ga/sol
Average Molecular Weight of Gasoline Vapors = 69.48 ga/sol

Molecular weight of Gasoline-Air Mixture = 43.58 ga/sol
Vapor Density of Gasoline-Air Mixture = 1811.60 ga/a^3
Weighted average air diffusion coefficient = 0.0726 ca^2/sec
Weighted average liquid density 0.7182 ga/ca^3
Weighted average gas density 1041.85 ga/a^3

Table 1–1B. (cont'd.)

Representative Chemical	Pure Chemical Vapor Density (GM/M³)	Concentration Over Liquid Gasoline (pps)	Boiling Point (deg. K)	Henry's Law Constant (dis.)	Gas Viscosity (μPoise)
Isobutane[a]	7162.14	96733.9	261.25	4304.39	70.01
n-Butane[a]	4944.83	33393.1	272.65	3404.37	70.90
Isopentane	2268.97	139200.7	300.95	4792.93	68.91
n-Pentane	1674.92	22019.2	309.15	4299.67	63.85
n-Octane	65.37	114.3	398.75	9330.24	61.06
Benzene	321.31	3604.0	353.25	18.07	72.22
Toluene	110.05	1478.6	383.75	21.38	74.01
Xylene (a)	35.78	506.8	466.25	21.60	53.72
n-Hexane	571.57	15900.0	341.85	5663.95	58.50
2-Methylpentane	808.51	19866.4	333.35	5724.28	59.40
Cyclohexane	357.04	3449.7	353.85	594.42	70.52
n-Heptane	194.86	664.1	371.55	7412.91	54.20
2-Methylhexane	284.50	3232.0	363.15	11282.83	55.00
Methylcyclohexane	194.50	460.2	374.05	1300.19	56.00
2,4-Dimethylhexane	145.75	2038.4	382.55	9839.65	52.20
Ethylbenzene	41.90	66.3	409.25	24.76	54.00
1-Pentene	2036.57	14165.0	303.05	1394.40	66.50
2,2,4-Trimethylhexane	79.30	219.9	399.65	10969.52	49.78
2,2,5,5-Tetramethylhexane	50.35	85.1	410.55	39373.43	48.11
1,4-Diethylbenzene	5.11	32.4	456.85	40.43	47.60
1-Hexane	690.40	3335.3	336.55	1401.18	60.50
1,3,5-Trimethylbenzene	11.36	89.7	437.85	17.84	50.20
Cl2-aliphatic	0.70	5.5	489.15	19822.78	41.40
		360660.5		5434.632	

a. Liquid properties are for the hypothetical superheated liquid.

1648 mm Hg to 2253 mm Hg as the temperature changes from 10°C to 20°C. Chemical engineering literature (e.g., Perry and Chilton, 1973) contains several pressure-temperature correlations. The Antoine equation is usually used, but correlations by Riedel, Kay, Reid, Miller, and Sherwood are also discussed by Perry and Chilton (1973).

Partial Pressure. For any mixture of compounds, the pressure exerted on the air by any one chemical component equals the product of the pure chemical vapor pressure and its liquid mole fraction:

$$P_i = x_i \cdot P_i^{\circ} \tag{1.1}$$

where P_i = partial pressure of contaminant i (mm Hg)
x_i = mole fraction of a contaminant i (dimensionless) $0 \leq x_i \leq 1$
P_i° = vapor pressure of pure contaminant i (mm Hg)

The total vapor pressure of a mixture is the sum of the partial pressures of the chemicals comprising the mixture:

$$P_t = \sum_{i=1}^{N} P_i \tag{1.2}$$

where P_t = Total pressure exerted on air by contaminant liquid (mm Hg)

At 20°C, the hypothetical gasoline exerts a pressure of 274 mm Hg on the air above the contaminant liquid, or 0.36 atmospheres (1 atm = 760 mm Hg). This may be expressed as 360,000 ppm on a volume per volume basis.

Volatilization Rate. The rate at which contaminants volatilize from the liquid phase is controlled by several physical and chemical parameters. Perhaps most important is the pure chemical vapor pressure. As described previously, this parameter is strongly temperature-dependent, and the rate of volatilization is much greater at higher temperatures. Soil temperature is fairly constant at any location throughout the year, but it varies from site to site (See Figure 8–5). Other important factors that affect volatilization in soil are described below.

Soil Moisture Content. The soil moisture content has two important effects on the volatilization of liquid product into the soil pores. Higher soil moisture contents reduce the air-filled pore space, which reduces volatilization (Farmer et al., 1980; Aurelius and Brown, 1987). The volatilization rate is also related to the water content through sorptive behavior. Lighty et al. (1988) and Houston et al. (1989) show that soil electrostatic force is greater for drier soils; as water content increases, water molecules displace contaminant molecules, and sorption decreases, increasing volatilization. Thus a plot of sorption vs. soil moisture content should show a maximum in sorption at some point. Figure 1-3 shows schematically the effect of soil moisture content on volatilization.

Changes in Physicochemical Properties ("Weathering"). Any product released into the subsurface environment is continually changing in nature. These changes, collectively referred to as "weathering", cause the remaining liquid contaminant to have different physical and chemical properties from fresh, or non-weathered, product. For example, when gasoline (a mixture of 200 or more compounds, each with varying vapor pressures) is released into the soil, the most volatile constituents are most likely to volatilize into the soil vapor and move away from the residual liquid. The constituents with the greatest solubilities are the most likely to dissolve into infiltrating precipitation and move toward the water table. The most degradable constituents are the most likely to be degraded. The effect of these changes is that the remaining liquid contaminant will have significantly different properties than the unweathered product. Johnson et al. (1989) present the following properties for "fresh" and "weathered" gasoline:

	Fresh	*Weathered*
Molecular Weight (g/mole)	95	111
Vapor Pressure (atm)	0.34	0.049
Vapor Density (mg/L)	1300	220

Figure 1-4 shows boiling point distribution curves for these two gasolines. From the above property data and Figure 1-4, it is apparent that the degree of weathering will be an important factor in the volatility of a released product.

1.2.2.2 Transport of the Mobile Phase

The mobilization/remobilization of the soil gas and vapor phase contaminant is controlled by physical properties and conditions of the soil, volatilized contaminant, and other immiscible phases present in locus no. 1. Three immiscible phases are considered—soil air, liquid contaminant, and soil water—which are in direct physical contact with one

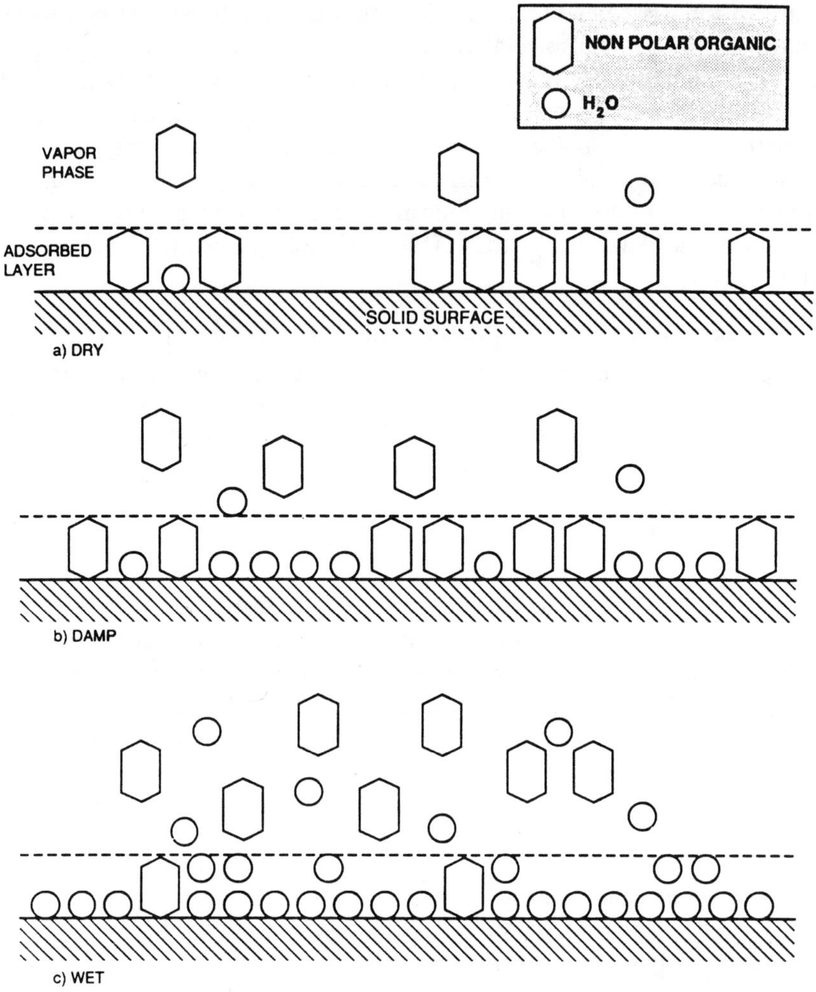

Figure 1–3. Illustration of VOC adsorption with three moisture regimes. *Source:* After Reible, 1989.

another. The sections below discuss the transport processes that move volatilized material through the soil air.

Diffusion. Diffusion is the movement of molecules from an area of high concentration to one of low concentration. The transport of mass in a direction of decreasing concentration is termed flux. Flux is proportional to the concentration gradient (the concentration difference divided by the distance); the higher the concentration gradient, the stronger the flux. This is stated in Fick's First Law (here shown for one-dimensional flux):

$$J = D_o \left(\frac{C_o - C_1}{\Delta x} \right) \tag{1.3}$$

where J = mass flux of volatilized contaminant (g/cm^2 sec)
D_o = air diffusion coefficient (cm^2/sec)
C_o = concentration of volatilized contaminant at X_o (g/cm^3)
C_1 = concentration of volatilized contaminant at X_1 (g/cm^3)
Δx = distance between X_o and X_1 (cm)

Tables 1–1A and 1–1B list values for the air diffusion coefficients of the constituents of the synthetic gasoline described previously. Values for D_o for other compounds are available in the literature.

Examination of Fick's First Law shows that a flux will continue to occur until the concentration equilibrates (i.e., until the gradient is zero). Thus volatilization will cease when the vapor phase concentration reaches the saturation value of the air in contact with the liquid contaminant. This law also shows that the rate of volatilization will be at a maximum when the vapor concentration at X_1 is zero. Remediation by soil venting operates by removing contaminant-saturated vapors and replacing them by clean air, which helps to maximize the rate of volatilization.

Fick's Second Law describes the change in concentration of volatilized contaminant in the soil air with time at a given location. Expressed in one-dimensional form:

$$\frac{dC}{dt} = D_o \cdot \frac{d^2C}{dx^2} \tag{1.4}$$

where x = distance (cm)
C = concentration of volatilized contaminant (g/cm^3)
D_o = air diffusion coefficient (cm^2/sec)
t = time (sec)

Analytical, closed-form solutions exist for these equations. The form of these solutions depends on the initial and boundary conditions on the system which the equation describes. An example of an initial condition is the specification of zero concentration in the system prior to initiation of diffusion. A boundary condition might require that a certain location in the system be always at some fixed concentration. For example, the ground surface could be set to equal zero concentration. Two simple

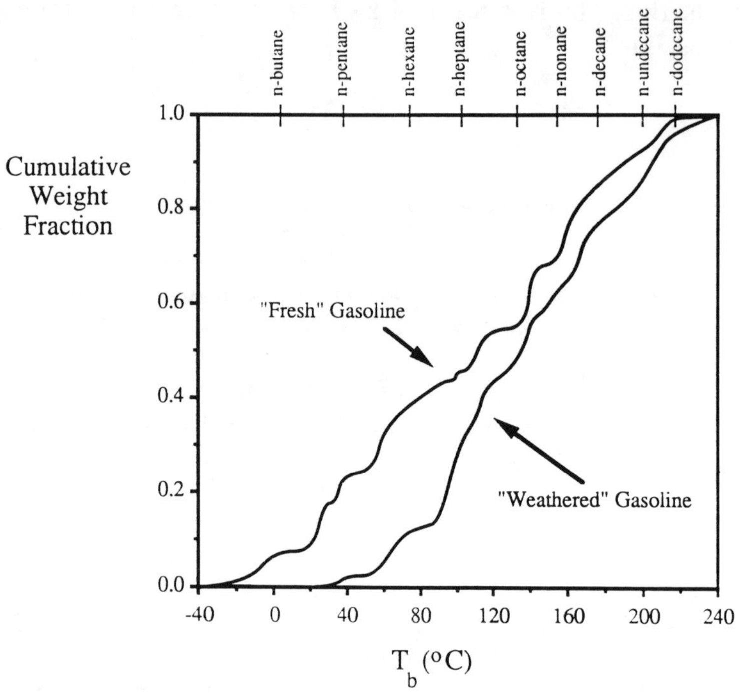

Figure 1–4. Boiling point distribution curves for samples of "fresh" and "weathered" gasolines. *Source:* Johnson et al., 1990. (Copyright Water Well Journal Publishing Co., 1990. Reprinted with permission.)

analytical solutions to the one-dimensional form of Fick's Second Law are presented below.

Impulse Input. Under this solution, a slug of contaminant enters the system instantaneously at one point and begins to diffuse. No further mass is added after the initial slug. It is assumed that there is no background contamination present in the system when the slug is added. The change in concentration with time and location is described by an exponential decay function:

$$C(x,t) = C_o \exp\left(\frac{-x^2}{4D_o t}\right) \tag{1.5}$$

where $C(x,t)$ = concentration at point and time t (mg/cm^3)
C_o = concentration at time zero (mg/cm^3)
x = distance from the source (cm)

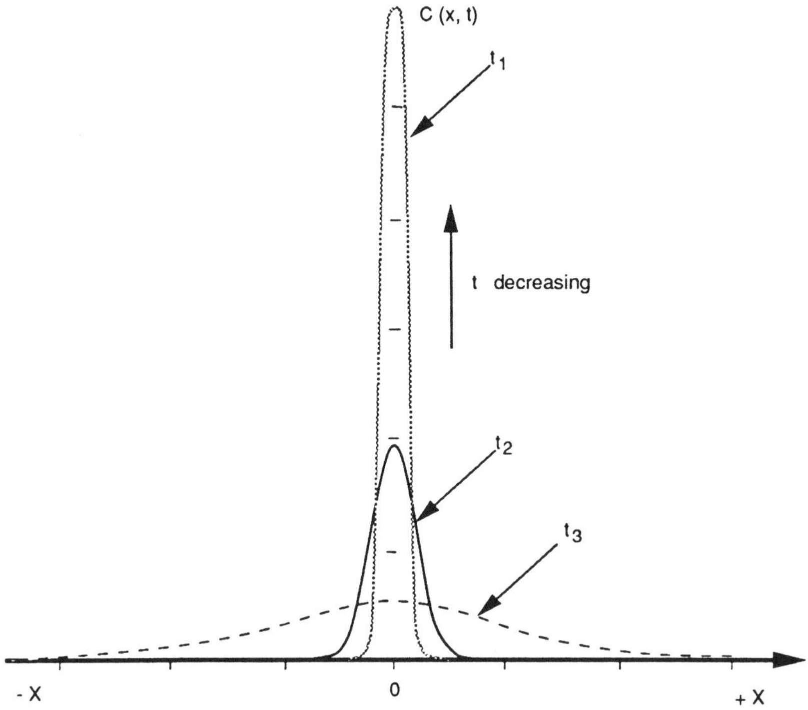

Figure 1–5. Decay through time of the impulse input. *Source:* Fischer et al., 1979. (Copyright Academic Press, Inc. 1979. Reprinted with permission.)

Figure 1–5 shows graphically the effects of diffusion upon a step input.

Step Input. A second solution for Fick's Second Law is for a continuous source of constant concentration (a "step input"). For this case, the initial condition is zero concentration everywhere at time $t=0$. The boundary conditions are $C(\infty, t)=0$ for $x=\infty$ and all t, and $C(0,t)=C_o$ for $x=0$ and all $t>0$. For these conditions, the solution to Fick's Second Law is:

$$C(x,t) = C_o \operatorname{erfc}\left(\frac{-x^2}{4D_o t}\right) \tag{1.6}$$

where erfc ( ) = complementary error function on the variable in ( )

$$\operatorname{erf c}(x) = 1 - \frac{2}{\sqrt{\pi}} \sum_{n=0}^{\infty} \frac{(-1)^n x^{2n+1}}{n!\,(2n+1)}$$

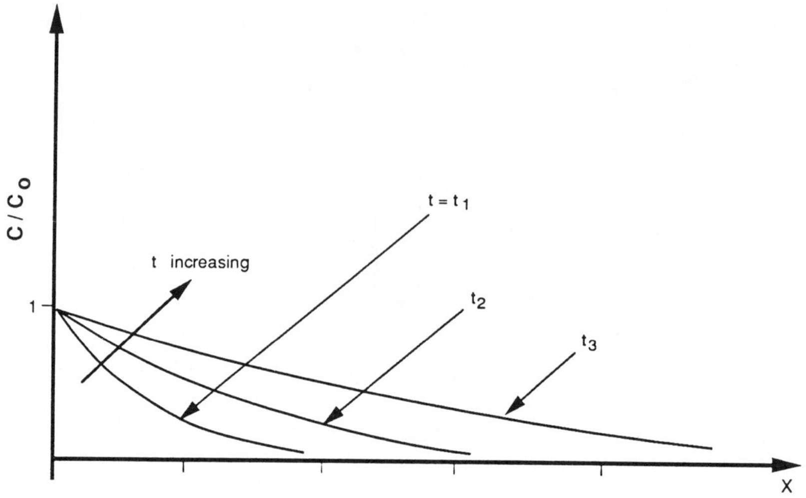

Figure 1–6. Description of the time-history of diffusion away from a constant source of contamination. *Source:* Fischer et al., 1979. (Copyright Academic Press, Inc. 1979. Reprinted with permission.)

Figure 1–6 shows the time history of diffusion away from a constant source of contamination. With time, the concentration C will be found throughout the immediate area of the source, an area that increases with time. The air diffusion coefficient, D_o, is a key variable that determines how quickly the concentration diffuses away from the source. A large diffusion coefficient would lead to a greater rate of diffusion away from the source.

Tortuosity. The air diffusion coefficient, D_o, should be adjusted to reflect subsurface conditions. For flow through a porous medium such as soil, the actual path length that the molecules diffuse across is dependent on the tortuosity of the medium. For highly tortuous paths, the actual length traveled by the molecules is greater. The more tortuous the path, the longer it takes for a molecule to diffuse through the soil, effectively reducing the air diffusion coefficient.

The air diffusion coefficient may be adjusted by a tortuosity factor. Various expressions formulated for this factor have been compiled by Weeks et al., 1982. A commonly used expression is presented by Millington and Quirk (1961):

$$D = D_o \frac{\Theta_a^{3.33}}{\Theta_t^{2}} \tag{1.7}$$

and

$$\Theta_t = \Theta_a + \Theta_w \tag{1.8}$$

where D = effective air diffusion coefficient (cm^2/sec)
D_o = diffusion coefficient in pure air (cm^2/sec)
Θ_a = air-filled porosity of soil
Θ_t = total porosity of soil
Θ_w = Water-filled porosity

This expression shows that the effective diffusion coefficient is substantially less than the air-diffusion coefficient even for dry soils. For example, if the following conditions were present for a dry, sandy soil:

$\Theta_t = 0.35$
$\Theta_a = 0.30$
$D_o = 0.07\ cm^2/sec$
then $D = 0.01\ cm^2/sec$

For the same soil with a higher water content (e.g., $\Theta_w = 0.2$) and a lower air-filled porosity ($\Theta_a = 0.15$), the effective diffusion coefficient is only about 0.015 D_o. Obviously then, diffusion through soils with significant water content will be greatly retarded.

Advection. Advection is the movement of the soil gas in response to a pressure gradient exerted on the soil gas. This pressure gradient may be caused by barometric pressure fluctuations, artificially induced gradients (for example, by vacuum extraction), gravity, water table fluctuations, or by wind-induced circulation. Gravity is discussed in the follow subsection.

Advection refers to the average speed of the bulk soil gas as it moves through the soil pores. The velocity of the soil gas is described by Darcy's law:

$$V_x = \frac{q_x}{\Theta_a} = -\left(\frac{k_a}{\mu_a}\right) \frac{d}{dx} (p_a + \rho_a gh)/\Theta_a \tag{1.9}$$

where V_x = gas velocity (cm/sec)
q_x = volume flow rate per unit area (cm/sec)
k_a = soil air permeability (cm^2)
μ_a = viscosity of air (g/cm sec)
d/dx = differential along x direction
P_a = pressure of air ($g/cm\ sec^2$)
ρ_a = density of air (g/cm^3)
g = acceleration of gravity (cm/sec^2)

h = elevation (cm)
Θ_a = air-filled porosity of soil

This form of Darcy's law accounts for both pressure-driven advection (e.g., due to barometric "pumping" or changes in the barometric pressure or from applied pressure gradients) and from gravity (discussed below). An examination of equation 1.9 shows that the air permeability is of great importance in determining the vapor flow. The air permeability depends on several factors including the grain size distribution, type and structure, porosity, and water content of the soil.

Advection and diffusion can be combined in the advection-diffusion equation, expressed here for three-dimensional flow:

$$\frac{dC}{dt} = \nabla^2 DC - \nabla VC \qquad (1.10)$$

where V is the average linear pore velocity and C is the solute concentration and ∇ is the three-dimensional gradient operator.

This equation must be solved numerically. Analytical solutions to the one-dimensional advection-diffusion equation exist that are similar in form to those presented in the diffusion subsection:

$$\text{impulse-input: } C(x,t) = C_o \exp\left[\frac{x - V_x t}{(4D_t)^{1/2}}\right] \qquad (1.11)$$

$$\text{step-input: } C(x,t) = C_o \operatorname{erfc}\left[\frac{x - V_x t}{(4D_t)^{1/2}}\right] \qquad (1.12)$$

Gravity-Driven Transport. As stated above, gravity can cause the downward migration of soil gas if the gas is denser than air. The vapor density of pure, dry air is about 1200 g/m³ at 20°C. The volatilization of the chemicals from the synthetic gasoline increases the vapor density of the soil gas, making it heavier than pure air, as demonstrated by the calculation of the vapor density of the gasoline vapor-soil air mixture:

$$\rho_v = \frac{P_a M}{RT} \qquad (1.13)$$

where ρ_v = vapor density of gasoline vapor-soil air mixture (gm/m³)
P_a = atmospheric pressure (mm Hg)
M = average molecular weight of gasoline vapor-soil air mixture (gm/mol)
R = universal gas constant, $R = 0.06236$ mm Hg m³/mol K)
T = temperature (K)

and

$$M = M_v(P_t/P_a) + M_a(P_a - P_t)/P_a \quad (1.14)$$

where M_a = average molecular weight of air, M_a = 28.96 g/mol
M_v = average molecular weight of gasoline vapors, M_v = 69.48 g/mol
P_t = equilibrium vapor pressure of liquid gasoline exerted on soil air, P_t = 274 mm Hg (@ 20°C. See Table 1–1B.)

Therefore, (69.48 g/mol · 274 mm Hg/760 mm Hg) + (28.96 g/mol)·(760 mm Hg−274 mm Hg)/760 mm Hg)
= 43.58 g/mol
ρ_v = 43.58 g/mol · 760 mm Hg/(293.15 K 0.06236 mm Hg m^3/mol · K)
ρ_v = 1800 g/m^3

This sample calculation (with values drawn from Table 1–1A) indicates that the gasoline vapor-soil air mixture is half again as heavy as air. Gravity will thus cause the downward migration of the contaminated soil air. The importance of gravity relative to advection and diffusion in the transport of vapors in the unsaturated zone is uncertain and represents a current area of much research.

Dispersion. As the contaminated soil gas moves through the soil pores it mixes with pure soil air, causing a spreading and reduction of the contaminant soil gas concentration. This mixing, known as hydrodynamic dispersion, is composed of molecular diffusion and mechanical dispersion:

$$D_l = \infty \cdot \bar{v} + D^* \quad (1.15)$$

where D_l = coefficient of hydrodynamic dispersion (cm^2/sec)
∞ = characteristic mixing length (cm)
$\bar{v}$ = average flow velocity (cm/sec)
D^* = effective air diffusion coefficient (cm^2/sec)

The relative contribution of each process depends on the flow conditions of the soil gas. Figure 1–7 depicts the ratio of mechanical dispersion to molecular diffusion with the Peclet number. The Peclet number is a dimensionless number and is defined as follows:

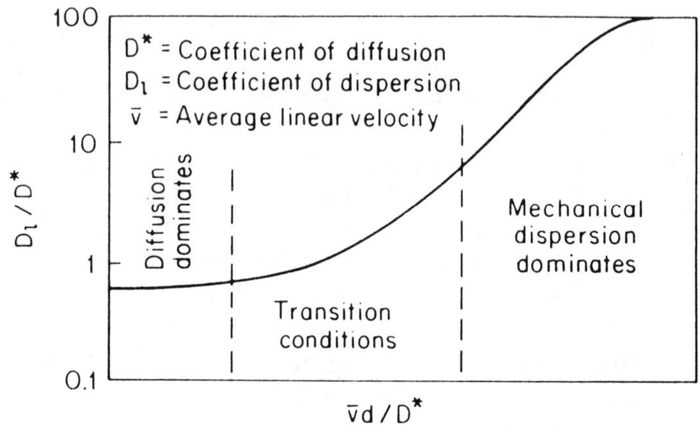

Figure 1–7. Relation between Peclet number and ratio of the longitudinal dispersion coefficient and the coefficient of molecular diffusion in a sand of uniform-sized grains. *Source:* Freeze and Cherry, 1979 (Copyright Prentice-Hall Inc., 1979. Reproduced with permission.)

$$P = \bar{v} \cdot d/D^* \tag{1.16}$$

where
P = Peclet number (dimensionless)
$\bar{v}$ = average flow velocity (cm/sec)
d = average pore diameter (cm)
D^* = effective molecular diffusion coefficient (cm^2/sec)

At low velocities (i.e., low Peclet numbers) molecular diffusion dominates over mechanical dispersion; higher velocities (i.e., higher Peclet numbers) indicate that mechanical dispersion dominates over diffusion.

The hydrodynamic dispersion causes some of the contaminant concentration to arrive before and after the average concentration traveling with the average flow velocity. This spreading about the average velocity is shown in Figure 1–8.

Dispersion properties of a porous medium are usually measured in field or laboratory studies. Much current research focuses on the determination of the characteristic mixing length, ∞. Dispersion is covered in more detail in locus no. 8. The analytical equations that include the effect of dispersion are similar to the equations used for the modeling of diffusion.

Enhancement of Mobilization. Little can be done to enhance the diffusive transport of vapor contaminant through soil pores. Advection of the soil gas, however, can be increased by applying a pressure gradient to the soil. This could most practically be accomplished by inducing a

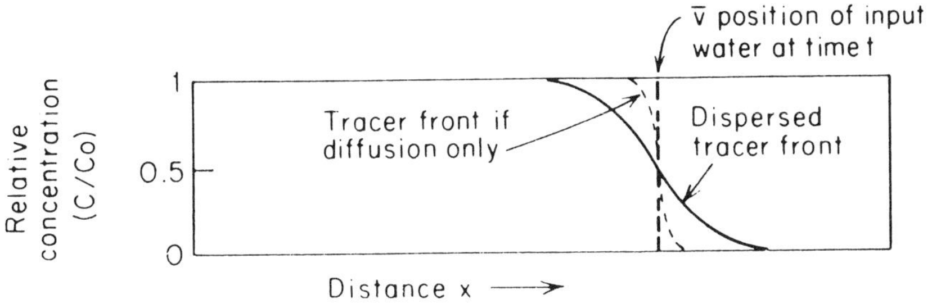

Figure 1–8. Schematic diagram showing the contribution of molecular diffusion and mechanical dispersion to the spread of a concentration front in a column with a step-function input. *Source:* Freeze and Cherry, 1979. (Copyright Prentice-Hall Inc., 1979. Reproduced with permission.)

negative pressure gradient, or vacuum, to the soil, as is done during vacuum extraction. The application of a vacuum causes advective flow through the soil pores.

Temperature plays a significant role in contaminant mobilization, mainly because of its effect on the contaminant's vapor pressure, but also through its effect on the viscosity and air diffusion coefficient. Higher temperatures result in higher vapor pressures, lower viscosities and higher air diffusion coefficients, all making contaminant vapors more mobile. Some research has been conducted (Hunt et al., 1986; Lord et al., 1987) regarding the use of steam injection as a means of increasing the soil temperature to increase contaminant mobility. The results are somewhat promising but this technique may not be applicable to a large area.

1.2.3 Fixation

1.2.3.1 Partitioning onto Immobile Phase

Vapors in locus no. 1 may be temporarily immobilized by absorption (dissolution) into pore water (locus no. 3 or 12) or by adsorption onto dry soils (part of locus no. 2). This distinction between "wet" absorption and dry adsorption is very important; the processes are significantly different, as are the resulting sorption capacities.

Absorption in Water (Details in Section 3). The process of absorption into water (or dissolution) is the reverse of volatilization from water, which process is described in detail in Section 3. The importance or degree of absorption in water is proportional to the water solubility of the vapors and the amount of water present. Although volatilization and absorption at the air/water interface is rapid, the overall rate of the processes may be controlled by the rate of diffusion of solutes in water, a much slower process.

Adsorption on Dry Soils (Details in Section 2). Adsorption onto dry soils is also considered to be a rapid process (Brunauer, 1945). Thus, any immobilization or fixation due to sorption is likely temporary and reversible unless diffusion into mineral grains or rocks (locus no. 10) or biodegradation (locus no. 11) is a significant process.

Water vapor and contaminant vapors compete for the same adsorption sites on soils. In general, water has the ability to dramatically reduce a soils' sorption capacity for vapors of organic chemicals. Water typically will preferentially occupy sorption sites or displace previously adsorbed organic molecules. Figure 1–9 shows the effect that increasing the relative water vapor humidity has on the soil sorption of benzene, dichlorobenzene and 1,2,4-trichlorobenzene. In the regions of low humidity, the adsorption isotherms are typically non-linear and reflect sorption—many molecular layers thick—on the soil surface. At high humidity, however, the sorption isotherms tend to be linear, as the contaminant vapors dissolve into water rather than sorb to soil.

Adsorption of Gases. A significant distinction exists between the adsorption of gases (temperature above critical point) and vapors (temperature below critical point). For gases, it is generally believed that only monomolecular layers can be sorbed onto the solid surface (Grain, 1987). Above the critical point of a gas, no further condensation from the gas to the liquid phase is possible.

A theoretical treatment of gas adsorption shows the amount adsorbed to be proportional to the vapor concentration (Grain, 1987):

$$\frac{x}{x_m} = \frac{K\,(P/P_o)}{1 + K\,(P/P_o)} \qquad (1.17)$$

where
- x = mass of gas adsorbed per weight of solid adsorbent (mg/kg)
- x_m = mass of gas adsorbed with a monolayer of coverage on adsorbent (mg/kg)
- K = a temperature-dependent empirical constant (dimensionless)
- P = vapor concentration of gas (atm)
- P_o = saturation vapor pressure of gas (atm)

There are many empirical correlations for gas adsorption on solids (Brunauer, 1945). One set of data showing the expected correlation be-

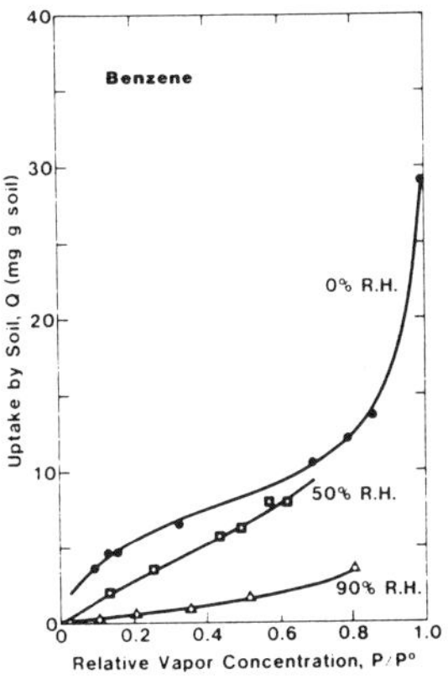

a) Vapor Uptake of Benzene by Dry Woodburn Soil at 20 °C as a Function of Relative Humidity.

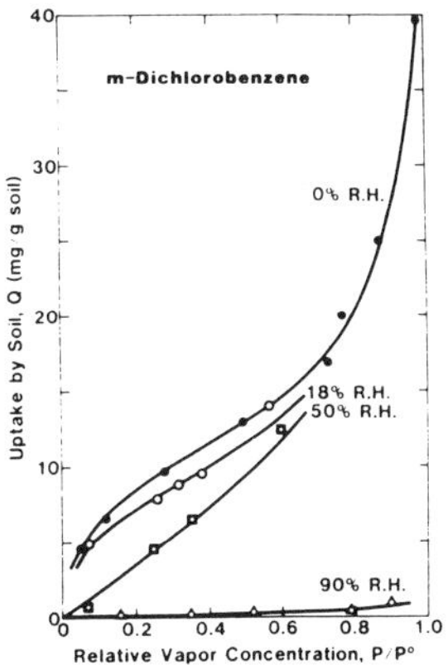

b) Vapor Uptake of m-Dichlorobenzene by Dry Woodburn Soil at 20 °C as a Function of Relative Humidity.

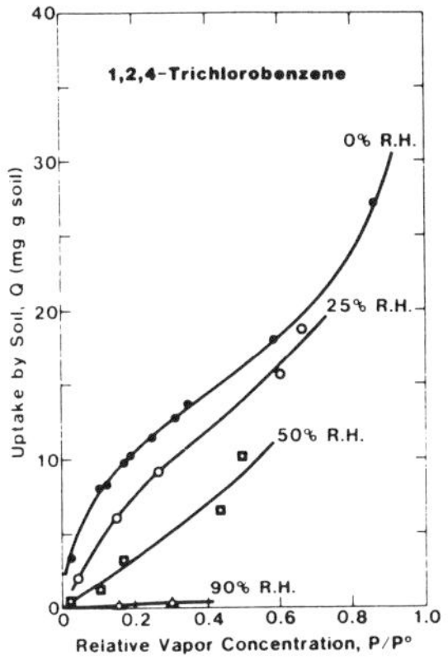

c) Vapor Uptake of 1,2,4-Trichlorobenzene by Dry Woodburn Soil at 20 °C as a Function of Relative Humidity.

Figure 1–9. Effect of relative humidity on soil sorption of organic chemical vapors. *Source:* Chiou and Shoup, 1985. (Copyright American Chemical Society. Reprinted with permission).

tween boiling point (a vapor pressure datum) and the amount adsorbed on charcoal is shown in Figure 1–10.

Many substances stored in USTs, such as gasoline, contain compounds that, in their pure state, are gases at ambient temperatures. Examples include n-butane and isobutane (boiling points of -0.5°C and -11.9°C, respectively). These highly volatile constituents sorb to soils much less than do compounds that have higher boiling points (which are

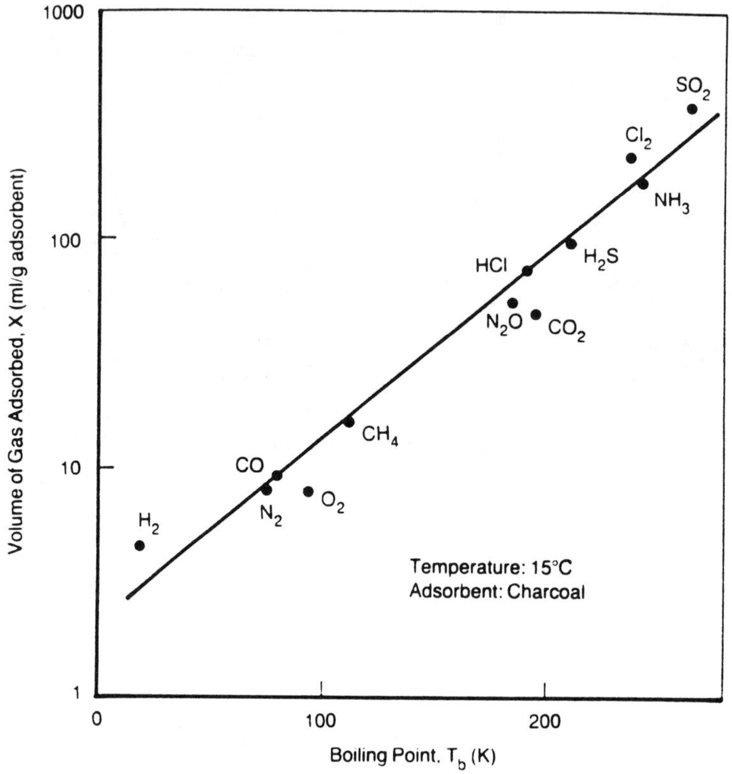

Figure 1–10. Adsorption of various gases as a function of their normal boiling points. *Source:* Grain, 1987. (Copyright Pergamon Press Inc., 1987. Reprinted with permission.)

vapors at ambient temperatures), due to the limitations discussed above regarding monomolecular layers. The lower sorption of the gases would result in higher mobility of these compounds. For gases, the extent of adsorption is likely to be directly controlled by the available surface area on the soils.

Adsorption of Vapors. The vapors of compounds that are liquids at ambient temperature can sorb onto soils in a process similar to condensation. The mass of vapors sorbed is not directly related to the surface area of the soil because vapors can sorb many layers thick. The sorption of vapors is limited by the accessible pore volume of the soil.

Sorption of vapors on soils is commonly analyzed via the Brunauer-Emmett-Teller (BET) isotherm equation for multilayer adsorption:

$$\frac{x}{x_m} = \frac{c\,(P/P_o)}{(1 - P/P_o)[1 + (c-1)\,(P/P_o)]} \tag{1.18}$$

where x, x_m, P and P_o are as defined in equation 1.17
c = temperature constant

The constant c is related to the net molar heat of adsorption at $x \leq x_m$ by:

$$-\ln c = (\Delta H_m + \Delta H_v)/RT \tag{1.19}$$

where ΔH_m = molar heat of adsorption (kcal/mol)
ΔH_v = molar heat of vaporization (kcal/mol)
R = gas constant, 0.001987 kcal/deg mol
T = temperature (K)

Examples of sorption isotherms for several chemicals are shown in Figure 1-11. The temperature effect on vapor sorption for two chemicals is shown in Figure 1-12. This negligible temperature effect is evidence that the heats of mineral adsorption (ΔH_m) and vapor condensation (ΔH_v) are essentially the same. Values of x_m, ΔH_v and ($\Delta H_m + \Delta H_v$) derived from these data are shown in Table 1-2.

Relatively few data are available showing the extent of soil sorption of organic vapors. Publications that contain data include Chiou and Shoup (1985), Isaacson and Sawheney (1984), Jurinak (1957) and Call (1957). Sorption of organic vapors onto solid particles in the atmosphere, a closely related process, is described in publications by Bidleman et al. (1986) and Cantrell et al. (1986).

The above discussion has indicated that modest soil sorption of organic vapors is possible, but that the presence or introduction of water vapor can lead to displacement of the sorbed vapors. Thus, the temporary fixation of organic vapors by soil sorption is enhanced by keeping the soils dry. Lowering the temperature would also increase the amount sorbed to soils.

1.2.3.2 Other Fixation Approaches

Other means of at least partially immobilizing vapors in soils include the blocking of soil pores (e.g., with water or other liquid agents) and the use of impermeable covers to prevent transport to the atmosphere.

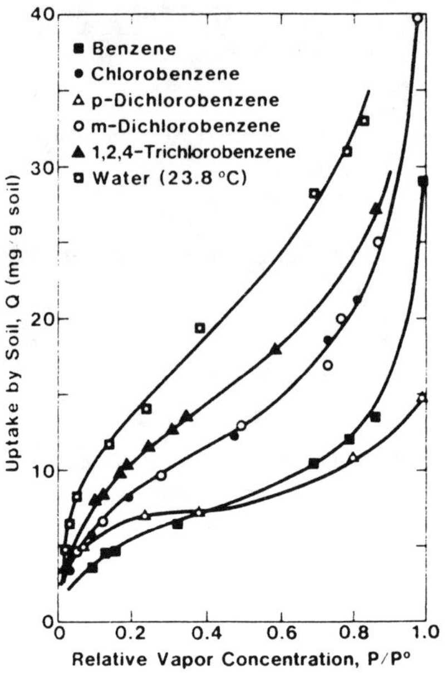

Figure 1–11. Uptake of organic vapors and moisture by dry woodburn soil at 20°C vs relative vapor concentration. *Source:* Chiou and Shoup, 1985. (Copyright American Chemical Society. Reprinted with permission.)

1.2.4 Transformation

1.2.4.1 Biodegradation (Details in Section 11)

Biodegradation in the subsurface occurs as soil microorganisms (mainly bacteria) metabolize contaminants. Oxygen, mineral nitrogen, and micronutrients are required in addition to the contaminants themselves. The optimum environment for this process is aqueous.

The role of soil gas as a transport medium of contaminants to microbes in liquid or solid phase loci may be more important than its role as a locus for microbial biodegradation.

1.2.4.2 Chemical Oxidation (Details in Section 3.2.4.2)

Gaseous components of petroleum hydrocarbons are subject to photo-oxidation when exposed to sunlight or within environments where free radicals exist. Within the soil environment, the oxidation of gaseous components is not expected to be significant.

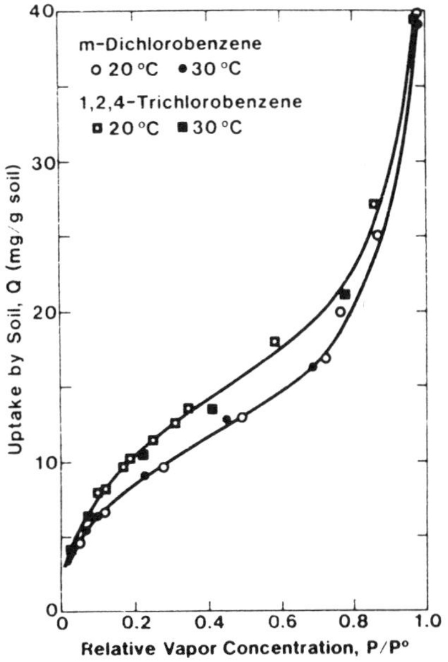

Figure 1–12. Vapor uptake of m-Dichlorobenzene and 1,2,4-Trichlorobenzene by dry woodburn soil at 20 and 30°C. *Source:* Chiou and Shoup, 1985. (Copyright American Chemical Society. Reprinted with permission.)

1.3 STORAGE CAPACITY IN LOCUS

1.3.1 Introduction and Basic Equations

The mass per unit volume of contaminant liquid volatilized into the vadose zone air may be calculated from:

Table 1–2. BET Monolayer Adsorption Capacities (x_m) and Associated Molar Adsorption Heats (ΔH_m) of Adsorbates on Dry Woodburn Soil at 20°C[a]

Compound	x_m (mg/g of soil)	H_v[b] (kcal/mol)	$-(\Delta H_m + \Delta H_v)$ (kcal/mol)
Benzene	5.57	8.09	1.52
Chlorobenzene	7.53	9.43	1.82
m-Dichlorobenzene	7.42	10.6	1.86
p-Dichlorobenzene	5.54	15.0	2.54
1,2,4-Trichlorobenzene	9.53	11.7	1.93
Water (23.8°C)	11.7	10.3	2.14

a. See equations 1.18 and 1.19 for definition of terms.
b. Molar heats of vaporization determined from vapor pressure data.
Source: Chiou and Shoup, 1985.

$$m_v = C_o \Theta_a \cdot \rho_v \qquad (1.20)$$

where m_v = mass of contaminant per unit volume (kg/m^3)
C_o = equilibrium vapor concentration of contaminant in soil air (m^3/m^3)
Θ_a = air-filled porosity (fraction)
ρ_v = contaminant vapor density (kg/m^3)

The air-filled porosity, Θ_a, is simply the difference between the total and water-filled porosity:

$$\Theta_a = \Theta_t - \Theta_w \qquad (1.21)$$

where Θ_t = total soil porosity, (fraction)
Θ_w = water-filled porosity, (fraction)

1.3.2 Guidance on Inputs for, and Calculations of, Maximum Value

The quantity of contaminant vapor is at a maximum when the air-filled porosity is at a maximum, or in other words, the water content is at a minimum. For this calculation, it is assumed that the water content in the soil is at the wilting point of the soil. The moisture content at the wilting point varies with the soil type, as shown in Figure 12–4.

Guidance on parameter values is as follows:

Total Porosity, Θ_t.
- Theoretical maximum porosity for sandy soil is 47.6% (thus, Θ_t = 0.476).
- Table 12–2 lists representative total porosity values for a variety of soils.
- Figure 12–4 shows total porosity in graphical format.

Water-Filled Porosity, Θ_w.
- Water-filled porosity can range from 0.05 (sand) to Θ_t.
- Figure 12–4 shows wilting point moisture content values for various soils.

Equilibrium Vapor Concentration C_o.
- User should supply equilibrium vapor concentration of contaminant.
- For synthetic gasoline in locus no. 1, recommend:

Temperature (°C)	C_o range (m^3/m^3)
0	0.11–0.28
10	0.17–0.42
20	0.24–0.61

- User should have information on contaminant composition, choosing high C_o value for mixtures with high fraction of C_4-C_6 alkane compounds.

Vapor Density, ρ_v.

- User should supply specific information on vapor density for contaminant of interest.
- Vapor density of gasoline vapors in Table 1–1A at 20°C is 1.042 kg/m^3.

1.3.3 Guidance on Inputs for, and Calculations of, Average Value

The average quantity of contaminant vapor held in locus no. 1 is assumed to occur when the water-filled porosity is at field capacity and the water is at a static condition.

Total Porosity, Θ_t
- See guidance given in Section 1.3.2.

Water-Filled Porosity, Θ_w
- Water-filled porosity should be chosen at field capacity, which may be obtained from Table 12–2.
- Recommend Θ_w = 0.09 at field capacity for a sandy soil.

Vapor Density
- See guidance given in Section 1.3.2.
- Suggest using half vapor density value given in Section 1.3.2 for average value calculation.

Equilibrium Vapor Concentration, C_o
- See guidance given in Section 1.3.2.
- Suggest using half equilibrium vapor concentration given in Section 1.3.2 for average value calculations.

1.4 EXAMPLE CALCULATIONS

1.4.1 Storage Capacity Calculations

1.4.1.1 Maximum Quantity

For this sample calculation, a sandy soil will be assumed. Since we wish to maximize the amount of volatilized contaminant liquid, we assume the water content in the soil is at the wilting point. For sandy soils, the water content at the wilting point is about 5% (Dunne and Leopold, 1978). The total porosity of the sandy soil is assumed to be 39.5% (Clapp and Hornberger, 1978). The air-filled porosity equals 34.5% (39.5% - 5%).

For gasoline, the vapor density (from Table 1-2) is 1.04 kg/m³. The maximum equilibrium concentration (from Table 1-1A) equals 0.61. Thus the maximum mass in locus no. 1 can be calculated:

$$m_V = C_o \cdot \rho_v \cdot \Theta_a$$
$$m_v = (0.61) \cdot (1.04 \text{ kg/m}^3) \cdot (0.345)$$
$$m_v = 0.22 \text{ kg/m}^3$$

The maximum mass per unit volume in locus no. 1 for the above conditions is 0.22 kg/m³.

1.4.1.2 Average Quantity

A sandy soil will also be assumed for the calculation of the average quantity of contaminant held in locus no. 1. It is assumed that the water-filled porosity at field capacity is about 9%. As stated in Section 1.4.1.1, the total porosity is 39.5%. Therefore, the air-filled porosity available at field capacity is 30.5%.

The equilibrium vapor concentration used to calculate the average mass of contaminant is 0.30 which is in the middle of the range of C_o reported in Section 1.3.2:

$$m_V = C_o \cdot \rho_v \cdot \Theta_a$$
$$m_v = (0.30) \cdot (1.04 \text{ kg/m}^3) \cdot (0.305)$$
$$m_v = 0.095 \text{ kg/m}^3$$

The average mass per unit volume in locus no. 1 for the above conditions is 0.095 kg/m³.

1.4.2 Transport Rate Calculations

Several examples of calculations for selected equations were provided in Section 1.2 following the discussion of the equation. One example calculation for vapor sorption is given below.

Vapor Sorption of Benzene. Estimate the amount of benzene vapors that have sorbed to a soil given x_m = 5.57 mg/g of soil (Table 1–2), $-(\Delta H_m + \Delta H_v)$ = 1.52 kcal/mole (Table 1–2), T = 20°C (293.15 K), P_o = 75.2 mm Hg (Table 1–1A), and P = 2.74 mm Hg (Table 1–1A).

From equation 1.19:

$$
\begin{aligned}
-\ln c &= (\Delta H_m + \Delta H_v)/RT \\
&= (-1.52)/[(1.987)(293.15)/(10^3)] \\
&= -2.61
\end{aligned}
$$

$$\text{Thus:} \quad c = 0.0736 \text{ (dimensionless)}$$

$$\text{Also,} \quad P/P_o = 2.74/75.2 = 0.0364$$

Substituting in equation 1.18:

$$x = (5.57)\frac{(0.0736)(0.0364)}{(1 - 0.0364)[1 + (0.0736-1)(0.0364)]}$$

$$= 0.016 \text{ mg/g of soil}$$

1.5 SUMMARY OF RELATIVE IMPORTANCE OF LOCUS

1.5.1 Remediation

The presence of contaminant vapors in the soil gas of the unsaturated zone provides a source of contamination of unsaturated and saturated zone water. The contaminant vapors are very mobile and are capable of migrating significant distances. The vapors may pose an explosion hazard, especially if they gather in sewer lines, basements, or other enclosed areas. Contaminant can sorb to soil particles or dissolve into subsurface water, resulting in further groundwater contamination. This migration pathway may be significant in view of the mobility of vapors and the mass of contaminant that may be held in locus no. 1 (Section 1–3).

The amount of contaminant vapors that reach the water table depends on the partitioning behavior of the contaminant. Partitioning between the air and water phases is governed by Henry's law, while partitioning between the contaminant liquid and soil air depends principally on the pure chemical vapor pressure of the contaminant. The equilibrium vapor pressure of the liquid and the volume of soil air determine the amount of mass held in locus no. 1.

The importance of locus no. 1 relative to the other loci depends then on the vapor pressure of the contaminant liquid, the air-filled porosity, and the Henry's law constants of the contaminant liquid chemicals. If locus no. 1 holds a maximum quantity of contaminant mass, as described in Section 1.4.1, it probably holds more mass per unit volume than loci nos. 3, 4, 8, 9, 10, 11 and 12.

Locus no. 1 provides an excellent medium by which to remediate unsaturated zone contamination for volatile compounds. Vacuum extraction or soil vapor extraction has been used extensively to remove volatile components from the unsaturated zone soils, while minimizing additional groundwater contamination by removing a source of recharge.

The success of vacuum extraction depends most heavily on the vapor pressure of the contaminants and the air permeability of the soil. Because only relatively volatile compounds will exist in the soil gas, the vapor pressure is the key contaminant variable with respect to the success of vacuum extraction. Bennedsen et al. (1985) state that compounds with a vapor pressure above 0.5 mm Hg have sufficient volatility for successful removal via vacuum extraction. The soil air permeability is the most important soil parameter. Because the removal of contaminant vapors depends on the ability to induce transport of the vapors through the unsaturated zone, the air permeability of the soil is important in determining whether this method can induce sufficient removal to make this method feasible.

1.5.2 Loci Interactions

Volatilization from liquid contaminant in loci nos. 5, 6, and 7 into the soil air represents the primary mechanism for partitioning into this locus. Concentrations in the soil gas are affected by advection and dispersion, gravity transport, and chemical retardation and sorption. Volatilized contaminant mass in the soil gas may partition to dry soil (locus no. 2) or wet soil (locus no. 4). Contaminant mass may also dissolve into the pore water of the unsaturated zone (loci nos. 3 and 12) or into the water of the saturated zone (locus no. 8). Contaminated soil gas is considered to be relatively mobile and venting to the atmosphere is the chief loss mechanism.

1.5.3 Information Gaps

Research into the following areas is important in better defining the transport of soil gas:

1. Importance of gravity transport as a transport mechanism relative to other transport mechanisms such as advection and diffusion.
2. Conditions under which volatilized contaminants partition to soil particles, including the influence of water content, soil type and organic carbon content.

1.6 LITERATURE CITED

Aurelius, M.W. and K.W. Brown. 1987. Fate of Spilled Xylene as Influenced by Soil Moisture Content. Water, Air and Soil Pollution. 36:23–31

Bennedsen, M.B., J.P. Scott, and J.D. Hartley. 1985. Use of Vapor Extraction Systems for In-situ Removal of Volatile Organic Compounds from Soil. Proceedings of the 5th National Conference on Hazardous Wastes and Hazardous Materials, HMCRI. pp. 92–95.

Bidleman, T.F., W.N. Billings and W.T. Foreman. 1986. Vapor-Particle Partitioning of Semi-volatile Organic Compounds: Estimates from Field Collections. Environ. Sci. Technol., 20(10):1038–1043.

Brunauer, S. 1945. The Adsorption of Gases and Vapors, Volume I: Physical Adsorption. Princeton University Press, reprinted 1967 by University Microfilms, Ann Arbor, MI.

Call, F. 1957. The Mechanism of Sorption of Ethylene Dibromide of Moist Soils. J. Sci. Food Agric., 8:630–636.

Camp, Dresser & McKee Inc. 1987. Modeling Vapor Phase Movement in Relation to UST Leak Detection. Interim Report. U.S. EPA, OUST. Washington, DC.

Cantrell, B.K., L.J. Salas, W.B. Johnson and J.C. Harper. 1986. Phase Distributions of Low Velocity Organics in Ambient Air. EPA/600/3-86/064, U.S. EPA, Research Triangle Park, NC.

Chiou, C.T. and T.D. Shoup. 1985. Soil Sorption of Organic Vapors and Effects of Humidity on Sorptive Mechanisms and Capacity. Environ. Sci. Technol., 19 (12):1196–1200.

Clapp, R.B. and G.M. Hornberger. 1978. Empirical Equations for Some soil Hydraulic Properties. Water Resources Research. Vol. 14, No. 4.

Dunne, T. and L.B. Leopold. 1978. Water in Environmental Planning. W.H. Freeman & Co.

Farmer, W.J., M.S. Yang, J. Letey, and W.F. Spencer. 1980. Land Disposal of Hexachlorobenzene Wastes: Controlling Vapor Movement in Soil. EPA/600/2-80/119.

Fischer, H.B., J. Imberger, R.C.Y. Koh, and N.H. Brooks. 1979. Mixing in Inland and Coastal Waters. Academic Press.

Freeze, R.A. and J.A. Cherry. 1979. Groundwater. Prentice-Hall, Inc.

Grain, C.F 1988. Gas Adsorption on Solids. In: Environmental Inorganic Chemistry. Pergamon Press, New York.

Houston, S.L., D.K. Kreamer, and R. Marwig. 1989. A Batch-Type Testing Method for Determination of Adsorption of Gaseous Compounds on Partially Saturated Soils. Journal of Testing and Evaluation, ASTM. 12(1):3–10.

Hunt, J.R., N. Sitar, and K.S. Udell. 1986. Organic Solvents and Petroleum Hydrocarbons in the Subsurface: Transport and Cleanup. Univ. of California at Berkeley. Sanitary Engineering and Environmental Health Research Laboratory. Report 86–11.

Isaacson, P.J. and B.L. Sawheney. 1984. Measurement of Vapor Sorption: Influence of Organic Matter on Clay Sorption of 2,6-Dimethylphenol. Soil Science 138(6):436–39.

Johnson, P.C., M.W. Kemblowski, J.D. Colthart, D.L. Byers, and C.C. Stanley. 1989. A Practical Approach to the Design, Operation, and Monitoring of In-situ Soil Venting Systems. Presented at the Soil Vapor Extraction Technology Workshop, Office of Research and Development, Edison, NJ. June 28–89. [Also published (1990): Ground Water Monitoring Review, 10(2):150–178].

Jurinak, J.J. 1957. The Effect of Clay Minerals and Exchangeable Cations on the Adsorption of Ethylene Dibromide Vapor. Soil Sci. Soc. Amer. Proc., 21:599–602.

Lighty, J.S., G.D. Silcox, D.W. Pershing, and V.A. Cundy. 1988. On the Fundamentals of Thermal Treatment for the Cleanup of Contaminated Soils. APCA 81st Annual Meeting, Dallas, TX. June 19–24.

Lord, A.E., Jr., R.M. Koerner, V.P. Murphy, and J.E. Brugger. 1987. In-situ Vacuum Assisted Steam Stripping of Contaminants from Soil. Proceedings of the 7th National Conference of Management of Uncontrolled Hazardous Wastes (Superfund '87). HMCRI. pp. 390–395.

Millington, R.J. and J.M. Quirk. 1961. Permeability of Porous Solids. Trans. Faraday Soc., 57:1200–1207.

Perry, R.H. and C.H. Chilton. 1973. Chemical Engineer's Handbook, 5th ed. McGraw-Hill, New York.

Reible, D.D. 1989. Introduction to Physico-Chemical Processes Influencing Enhanced Volatilization. Presented at the workshop on Soil

Vacuum Extraction. R.S. Kerr Environmental Research Laboratory, Ada, OK. April 27–28.

Weeks, E.P. 1982. Unsaturated Zone Diffusion Parameters. Water Resources Research. 18(5).

SECTION 2

Liquid Contaminants Adhering to "Water-Dry" Soil Particles in the Unsaturated Zone

CONTENTS

List of Tables 50
List of Figures 50
2.1 Locus Description 51
  2.1.1. Short Definition 51
  2.1.2 Expanded Definition and Comments 51
2.2 Evaluation of Criteria for Remediation 52
  2.2.1 Introduction 52
  2.2.2 Mobilization/Remobilization 55
    2.2.2.1 Partitioning into Mobile Phase 55
      Displacement of NAPL by Water 55
    2.2.2.2 Transport of Mobile Phase 56
      Soil Permeability 57
      Viscosity 59
      Pressure Drops 61
  2.2.3 Fixation 65
    2.2.3.1 Partitioning onto Immobile (Stationary) Phase 65
  2.2.4 Transformation 66
    2.2.4.1 Biodegradation 66
    2.2.4.2 Chemical Oxidation 66
2.3 Storage Capacity in Locus 66
  2.3.1 Introduction and Basic Equations 66
  2.3.2 Guidance on Inputs for, and Calculation of, Maximum Value 67
  2.3.3 Guidance on Inputs for, and Calculation of, Average Values 67
2.4 Example Calculations 68
  2.4.1 Storage Capacity Calculations 68
    2.4.1.1 Maximum Value 68
    2.4.1.2 Average Value 69

2.4.2 Transport Rate Calculations 69
2.5 Summary of Relative Importance of Locus 70
2.5.1 Remediation 70
2.5.2 Loci Interactions 70
2.5.3 Information Gaps 70
2.6 Literature Cited 71

TABLES

2–1 Relationship Between Diameter of Particles and Surface Area 59
2–2 Fluid Viscosities at 20°C 60
2–3 Surface Tension Values, Liquid-Vapor Interfaces 64
2–4 Retention of Gasoline in Soil 68

FIGURES

2–1 Schematic Cross-Section Diagram of Locus No. 2—Liquid Contaminants Adhering to "Water-Dry" Soil Particles or Rock Surfaces in the Unsaturated Zone ... 52
2–2 Schematic Representation of Important Transformation and Transport Processes Affecting Other Loci 53
2–3 Schematic Cross-Section of Locus No. 2 Contaminants Before and After Draining by Gravity 54
2–4 Capillary Pressure Effects on Variation of Water Content and Elevation 58
2–5 Schematic Cross-Section of Soil Particles Showing Liquid Held by Capillary Forces 63
2–6 Capillary Rise 64
2–7 Interfacial Surface Tensions and Contact Angle 65

SECTION 2

Liquid Contaminants Adhering to "Water-Dry" Soil Particles in the Unsaturated Zone

2.1 LOCUS DESCRIPTION

2.1.1 Short Definition

Liquid contaminants adhering to "water-dry" soil particles in the unsaturated zone.

2.1.2 Expanded Definition and Comments

The principal feature of this locus is the presence of liquid contaminants, either as a continuous phase (e.g., in a spill front) or as a discontinuous phase (e.g., separate droplets), adhering directly to soil surfaces. In either case, the liquid contaminants are somewhat mobile, the movement being due to gravity, the pressure (from above) of infiltrating water, and capillary tension. Such liquid contaminants, whether adhering directly to soil surfaces or not, are often referred to as "non-aqueous phase liquids", or NAPL.

This locus is intended to include the so-called "oil-wet" case where the liquid contaminants preferentially wet the soil surface, displacing water. It also obviously includes the situation where the soils are dry (thus there is no water to displace), so that the liquid contaminants come into direct contact with the soil. This locus does not include either liquid contaminants in the pores where water is covering the soil surfaces (this "water-wet" scenario is covered by locus no. 6) nor the presence of the "oil-wet" case in the saturated zone which is included in the locus no. 5 definition. Contaminants that sorb out of the vapor phase onto dry soils are considered a part of this locus. The sorption mechanism in this case is analogous to condensation and the formation of an ultra-thin liquid film on the surface of the particles. Figures 2–1 and 2–2 present a schematic cross-sectional diagram of locus no. 2 and a schematic representation of

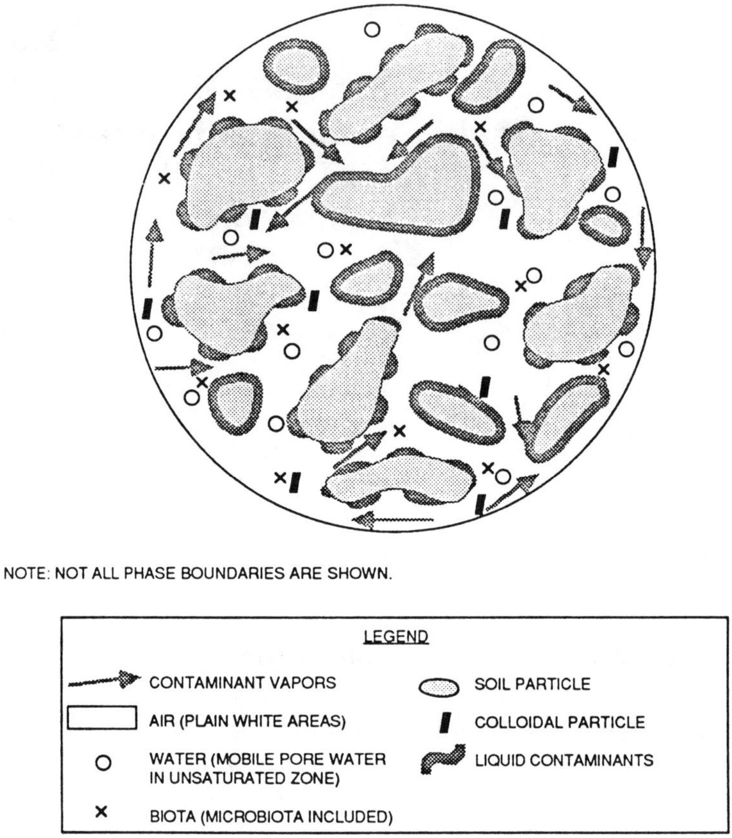

Figure 2–1. Schematic cross-sectional diagram of locus no. 2—liquid contaminants adhering to "water-dry" soil particles or rock surfaces in the unsaturated zone.

the transformation and transport processes affecting other loci, respectively.

2.2 EVALUATION OF CRITERIA FOR REMEDIATION

2.2.1 Introduction

Liquid contaminants in pore spaces between soil particles in the unsaturated zone may exist either as a continuous bulk liquid phase, or as discontinuous films and rings. Initially, after a relatively large release, the entire pore volume may be filled with organic liquid and any residual water not displaced by the descending organic plume. Over time, much

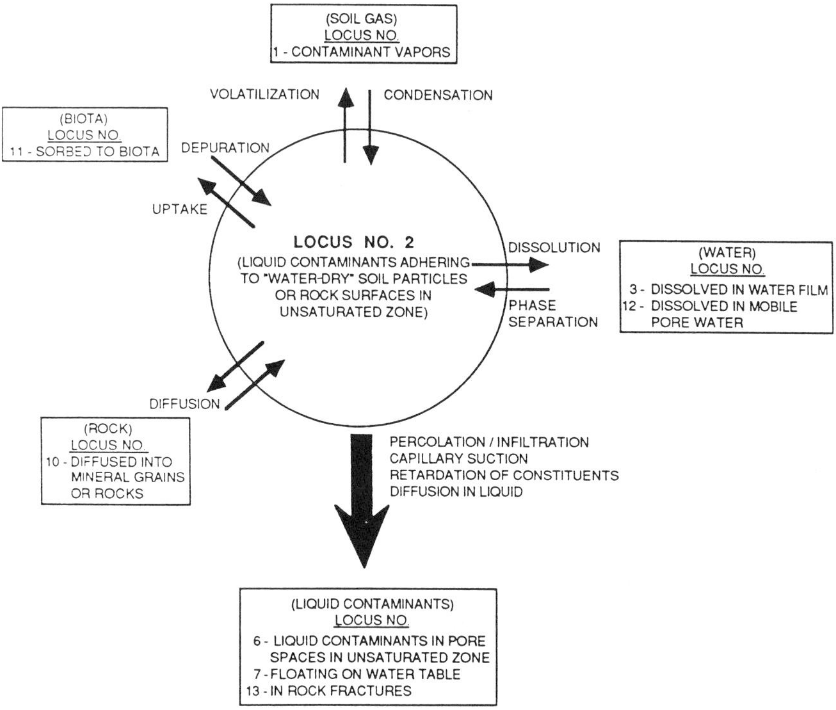

Figure 2–2. Schematic representation of important transformation and transport processes affecting other loci.

of the liquid contaminant will drain from the pores by gravity, leaving the residual "attached" to the soil particles. Figure 2–3 shows the saturated and drained states of locus 2.

Contaminants may also enter this locus through condensation of vapors. English and Loehr (1989) found that hydrocarbon vapors partition onto the solid phase in the unsaturated zone, and that sorption is greatest at low moisture content where organic vapors find less competition with water vapors for sorption sites. Detailed discussion of contaminant vapors is presented in Section 1.

The locus preferentially wets the soil particles, displacing residual water from the surface into the pore space and creating an "oil-wet" scenario. For the purposes of this discussion, it is assumed that residual water can be trapped in the pore space and not be transported out of the locus, as shown in Figure 2–3.

Mobilization of the contaminants in locus no. 2 can be accomplished by volatilization, dissolution, or pressure gradients arising from gravity and liquid surface tension. Transport of contaminants through the locus

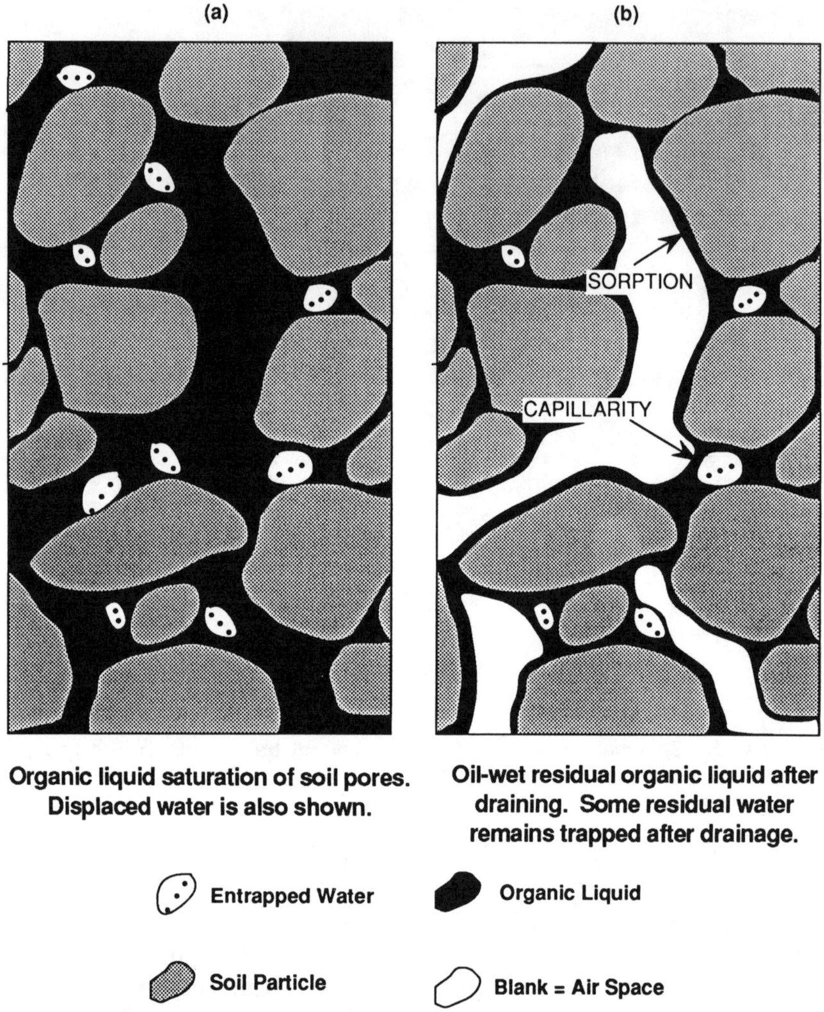

Figure 2–3. Schematic cross-section of locus no. 2 contaminants before and after draining by gravity.

will be a function of the liquid properties of the contaminant, such as viscosity, and the bulk properties of the soil, such as porosity and surface area. Transport can be increased by affecting soil properties, liquid properties, or a combination of both.

Immobilization of residual contaminants can be enhanced by minimizing the parameters that increase contaminant mobilization and transport. In addition, partitioning within the contaminant phase itself will increase the potential for immobilization.

2.2.2 Mobilization/Immobilization

2.2.2.1 Partitioning into Mobile Phase

Contaminants enter locus no. 2 as a mobile phase. The phase may be a liquid such as continuous liquid contaminant, or a vapor such as the volatile components ahead of a liquid front. In the former case, the liquid contaminant wets the soil particles, displacing any residual water from the surface. In the case of contaminated infiltrating water entering the locus, the oil-wet nature of locus no. 2 assumes a preferential sorption or partitioning of the contaminant to the soil surface. (Water-wet partitioning of contaminants is examined in locus no. 5). In the case of contaminants entering locus no. 2 as vapors, the vapors may sorb and condense on the "dry" soil.

Because contaminants enter locus no. 2 as a mobile phase, the contaminants in the locus may generally be viewed as being somewhat mobile, with the exception of the first molecular layer(s) of contaminant. Contaminants sorbed to (or directly contacting) the soil particles will be somewhat less mobile because the proximity of the liquid and solid enhance the attractive forces between them. This phenomenon has implications regarding fixation as described in Section 2.2.3 below.

Mobilization of contaminants in locus no. 2 is accomplished through three basic mechanisms for either the continuous liquid phase or residual saturation: pressure gradients, volatilization, and dissolution. Pressure gradients arise from gravity, including the weight of infiltrating liquid as well as liquid surface tension. They also affect the transport of the mobile phase (Sec. 2.2.2.2). Volatilization involves a contaminant phase change from liquid to vapor and is a function of the vapor pressure of the contaminant, the temperature, and vapor space saturation of the soil gas. Conversely, contaminant vapors can condense, partitioning into locus no. 2 from the soil gas. Locus no. 1 describes in detail the factors affecting volatilization of contaminants. Dissolution involves dissolving contaminants into water solution and is treated in detail in locus no. 5.

Displacement of NAPL by Water. Whenever a porous medium is saturated with more than one fluid (e.g., NAPL and water), it will demonstrate a capacity to retain both fluids (Sale and Piontek, 1989). Locus No. 2 contaminants can be displaced by infiltrating rain water or through a remedial flushing process. The degree to which displacement occurs depends on the fluid viscosities and the effective permeabilities of the soil for the fluids. The mobility ratio is a measure of the ability of one fluid (e.g., water) to displace another (e.g., NAPL), and is defined by:

$$M = (k_{ew} \cdot \mu_o)/(k_{eo} \cdot \mu_w) \qquad (2.1)$$

where M = mobility ratio (dimensionless)
k_{ew} = effective permeability of water at residual NAPL saturation (cm^2)
k_{eo} = effective permeability of NAPL at residual water saturation (cm^2)
μ_w = viscosity of water (g/cm-sec)
μ_o = viscosity of NAPL (g/cm-sec)

Effective permeability is a measure of the soil's ability to conduct a particular fluid. In a two-fluid system, the effective permeability of one fluid will increase as the fraction of pore volume it occupies increases, while at the same time the effective permeability of the other fluid will decrease. As water is passed through the system, NAPL is displaced until the effective permeability of NAPL approaches zero and the remaining NAPL is bound to the soil too tightly for further displacement.

The addition of surfactants (e.g., alkalis or polymers) to water can enhance displacement of NAPL. Surfactants reduce the interfacial tension (ITF) between NAPL and the aqueous phase, reducing the forces binding the contaminant to the soil surface. Alkalis can also reduce ITFs between NAPL and the aqueous phase by forming surfactants. In addition, alkaline agents can neutralize soil surfaces, reducing the soil/NAPL affinity. Addition of a polymer to water increases its viscosity. This will reduce the preferential flow of water through the larger pores and result in more thorough contact between the two fluids.

2.2.2.2 Transport of Mobile Phase

Transport of contaminants within locus no. 2 is primarily restricted to the continuous liquid phase (as shown in Figure 2–3a). Transport is a function of liquid pressure gradients due to gravity, viscous forces, and liquid surface tension. Transport in or out of the locus is described by volatilization (locus no. 1) and dissolution (loci no. 3 and 12). In as much as condensation from the vapor phase (or sorption from solution) represents transport of contaminant to locus no. 2, this section will be limited to transport of the condensed or liquid phase. In addition, although the liquid contaminant phase may be in contact with air or water, this section will treat the phases as essentially immiscible.

Contaminant flow as a continuous phase through the porous medium of locus no. 2 is inversely proportional to the viscosity of the liquid contaminant, and proportional to the permeability of the soil and the force (pressure) acting on (and by) the liquid. Darcy's law states that the volume flow rate per unit area through a porous medium treated as a bundle of capillaries is:

$$Q = \frac{r^2\,\Delta P}{8u} \times \frac{A}{l} = \frac{k\,\Delta P}{u} \times \frac{A}{l} \qquad (2.2)$$

where Q = volume flowrate (cm^3/sec)
r = radius of capillaries (cm)
k = soil permeability (cm^2)
u = dynamic viscosity (g/cm.sec)
ΔP = pressure drop of the fluid (g/cm.sec^2)
A = cross-sectional area of porous medium (cm^2)
l = length of the porous medium (cm)

Equation 2.2 applies only to conditions where soil pores are fully saturated with liquid. An example calculation is provided in Section 2.4.2. The permeability of various soils can be estimated from Table 12–2. Table 2–2 (in the subsection on Viscosity) lists viscosities for some common fluids. Equation 2.5 can be used to obtain the fluid pressure drop. For a unit area or length, A or l can be specified accordingly.

The capillaries consist of the interconnecting pores between the particles, where the radii vary with the grain-size distribution in the unsaturated zone. For example, Figure 2–4 illustrates two representations of the capillaries just above the saturated zone termed the "capillary fringe." (Water remains/flows above the saturated zone in the capillary fringe due to interfacial surface tension. Contaminant capillary flow is discussed further below.)

To increase the volume flow rate of liquid contaminant in locus no. 2, the pressure on the contaminant, the capillary radii, or the soil permeability must be increased, or contaminant viscosity must be decreased. Conversely, decreasing r, P, or k, or increasing u will reduce contaminant flow, effectively immobilizing contaminants in the locus. Changing the size of the capillary radii is not considered feasible.

Soil Permeability. As illustrated in equation 2.2, soil permeability (k) is a fundamental parameter governing contaminant flow in a porous medium. Essentially, soil permeability is a measure of resistance to liquid flow, and is independent of the physical properties of the liquid. (It is therefore an important parameter for any liquid flow in the subsurface, including air and vapors.) Soil permeability is generally related to

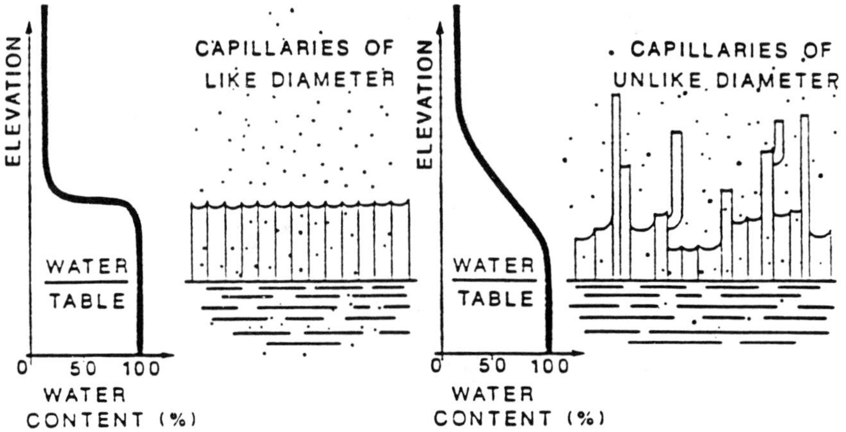

Figure 2–4. Capillary pressure effects on variation of water content and elevation. *Source*: Schwille, 1987.

soil grain-size distribution, and is directly proportional to soil porosity and inversely proportional to soil surface area. Relatively large-grained soils such as sands are generally more permeable than small-grained materials, such as clays.

The surface area of the soil particles plays an important role in intrinsic soil permeability. This is because surface features of the soil particles, such as depressions or micropores, produce resistance to liquid flow in the unsaturated zone in addition to that produced by interparticle capillaries. Altering soil surface area may have implications for mobilization or immobilization of contaminants in the unsaturated zone, but its practical application in remediation may be less feasible than other techniques.

Table 2–1 shows typical surface areas and particle diameters for various soils. As indicated, heat activated (sintered) clays have a very large surface area compared to other materials. Increasing the surface area of the soil in locus 2 will markedly decrease permeability. A decrease in permeability will have a corresponding decrease on contaminant flow rate in equation 2.2.

If lowering the mobility of a contaminant in locus no. 2 is desired, heat activating the soil (if sinterable) will greatly increase the area available for sorption or adhesion, and reduce liquid transport out of the locus. Liquid transport will occur on a microscale, however, as the contaminant spreads over the new available surface area. (As a practical matter, supplying heat to locus no. 2 will have competing effects, even

Table 2–1. Relationship Between Diameter of Particles and Surface Area

Diameter of Particles (mm)	Description	Approximate Surface Area (m^2/g)
1.0–2.0	Very Coarse Sand	0.001–0.003
0.5–1.0	Coarse Sand	0.003–0.005
0.25–0.5	Medium Sand	0.005–0.01
0.1–0.25	Fine Sand	0.01–0.03
0.05–0.1	Very Fine Sand	0.03–0.1
0.002–0.05	Silt	0.1–1
<0.002	Clay	>1
<0.002	Activated Clay	1.50–2.25

Source: Adapted from Hillel, 1980.

for activated soils. Heat will increase volatilization as described in locus no. 1, and will lower viscosity; see below).

In addition, heat treating soils that can be activated beneath UST installations may serve as a useful in-situ preventative measure, providing a sub-bedding of low liquid permeability and high adsorbent capacity for soil vapors. Alternatively, if native soils can not be heat activated, a sub-bedding of activated adsorbent may be desirable.

To increase permeability and contaminant flow rate, soil porosity may be increased by physically increasing the void space per unit volume of soil. This can be accomplished by plowing or mechanically breaking the soil, increasing permeability in equation 2.2. Increasing porosity effectively increases capillary radii in equation 2.2 as well. Conversely, soil compaction reduces the interparticle radii and soil porosity, reducing contaminant flowrate. Again, the practical application of these techniques at UST sites, where the tank may be several meters below ground surface, may be limited.

Soil permeability may be determined indirectly from hydraulic conductivity measurements, since hydraulic conductivity is a function of soil permeability, liquid density, and liquid viscosity. (See Section 6.2.2.2). Estimates of soil permeability can also be found for various soil types in Table 12-2. Other liquid phases produce resistance to liquid flow in addition to resistance caused by the soil particles, and also compete for available flow area. Section 6 discusses relative permeability factors that account for the presence of additional immiscible liquid phases (such as water) and can be used to adjust soil permeability. However, increasing porosity or interparticle dimensions still increases transport, regardless of the presence of other phases.

Viscosity. Whereas permeability is a measure of external resistance to fluid flow, viscosity is a measure of the internal resistance to flow exhib-

Table 2–2. Fluid Viscosities at 20°C

Fluid	Viscosity (cP)
Air	0.018
Gasoline	0.4–0.6
Benzene	0.64
Water	1.0
Kerosene	2.0
Crude oil	7.2

Source: Nash, 1987.

ited by a fluid. (Specifically, it is the ratio of shearing stress to rate of shear.) Table 2–2 compares viscosities of some common fluids.

The viscosities listed are dynamic viscosities and may be used directly in equation 2.2. However, fluid flow rate is not linearly dependent on dynamic viscosities because different liquids exhibit different pressure drops for a given porous bed. Kinematic viscosity, which is the dynamic viscosity divided by fluid density, produces a direct comparison of fluid flow rate. For example, "thinner" water (less viscous by a factor of 2) does not flow twice as fast as kerosene in a porous medium, because kerosene has a density of 0.8 g/cm^3; water flows 2.5 times faster (by volume) than kerosene.

Whether comparing dynamic or kinematic fluid viscosities, decreasing viscosity increases contaminant flow rate. One way to change liquid viscosity is to change the liquid temperature. Thus, one way to increase contaminant transport is to raise the temperature. Because soil temperature remains fairly constant on a daily and seasonal basis below a depth of 2 meters, heat will typically be applied by artificial means. Two common methods used for tertiary recovery in oil fields are steam and fire. As mentioned above with respect to soil permeability, high heat may cause localized activation of the soil, potentially immobilizing or decreasing the transport of liquid contaminants. (Steam is typically used to activate carbon, clays, alumina and other adsorbents).

Another method of changing liquid viscosity is by chemical addition. For example, carbon dioxide is often employed by oil companies for tertiary recovery. Chemical additives such as viscosity index improvers or viscosity extenders are added to motor oil to lower the viscosity at low temperatures and to raise it at high temperatures. Agents used for this purpose are polymers of alkyl esters of methacrylic acid, polyisobutylenes, and others. Two drawbacks to the use of viscosity extenders for UST leak remediation are in situ mixing, and the addition of hydrocarbons to the subsurface environment.

The mobility of contaminants in locus no. 6 can also be reduced by emulsification, which involves the addition of emulsifiers (surfactants) to the locus to increase viscosity. There are drawbacks to emulsification because it requires the presence of considerable amounts of water to convey the emulsifier to the locus. Dissolution of contaminants into the aqueous phase may occur. Care in emulsifier selection is also required to avoid unintended consequences. For example, while emulsifiers increase viscosity, slowing transport, they may not completely immobilize contaminants. Furthermore, emulsions can break (separate), particularly if the dispersed (i.e., dilute) phase was not completely emulsified. An unemulsified portion of the dispersed phase will serve as a "seed" to break the emulsion.

Pressure Drops. Pressure gradients in locus no. 2 arise from gravity (including hydrostatic pressure) and surface tension. Gravity produces a downward force on the contaminant, pulling the liquid through locus no. 2. An additional force is the weight (due to gravity) of liquid at a higher elevation acting on liquid beneath it. This hydrostatic pressure (or "head") can be described by

$$\Delta P = \Delta\rho\, gh \tag{2.3}$$

where $\Delta\rho$ = difference in density between the liquid and the gas phase (g/cm^3)
g = acceleration due to gravity (980 cm/sec^2)
h = height of liquid column (cm)

As can be seen by substituting ΔP of equation 2.3 into equation 2.2, increasing the density of a liquid contaminant or increasing the height of the liquid column by addition of liquid will increase the flow rate of a contaminant in locus no. 2. However, the former is impractical, while the latter is generally unacceptable (except for water flushing) and in an oil-wet regime possibly ineffective. Because gravity varies so little with elevation (0.05 percent per kilometer, ignoring tidal effects), and hydrostatic pressure is normally localized and of small magnitude, hydrostatic pressure is normally ignored in the subsurface environment. For a release from an UST, there is an instantaneous pressure from the height of the liquid in the tank. However, the pressure quickly dissipates laterally. Even for no lateral dissipation, as in the case of rainwater infiltration, the static pressure is minimal. For example, an instantaneous rainfall of one inch would result in an increased pressure of only 0.035 psi in the locus. Also, density is a stronger function of temperature, resulting in a net reduction of flow rate with reduced temperature.

Thus, for the continuous phase (Figure 2–3a), gravity is the principal pressure gradient in locus no. 2. (This holds true for flushing with a continuous water phase as well). If the height of liquid is comparable to the pore dimensions, and discontinuous films, rings, and blobs form residual saturation (Figure 2–3b), then static pressure approaches zero and surface tension and capillary forces dominate. Under these conditions, contaminants are effectively immobilized.

A curved interface between two liquids or between a liquid and a gas indicates a difference in pressure exists between the two (Hoag and Marley, 1986). Capillarity concerns interfaces that are sufficiently mobile to assume an equilibrium shape. The most common examples are meniscuses and drops formed by liquid in air. The fundamental equation of capillarity was given in 1805 by Young and Laplace (Adamson, 1981):

$$\Delta P = \gamma\left(\frac{1}{R_1} + \frac{1}{R_2}\right) \qquad (2.4)$$

where ΔP = capillary pressure or tension ($dyne/cm^2$)
γ = liquid surface tension (dyne/cm)
R_1 and R_2 = radii of curvature for the solid surface (cm)

Figure 2–5 shows a cross-section of locus 2 contaminants held by capillary forces. One of the two radii of curvature is identified. In the case of a cylinder, $R_1 = R_2$ and equation 2.4 reduces to:

$$\Delta P = \frac{2\gamma}{r} \qquad (2.5)$$

where r is the radius of the cylinder. Assuming the pores in locus no. 2 are cylindrical, and the liquid wets the soil, equation 2.5 can be used to show the pressure exerted by the liquid contaminant on the pores within the locus.

If the radius is not too large, the surface tension of the liquid will cause the liquid to spread across the particle surface in all directions, thereby "wetting" the soil particles. The vertical rise (up or down) from a flat liquid surface as shown in Figure 2–6, (for which ΔP must be zero) can be found by combining equations 2.3 and 2.5

$$h = \frac{2\gamma}{r\Delta p\ g} \qquad (2.6)$$

As can be seen from equation 2.5, capillary tension will be the greatest for the smallest pores, regardless of saturation. Indeed, for water in

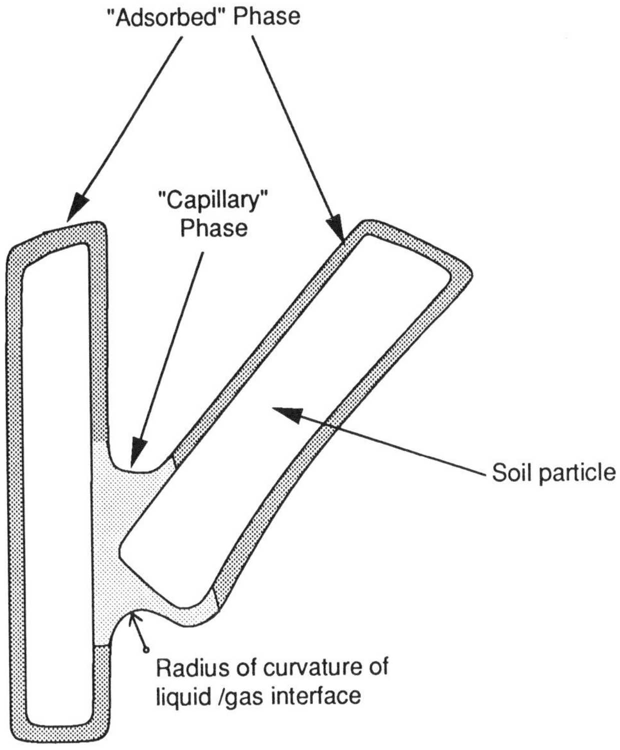

Figure 2–5. Schematic cross-section of soil particles showing liquid held by capillary forces. *Source*: Adapted from Hillel, 1980.

loam the capillary pressure is in excess of 4 atmospheres (Brooks and Corey, 1964). Table 2–3 gives surface tensions for some liquids.

Thus, it is inherently exceedingly difficult to transport the entire residual saturation in locus no. 2. Equation 2.5 indicates that increasing pore radius reduces capillary tension, resulting in increased hydrostatic pressure differential in the locus. As shown in equation 2.2, an increase in pressure differential will cause an increase in contaminant flow rate. This corresponds to increasing porosity by mechanical plowing. Alternatively, liquid surface tension is a weak function of liquid temperature, decreasing slightly with increased liquid temperature. Increasing temperature will reduce capillary pressure (and reduce viscosity), increasing contaminant transport. Conversely, compaction will reduce pore radii, increasing capillary pressure and decreasing contaminant flow rate.

Discontinuous rings and films adhering to soil particles can also be

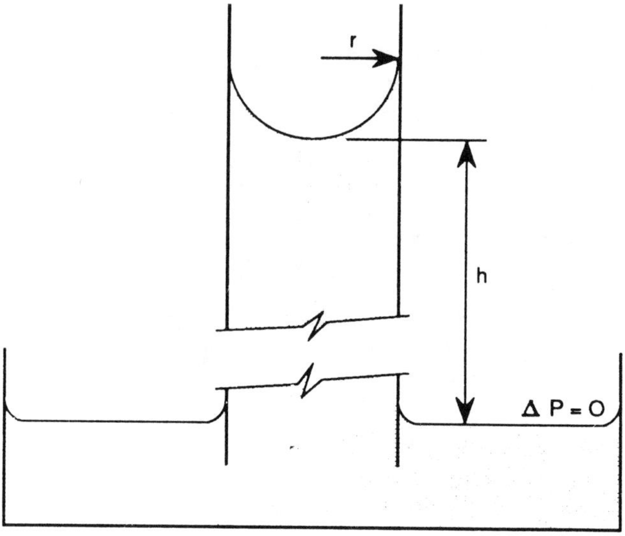

Figure 2–6. Capillary rise (capillary much magnified in relation to dish).

mobilized by increasing the contact angle between the liquid and the soil. The contact angle is defined by

$$\Theta = \cos^{-1}\left(\frac{\gamma_{pw} - \gamma_{pc}}{\gamma_{cw}}\right) \tag{2.7}$$

Where γ_{pw}, γ_{pc} and γ_{cw} are the interfacial surface tensions for the particle and water (pw), particle and contaminant (pc), and contaminant and water (cw). Figure 2–7 illustrates the relation between interfacial tension forces and contact angle for a globule of organic liquid in a water-saturated pore space. The magnitude and direction of the ITFs at the soil/water/organic liquid interface are shown. Decreasing the surface tension between the particle and water with a surfactant will permit the water to wet the soil, mobilizing the contaminant.

However, choice of surfactant is crucial. In an oil-wet regime where the soil substrate has a surface electronegativity from hydrated silicates,

Table 2–3. Surface Tension Values, Liquid-Vapor Interfaces

Liquid	Temp. (°C)	Surface Tension; (dyne/cm)
Water	20	72.88
Benzene	20	28.88
Toluene	20	28.52
Octane	20	21.62

Source: Adamson, 1981.

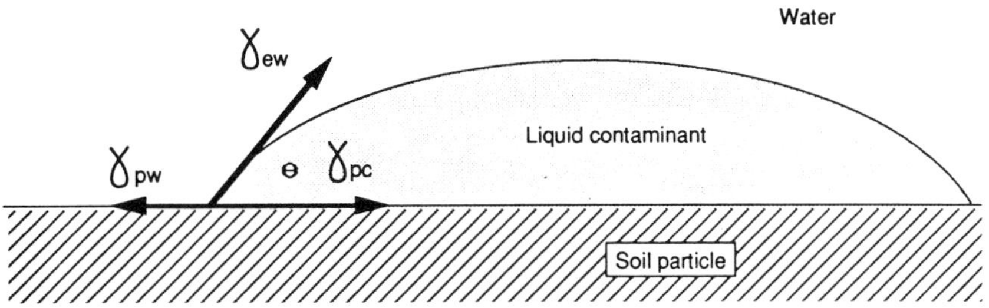

Figure 2–7. Interfacial surface tensions and contact angle.

an anionic surfactant will help to immobilize the contaminant. To transport the contaminant out of the locus, a cation surfactant should be used.

2.2.3 Fixation

2.2.3.1 Partitioning onto Immobile (Stationary) Phase

As mentioned in Section 2.2.1, contaminant enters locus no. 2 as a mobile liquid or condensing vapor phase. The contaminant adheres to and spreads on the soil particles by surface tension and moves through the locus by gravity and capillary pressure gradients. Equation 2.2 described the parameters governing mobility in the locus.

Just as contaminant transport can be increased by increasing (or decreasing) relevant parameters of flow rate, fixing contaminants (or greatly reducing flow rate) can be achieved by decreasing the parameters that favor mobility or increasing those that inhibit mobility. Flow rate, or mobility, varies with soil permeability, liquid viscosity, and pressure differentials.

Soil permeability can be reduced by heat activation of the soil, effectively increasing soil surface area and reducing mobility. Viscosity of the liquid can be altered with polymers and surfactants. Pressure differentials are influenced by soil/liquid interfacial tension forces and changes in liquid column height. The specific physical and chemical principles that influence these parameters as they relate to fixation of liquid contaminant were discussed in detail earlier in this section.

A likely first step in a remedial fixation approach would be to minimize pressure differentials caused by infiltrating rain water. Other techniques are more complex and require careful preparation and execution. To increase liquid viscosity, water may be required to convey the viscosity enhancers to the locus. This may result in displacement and/or dissolution of contaminant. Also, the chemical enhancers must be evenly distributed through the contaminated zone at the proper concentration or results will be less than optimal.

2.2.4 Transformation

2.2.4.1 Biodegradation (Details in Section 11)

Biodegradation is most likely to take place in an aerated aqueous environment in the unsaturated zone. Contaminant liquids in locus no. 2 exist in a predominantly water-free environment and while some microbes can live suspended in pure hydrocarbon, these are rare and metabolize oil slowly. Therefore, locus no. 2 contaminants would likely diffuse into other loci before they could be biodegraded.

2.2.4.2 Chemical Oxidation (Details in Section 3.2.4.2)

Liquid contaminants in locus no. 2 are in contact with mineral surfaces and may undergo catalytic transformations. Chemical oxidation in this locus may also be enhanced by introducing an oxidant such as ozone or peroxide.

2.3 STORAGE CAPACITY IN LOCUS

2.3.1 Introduction and Basic Equations

The bulk or gross storage capacity of contaminant in locus no. 2 is a function of soil porosity. Soil with high void space per unit volume can store more contaminant than soil with low porosity. The soil void space available for storage is

$$\Theta_a = \Theta_t - \Theta_w \qquad (2.8)$$

where Θ_t is the total porosity of the soil, Θ_w is the porosity filled by residual water saturation, and Θ_a is the resultant porosity of the air space (i.e., not soil or water). As stated in the Locus Description, although water may be displaced by the contaminant, it is assumed that it does not leave the pore space.

To calculate the maximum mass of contaminant per unit volume of soil, the air space porosity is multiplied by the contaminant density to obtain

$$m = \rho\ \Theta_a \qquad (2.9)$$

where m = maximum storage capacity (g/cm^3)
ρ = contaminant density (g/cm^3)

After the contaminant has flowed or drained through the locus, residual films and rings remain. The mass of residual saturation per unit volume is also a function of contaminant density and porosity.

$$m_r = \rho\Theta_a R \qquad (2.10)$$

Here R is retention capacity, the fraction of total soil porosity storing residual contaminants. Table 2–4 lists experimentally derived retention capacity values for gasoline. "Mixed" sand type contains equal parts of fine, medium and coarse sand.

2.3.2 Guidance on Inputs for, and Calculation of, Maximum Value

Maximum storage capacity in locus no. 2 is a function of porosity and liquid density, and may be calculated from equations 2.8 and 2.9.

- Values of porosity for various soils can be estimated from Figure 12–4. If dry soils are assumed, the total porosity is used. If some water is assumed to be retained as residual saturation, air-filled porosity (Θ_a) can be estimated from equation 2.8 by substituting field capacity or wilting point values from Figure 12–4 for Θ_w.
- The density of liquid contaminants can be found in various literature sources. Liquid density of gasoline (0.74 g/cm^3) is shown in Table 1–1A.

2.3.3 Guidance on Inputs for, and Calculations of, Average Values

For residual contaminant saturation, equation 2.10 may be used to calculate average storage values in locus no. 2. Porosity and liquid density values are as discussed in Section 2.3.2. Retention capacity (R) for gasoline can be estimated from Table 2–4.

Table 2–4. Retention of Gasoline in Soil

Run Number	Sand Type	Dry Density, g/cm³	Moisture State	Mass of Gasoline Retained, g[a]	(R) Degree of Saturation of gasoline, percentage
1	Medium	1.44	Dry	123.0	19.6
2	Medium	1.44	Dry	93.0	14.8
3	Medium	1.44	Dry	112.0	17.8
4	Coarse	1.47	Dry	90.5	14.9
5	Fine	1.47	Dry	330.0	54.0
6	Medium	1.55	Dry	117.0	20.5
7	Mixed	1.74	Dry	218.0	46.3
8	Medium	1.55	Dry	118.0	20.7
9	Medium	1.55	Dry	120.0	21.0
10	Coarse	1.57	Dry	100.0	17.8
11	Fine	1.62	Dry	290.0	54.3
12	Medium	1.71	Dry	128.0	26.3
13	Mixed	1.85	Dry	218.0	52.6
14	Medium	1.71	Dry	132.0	27.1
15	Medium	1.68	Dry	117.0	24.0
16	Coarse	1.72	Dry	98.0	19.3
17	Fine	1.72	Dry	290.0	60.2
18	Medium	1.48	Field	100.0	16.4
19	Mixed	2.00	Dry	200.0	59.5
20	Medium	1.48	Field	77.0	12.7
21	Medium	1.47	Field	84.0	13.8
22	Medium	1.57	Field	81.0	14.5
23	Medium	1.56	Field	69.0	12.3
24	Medium	1.71	Field	88.0	18.0
25	Medium	1.55	Field	85.0	15.2
26	Medium	1.72	Field	83.0	16.9
27	Medium	1.71	Field	85.0	17.4
28	Fine	1.47	Field	121.4	19.9
29	Fine	1.62	Field	100.0	18.7
30	Fine	1.72	Field	124.3	25.8

a. Volume of soil contacted = 1,831 cm^3
Source: Hoag and Marley, 1986.

2.4 EXAMPLE CALCULATIONS

2.4.1 Storage Capacity Calculations

Assume a release of gasoline in a medium sand with moisture content at field capacity.

2.4.1.1 Maximum Value

- Liquid density of gasoline is 0.74 g/cm^3 (Table 1–1A).
- Soil porosity for medium sand is 0.42 and field capacity is 0.10 (Figure 12–4).

Using equation 2.8, air-filled porosity is:

$$\Theta_a = 0.42 - 0.10 = 0.32$$

Using equation 2.9, maximum storage capacity is:

$$m = 0.74\ g/cm^3\ (0.32) = 0.24\ g/cm^3$$

2.4.1.2 Average Value

- Liquid density of gasoline is 0.74 g/cm^3.
- Air-filled soil porosity is 0.32.
- The range of retention capacity values from Table 2–4 for medium sand at field capacity moisture content is 0.123 to 0.18. The average R is 0.15.

Using equation 2.10, average storage capacity is:

$$m_r = 0.74\ g/cm^3\ (0.32)(0.15) = 0.036\ g/cm^3$$

2.4.2 Transport Rate Calculations

As mentioned in Section 2.2.2.1, contaminant transport out of or through the locus can occur through volatilization, dissolution, and/or pressure gradients (neglecting biological and chemical oxidation processes). Rate of volatilization, and dissolution are examined in loci nos. 1 and 5 respectively. Rate of liquid contaminant transport in the locus can be calculated from equation 2.2.

If gasoline is flowing through locus no. 2, the flow rate can be calculated by equation 2.2.

- Average pore diameter is assumed to be 0.01 cm (0.005 cm in radius).
- Liquid viscosity is 0.005 poise (Table 2–2).
- Surface tension of gasoline (taken as octane) is 21.62 dyne/cm (Table 2–3).
- ΔP is determined by equation 2.5.
  $\Delta P = 2(21.62)/0.005 = 8648\ dyne/cm^2$
- Unit area (A) and length (l) are assumed to be 1 cm^2 and 1 cm, respectively.

Then, using equation 2.2, flow rate is:

$$Q = 0.005^2\ (8648)/(8)(0.005) = 5.4\ cm^3/sec$$

The flow rate is a linear velocity of fluid volume per unit time per area ($cm^3/sec/cm^2$). This flow rate (5.4 cm/sec) is very high because flow length (l) was assumed to be very short. Increasing the length of the medium increases surface area and resistance to flow. Thus, for the same release, the flow rate per cm^2 for a porous bed 1 m long would be:

$$Q = 5.4/100 = 0.054 \text{ cm}^3/\text{sec}$$

2.5 SUMMARY OF RELATIVE IMPORTANCE OF LOCUS

2.5.1 Remediation

Residual contaminant in the oil-wet unsaturated zone can provide a continual source of dissolved contamination via infiltrating water as well as vapor phase contamination through volatilization. Active venting of contaminated soils to enhance removal by volatilization is an increasing popular remedial approach. Decreasing viscosity through heating or chemical additives can also be used to remove residual contaminants, although small capillaries may remain unaffected. (Heating the locus will also facilitate volatilization). Alternatively, the residual saturation can be immobilized by emulsification, although mixing is a problem. Contaminants can also be immobilized by reducing soil porosity, either by compaction or filling the unsaturated porosity with an immiscible, immobile phase.

2.5.2 Loci Interactions

Because residual saturation of liquid contaminant can be significant (greater than 50 percent of porosity), locus no. 2 can be a source of significant contamination for other loci. For residual saturation, the principal partitioning processes are volatilization into locus no. 1, and dissolution into residual or infiltrating water (loci nos. 3 and 12, respectively).

2.5.3 Information Gaps

The principal information gaps in understanding mobilization or immobilization of oil-wet contaminants in the unsaturated zone include:

(1) Fundamental data on multi-component contaminant surface tension
(2) Distribution of pore geometries and radii
(3) Data on emulsifier selection and in situ influence on contaminant viscosity

(4) Data on surfactant selection and in situ influence on contaminant contact angle and mobilization
(5) Understanding of extent of in situ heat activation of soils
(6) Biodegradation and chemical oxidation in the locus.

2.6 LITERATURE CITED

Adamson, A.W. 1981. Physical Chemistry of Surfaces. John Wiley & Sons, New York.

Brooks, R.H. and Cory, A.T. 1964. Hydraulic Properties of Porous Media, Hydrology Papers, Colorado State University, #3, Ft. Collins.

English, C.W. and R.C. Loehr. 1989. Sorption of Volatile Organic Compounds in Soil. Proceedings of Petroleum Hydrocarbons and Organic Chemicals in Groundwater. NWWA. Houston. 383-391.

Hillel, D. 1980. Fundamentals of Soil Physics. Harcourt, Brace, and Jovanovich. New York.

Hoag, G.E. and M.C. Marley. 1986. Gasoline Residual Saturation in Unsaturated Uniform Aquifer Materials. Journal of Environmental Engineering. Vol. 112, No. 3. 586-604.

Nash, J.H. 1987. Chemical and Physical Properties of Gasoline. Proceedings of Underground Environment of an UST Motor Fuel Release. EPA Office of Research and Development. Edison, NJ.

Sale, T. and K. Piontek. 1989. Chemically Enhanced In Situ Soil Washing. Procedures of Petroleum Hydrocarbons and Organic Chemicals in Groundwater. NWWA. Houston, 487-504.

Schwille, F. 1967. Petroleum Contamination of the Subsoil—A Hydrological Problem. In: Joint Problems of Oil and Water Industries Institute of Petroleum, pp. 23-54. London, England.

SECTION 3

Contaminants Dissolved in the Water Film Surrounding Soil Particles in the Unsaturated Zone

CONTENTS

List of Tables 74
List of Figures 75
3.1 Locus Description 77
3.1.1 Short Definition 77
3.1.2 Expanded Definition and Comments 77
3.2 Evaluation of Criteria for Remediation 78
3.2.1 Introduction 78
3.2.2 Mobilization/Remobilization 80
3.2.2.1 Transfer to Mobile Pore Water 80
Advection (Mixing) 80
Diffusion in Water 81
3.2.2.2 Transfer to Soil Gas (Volatilization from Solution) 82
Equilibrium Partitioning 82
Rate of Volatilization 83
3.2.2.3 Transport of/with Mobile Phase 87
3.2.3 Fixation 87
3.2.3.1 Partitioning onto Immobile (Stationary) Phase 87
Sorption 87
3.2.3.2 Other Fixation Approaches 87
3.2.4 Transformation 87
3.2.4.1 Biodegradation 87
3.2.4.2 Abiotic Transformation of UST Contaminants 88
Complexation 88
Photo-oxidation 89
Hydrolysis 89
Elimination 89
Dehydrogenation 90

Redox 90
Polymerization 93
Summary 98
Applicability to Locus 3 98
3.3 Storage Capacity in Locus 98
3.3.1 Introduction and Basic Equations 98
3.3.2 Guidance on Inputs for, and Calculations of, Maximum Values 99
3.3.2.1 Concentration of Contaminants in Water 99
3.3.2.2 Soil Moisture Content: Water-Filled Porosity 100
3.3.2.3 Soil Bulk Density 100
3.3.3 Guidance on Inputs for, and Calculations of, Average Values 100
3.3.3.1 Concentration of Contaminants in Water 100
3.3.3.2 Soil Moisture Content: Water-Filled Porosity 100
3.3.3.3 Soil Bulk Density 100
3.4 Example Calculations 100
3.4.1 Storage Capacity Calculations 100
3.4.1.1 Maximum Value 100
3.4.1.2 Average Value 101
3.4.2 Transport Rate Calculations 101
3.4.2.1 Diffusion Flux 101
3.4.2.2 Air/Water Concentrations 102
3.5 Summary of Relative Importance of Locus 102
3.5.1 Remediation 102
3.5.2 Loci Interaction 103
3.5.3 Information Gaps 103
3.6 Literature Cited 103

TABLES

3–1 Diffusion Coefficients of Selected Organic Chemicals in Water 82
3–2 Estimated Concentrations of Gasoline Contaminants in Locus After Losses to Air and Soil 86
3–3 Oxidation Processes 91
3–4 Selected Values for Soil Bulk Density 99

FIGURES

3-1 Schematic Cross-Sectional Diagram of Locus No. 3 – Contaminants Dissolved in the Water Film Surrounding Soil Particles or on Rock Surfaces in the Unsaturated Zone 78
3-2 Schematic Representation of Important Transformation and Transport Processes Affecting Other Loci 79
3-3 Henry's Law Constants for Gasoline Components 84
3-4 Humic Substances Component Structures 94
3-5 Abiotic/Biotic Humic Substance Formation Pathways .. 96

SECTION 3

Contaminants Dissolved in the Water Film Surrounding Soil Particles in the Unsaturated Zone

3.1 LOCUS DESCRIPTION

3.1.1 Short Definition

Contaminants dissolved in the water film surrounding soil particles in the unsaturated zone.

3.1.2. Expanded Definition and Comments

This locus includes only the water (with its dissolved contaminants) that exists as a thin film around, or in narrow pores between, soil particles. As such, it represents the relatively immobile water retained in a well-drained soil.

This locus does not include pore water in the unsaturated zone that is not in close proximity to soil surfaces. This water, which is covered in the definition of locus no. 12, is quite mobile and may, at times, completely fill the available soil pores.

The locus no. 3 water will exist in times and places of dryness and/or low infiltration. It could thus include well-drained soils under an impervious cover (e.g., paved road), or slightly moist soils in unpaved areas where infiltration is minimal. The thickness of the film is not precisely defined, and there is clearly some gradual change between water (and its dissolved contaminants) that will be considered in locus no. 3 vs. that which is considered to be in locus no. 12. It will comprise only a fraction (e.g., 1 to 10 percent) of the total soil porosity. Figures 3–1 and 3–2 present a schematic cross-sectional diagram of locus no. 3 and a schematic representation of the transformation and transport processes affecting other loci, respectively.

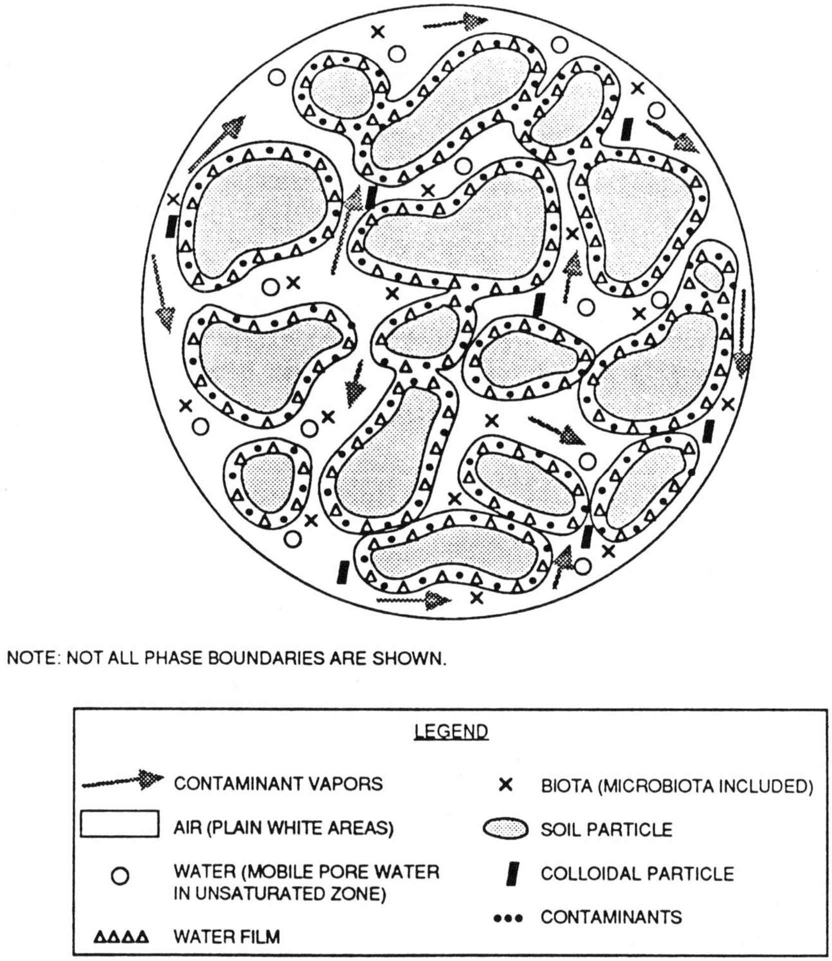

Figure 3–1. Schematic cross-sectional diagram of locus no. 3–contaminants dissolved in the water film surrounding soil particles or on rock surfaces in the unsaturated zone.

3.2 EVALUATION OF CRITERIA FOR REMEDIATION

3.2.1 Introduction

As stated above, locus no. 3 consists of contaminants dissolved in a thin, immobile film of water surrounding soil particles in the unsaturated zone. During times of rainfall infiltration, percolating water (locus no. 12) is in intimate contact with, and partially mixes with, some of the

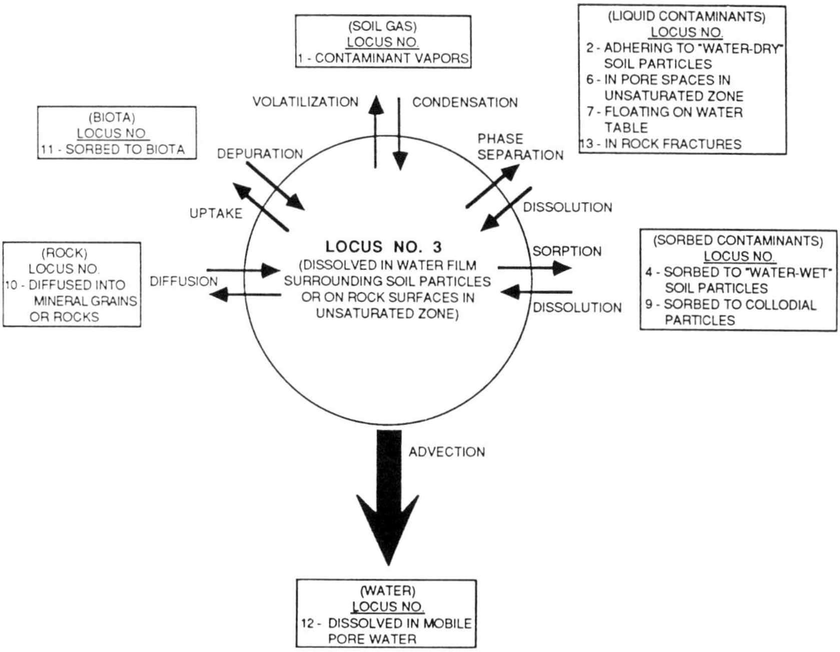

Figure 3–2. Schematic representation of important transformation and transport processes affecting other loci.

water of locus no. 3. However, a significant fraction of the locus 3 water may be held in small pores and fractures that play little or no role in infiltration and are thus bypassed by the percolating water.

The distinction between loci 3 and 12 is discussed in more detail in Section 12.2.2.2. The point is made there that locus 12 (mobile) water might be considered to certainly include water held in excess of a soil's field capacity, and, possibly, also the less mobile water that is in excess of the wilting point but less than the field capacity. Figure 12–4 (in Section 12) provides data on the moisture content of various soils at field capacity and at the wilting point. Wilting points occur at about 5 percent moisture content in sandy soils and about 20 percent in clayey soils. It should be understood, however, that soil moisture content cannot be used to classify all of the moisture in a soil as being either locus no. 3 or 12. In many (perhaps most) soils, both will be present.

The properties of the water held in locus no. 3 are assumed here to be the same as for bulk water. It is known, however, that water within a few molecular diameters of a solid phase, such as a clay particle, may have a more rigid structure and be more immobile than is water at greater distances. This structure is induced by electronic interactions between surface charges on the soil and the dipole of the water molecule. Some hydrogen bonding may also be involved. Such interactions are the molecular basis for interfacial tension forces which, in turn, are the basis for capillary suction.

3.2.2 Mobilization/Remobilization

Contaminants held in locus no. 3 are, by definition, in an immobile phase. The only two approaches available to mobilize these contaminants are by transfer to mobile pore water (locus no. 12) or by volatilization into the mobile soil gas (locus no. 1). Each process is discussed below.

3.2.2.1 Transfer to Mobile Pore Water

Two mechanisms are available to transfer contaminants into mobile pore water (locus no. 12): advection (or mixing) and diffusion.

Advection (Mixing). Although locus no. 3 is defined as being an immobile water film, there is clearly a continuum of mobility between the strongly held water (a few molecules thick) essentially touching the soil surfaces and the free mobility of locus no. 12 waters. When both loci are present (i.e., the moisture content is above the wilting point), then gravitational flow—and other pressure gradients—will cause some intermingling of the two.

Because the distinction between loci 3 and 12 is an invention of this report, there are—presumably—no published reports which directly address the extent of mixing during periods of infiltration. It would appear obvious, however, that mixing would be facilitated:

- During periods of rapid infiltration associated with saturated soils and high hydraulic heads;
- In high porosity soils;
- In homogeneous soils; and
- With more soluble contaminants (since sorption is less important for them).

It cannot be assumed that advection and mixing would increase in direct proportion to the amount of infiltrating water. As is discussed in Section 12.2.2.2, the presence of macropores often leads to a situation where a small fraction of a soil's porosity is responsible for a large fraction of the total water flow during periods of rapid infiltration. For example, 10 percent of the porosity—in the form of macropores—might be responsible for 80 percent of the flow. The water held in the remaining 90 percent of the porosity moves at a much slower rate and is mostly bypassed. In such situations, one might anticipate having to flush a soil with the equivalent of several pore volumes to achieve a degree of mixing and dilution that a well-mixed system (e.g., a high porosity soil) would give with a single pore volume.

When advection and mixing do occur, solute retardation due to soil sorption will occur. Detailed discussions of sorption and retardation are given in Sections 4 and 8, respectively.

Diffusion in Water (Additional details in Section 7.2.2) When both loci 3 and 12 are present in a soil, diffusion of contaminants through the aqueous phase can lead to some mobilization even if the water is stagnant. Diffusion is driven by concentration gradients. The constant of proportionality between the concentration gradient (ΔC) and the induced flux (J) is called the diffusion coefficient (D), i.e.:

$$J = D\ \Delta C/\Delta x \tag{3.1}$$

where J = net molal flux of contaminant across a hypothetical plane ($mol/cm^2.s$)
D = diffusion coefficient of contaminant in water (cm^2/s)
ΔC = concentration gradient of contaminant at the hypothetical plane (mol/cm^3)
Δx = distance over which concentration gradient is measured (cm)

Selected values of D are provided in Table 3-1. As can be seen from this table, most organic chemicals—especially neutral ones of low molecular weight—diffuse at about the same rate in water. Also, diffusion rates decrease with decreasing temperature.

For diffusion on a macroscale, the blocking action of the soil particles must be accounted for; this is usually done by including a dimensionless tortuosity factor ($O < \Upsilon < 1$) and the calculation of an effective diffusion coefficient:

$$D_{eff} = \Upsilon\ D \tag{3.2}$$

Soils with a high degree of tortuosity in the water-filled pores have lower values of Υ and, thus, lower effective diffusion rates. Values of Υ are typically in the range of 0.01 to 0.5 (Tucker and Nelken, 1981). Although expressions have been given for the estimation of Υ for air-

Table 3–1. Diffusion Coefficients of Selected Organic Chemicals in Water

	Diffusion Coefficient × 10^5 (cm^2/s)			
Chemical	**25°C**	**20°C**	**2°C**	**Source**
HYDROCARBONS				
n-Butane	—	0.89	0.50 (4°C)	a
n-Pentane	—	0.84	0.46 (4°C)	a
Cyclopentane	—	0.93	0.56	a
Cyclohexane	—	0.84	0.46	a
Methylcyclopentane	—	0.85	0.48	a
Benzene	—	1.02	0.58	a
Toluene	—	0.85	0.45	a
Ethylbenzene	—	0.81	0.44	a
OXYGEN CONTAINING				
Methanol	1.66	—	—	b
Ethanol	1.24	—	—	b
Formic acid	1.52	—	—	b
Acetic acid	1.19	—	—	b
Valeric acid	0.82	—	—	b
Acetone	1.28	—	—	b
Ethyl acetate	1.12	—	—	b
Benzyl alcohol	0.93	—	—	b

a. Witherspoon and Bonoli, 1969.
b. Hayduk and Laudie, 1974.

filled pores—and diffusion in air (see Section 1.2.2)—it is not clear if the same expressions can be used for diffusion in soil water.

From the above it is seen that mobilization via diffusion is enhanced by increasing temperature, increasing porosity and increasing concentration gradient. While remedial actions can be taken to provide some increases in temperature and concentration gradient (the latter, for example, by vacuum extraction of gases), little can be done to change the porosity. Tucker and Nelken (1982) suggest that diffusion can probably be ignored if pore water velocities exceed 0.002 cm/s.

3.2.2.2 Transfer to Soil Gas (Volatilization from Solution)

Equilibrium Partitioning. Contaminants held in locus no. 3 can be mobilized by transfer into the soil gas phase. This air-water transfer is a natural process. When contaminants such as petroleum hydrocarbons are present in a system containing both water and air, they will partition themselves between the two phases in proportion to their value of Henry's law constant:

$$C_a = H\,C_w \tag{3.3}$$

where H = Henry's law constant (atm m^3/mol)
C_a = concentration in air (atm)
C_w = concentration in water (mol/m^3)

Values of H for selected gasoline constituents and additives are provided in Figure 3–3. (Numeric tabulations were provided in Tables 1–1A and 1B.) As can be seen from these data, values of H range over eight orders of magnitude. Hydrocarbons range over three orders of magnitude, with the saturated aliphatics being the most volatile (H ~ 1–10 atm m^3/mol @ 25°C), unsaturated and cyclo-aliphatics slightly lower in volatility (H ~ 0.1–1 atm m^3/mol @ 25°C), and light aromatics being the least volatile (H ~ 0.007 - 0.02 atm m^3/mol @ 25°C). Some additives (e.g., ethanol) are of such low volatility that mobilization into the soil gas will be extremely difficult unless the soil is taken to dryness.

Values of H for hydrophobic solutes may be estimated from the ratio of vapor pressure to water solubility (i.e., H = P/S). The influence of environmental variables on H can thus often be estimated from the corresponding influence on vapor pressure (P) and solubility (S). For hydrocarbons, we can expect H to: increase significantly with increasing temperature; increase slightly with increasing water salinity; and decrease slightly with increasing concentration of dissolved organic carbon (DOC) in water. Some data that are specific to hydrocarbons are provided by Mackay and Shiu (1981), Leighton and Calo (1981), Gossett and Lincoff (1981), Mackay et al. (1979), and Lion and Garbarini (1983). Additional information on the variability of vapor pressure and solubility with various environmental parameters is provided in Sections 1.2.2 and 7.2.2, respectively.

It is interesting to note that the combined effects of volatilization and sorption lead, for a typical gasoline, to a very strong (relative) enrichment in the concentration of light aromatics: benzene, toluene, ethylbenzene and xylene (BTEX). In one model calculation (W. Lyman, unpublished data), a hypothetical gasoline with 23 constituents was considered to have contacted pore water in the unsaturated zone until dissolution had reached equilibrium. The dissolved constituents were then allowed to partition into the soil gas and, via sorption, to the soil. The model results are shown in Table 3–2.

Rate of Volatilization. In instances when the concentration in water exceeds the equilibrium concentration, volatilization will occur. The rate of volatilization can be approximated from an approximation to Ficks' diffusion law (c.f. eqn. 3.1) (Thomas, 1982):

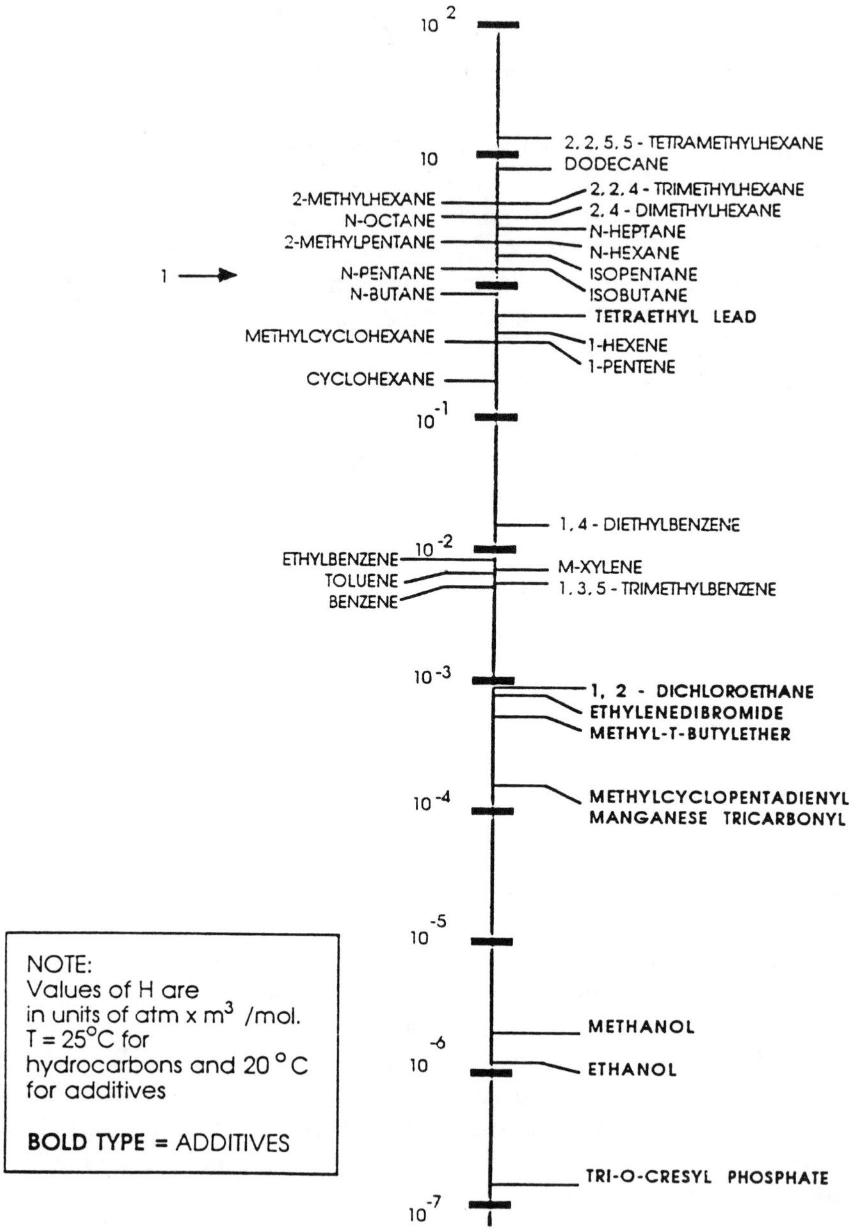

Figure 3–3. Henry's law constants for gasoline components. *Source:* Most hydrocarbon data are from Mackay and Shiu, 1981. Remainder are estimates by Camp, Dresser & McKee, Inc., 1987.

$$J_{vl} = k \, \Delta C \tag{3.4}$$

where J_{vl} = flux of contaminant across air-water interface (g/cm^{2}s)
k = a first-order exchange constant (cm/s)
ΔC = concentration difference across interface (g/cm^3)

In practice, k is usually expressed in terms of a gas-phase (k_g) and a liquid-phase (k_l) mass transfer coefficient, expressions which always reflect the movement (i.e., degree of mixing as reflected in a velocity term) of the fluid phase. Several expressions for k_g and k_l have been derived for surface waters and impoundments (Thomas, 1982), and they show the value of k going to zero as the fluid velocity (and thus mixing) goes to zero. Thus, in quiescent environments—such as the stagnant water of locus no. 3—the rate of volatilization may reach very low levels and be controlled by the rate of diffusion of contaminants in the air and water phases on either side of the interface.

In cases where vacuum extraction is implemented, the air phase may be relatively well mixed, but the locus 3 water may not be. In any case, where H is high ($> 10^{-3}$ atm m^3/mol), as it is for most petroleum hydrocarbons, the resistance of the water film dominates by a factor of ten at least, i.e., k_l is at least ten times less than k_g. The volatilization transfer is liquid-phase-controlled and equation 3.4 can be written as:

$$J_{vl} = K_l \, (C_w - C_a/H) \tag{3.5}$$

where J_{vl} = flux of contaminant (mol/m^{2}s)
K_l = overall liquid phase mass transfer coefficient (m/s)
C_w = concentration of contaminant in water (mol/m^3)
C_a = concentration of contaminant in air (atm)
H = Henry's law constant (atm m^3/mol)

Again, however, K_l is directly related to water velocity (and air velocity, too); and when either of these go to zero, so does K_l. Thus, it would appear that, in vacuum extraction remediation, a state would soon be reached where the rate of movement of (and diffusion in) water would be the rate-limiting step for mass recovery. Any attempts to increase mass recovery by changing only the air velocity (and air concentration) will be futile until the soils have dried out.

Once the contaminants have entered the soil gas (locus no. 1) their movement in, and transport with, soil gas is controlled by a variety of factors. Such transport is described in Section 1.

Table 3–2. Estimated Concentration of Gasoline Contaminants in Locus after Losses to Air and Soil

	Estimated Partitioning in Top Soil (% Mass)			C	C × % Water × 10
Gasoline Component	% in Air	% in Water	% in Soil	Estimated Concentration in Gasoline-Saturated Water (mg/L)	Residual Concentration in Vadose Pore Water After Gasoline-Saturated Water Loses Fraction to Air and Soil by Partitioning (μg/L)
n-Butane	71.2	0.61	28.2	1.00	6.1
Isobutane	73.3	0.51	26.2	1.59	8.1
n-Pentane	70.8	0.47	28.7	1.62	7.6
Isopentane	74.0	0.44	25.5	8.92	39.2
I-Pentene	57.9	1.2	40.9	3.00	36.
n-Hexane	64.0	0.31	35.7	1.24	3.8
1-Hexene	44.8	0.88	30.5	0.84	7.4
2-Methylpentane	65.6	0.32	34.1	1.25	4.0
Cyclohexane	27.9	1.3	70.9	2.02	26.3
Benzene	4.9	7.4	87.7	64.80	4800.
n-Heptane	52.7	0.19	47.1	0.038	0.072
2-Methylhexane	62.7	0.15	37.2	0.12	0.18
Methylcyclohexane	30.2	0.62	69.2	0.14	0.87
Toluene	3.5	4.4	92.1	27.6	1210.
n-Octane	42.1	0.12	57.8	0.0055	0.0066
2,4-Dimethylhexane	51.9	0.14	48.0	0.10	0.14
Ethylbenzene	2.4	2.5	95.1	2.99	74.8
m-Xylene	2.0	2.4	95.6	10.10	240.
2,2,4-Trimethylhexane	46.9	0.11	53.0	0.012	0.013
1,3,5-Trimethylbenzene	1.2	1.6	97.3	2.86	46.
2,2,5,5-Tetramethylhexane	57.8	0.04	42.2	0.0013	0.00052
1,4-Diethylbenzene	1.3	0.76	97.9	0.53	4.0
Dodecane	14.5	0.02	85.5	3×10^{-4}	0.00006
					6514. μg/L
				TOTAL = 131 mg/L	6.5 mg/L

Source: W. Lyman, unpublished data, 1987.

3.2.2.3 Transport of/with Mobile Phase

Transport of contaminants in soil gas is described in Section 1. Transport of contaminants in mobile pore water within the unsaturated zone is described in Section 12.

3.2.3 Fixation

3.2.3.1 Partitioning onto Immobile (Stationary) Phase

Sorption (Details in Section 4.2.2). The contaminants in locus no. 3 are already in a relatively immobile phase. They may be made more immobile by sorption to soil. Sorption will be most important for the constituents of low water solubility and low vapor pressure. For petroleum hydrocarbons, this generally means saturated (and some unsaturated) compounds having roughly nine or more carbons. Soil sorption is not very effective at immobilizing the light aromatics (BTEX fraction).

The sorption equilibrium is not easily changed by parameters that can be adjusted during common remediation activities. It is, in fact, easier to enhance the desorption process via addition of water, detergents, solvents or—in some cases—heat. Only if the soils are taken to dryness will the higher molecular weight fraction of contaminants, especially those that are solids at the system temperature, become fairly well immobilized.

3.2.3.2 Other Fixation Approaches

Isolation from infiltration (e.g., via the use of impermeable covers) will prevent transfer to locus no. 12 (mobile pore water). An impermeable cover will also reduce volatilization to the air.

3.2.4 Transformation

3.2.4.1 Biodegradation (Details in Section 11)

Biodegradation by microbes occurs most readily when microbes, organic contaminants and other nutrient substrates (nitrogen, oxygen, phosphorus, potassium) come in close physical contact. This may well be the case for locus no. 3. However, the discontinuous nature of the water film precludes significant communication with other sources of substrate. Therefore, biodegradation may be high initially but drop off significantly when any of the above named substrates, including the organic contaminant becomes limiting. A full treatment of biodegradation kinetics is found in the locus no. 11 section.

3.2.4.2 Abiotic Transformation of UST Contaminants

Contaminants released to the environment from an UST will be subject to biotic and abiotic "weathering" reactions in the soil/groundwater media. The rate and degree of weathering will be related to the characteristics of the contaminant, whether it be gasoline or no. 2 fuel oil, for example, mediated by environmental factors including temperature, moisture content and soil chemical, physical and microbial properties.

Abiotic and biotic properties operate in concert to weather hydrocarbons introduced to the soil environment. Although mineralization may be envisioned as the ideal end point for remediation of hydrocarbon contaminated media, certain degradation products may remain as stable components in soils for extended periods of time. These degradation products include aliphatic acids and aromatics that are integral components of soil humus (Alexander, 1977).

Humic substances are brown to black colored organic acids which have high molecular weight and are structurally complex. Humic substances play a significant role in both the biotic and abiotic transformation of introduced hydrocarbons as well as being formed from the degradation of the hydrocarbons. Degradation of petroleum hydrocarbons to humic substances rather than to the mineralization products of CO_2 and H_2O could therefore be expected to beneficially impact soil physical and chemical properties.

Abiotic processes which are involved in the transformation of UST contaminants include:

- complexation
- photo-oxidation
- hydrolysis
- elimination
- dehydrogenation
- redox and
- polymerization

Complexation. Clays have been shown to be of significance in the stabilization of organic carbon by complex formation with active aluminum groups or sesquioxides. The silica-alumina humus complex may also increase the resistance of bound humic substances to degradation (Huang, 1987).

Photo-oxidation. Photo-oxidation is an important abiotic transformation process when petroleum is spilled on soil surfaces or in aquatic systems. The excited state brought about by sunlight that leads to photo-oxidation does not occur in the subsurface soil system and is, therefore, not considered a significant factor in abiotic transformations for USTs. However, photo-oxidation could potentially be incorporated in certain pump and treat schemes.

Hydrolysis. Hydrolysis is the reaction of a compound with water, hydronium or hydroxide ions resulting in bond cleavage. Equation 3.6 is an example of an hydrolysis reaction transforming an epoxide to a diol (Valentine, 1987):

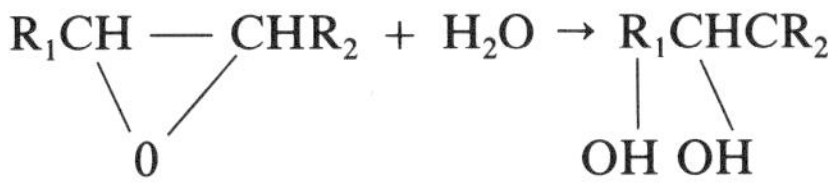

(3.6)

At a constant pH, hydrolysis appears to occur as a first order or "pseudo-first order" reaction (Amdurer et al., 1986). The rate of disappearance of the substrate (Rx) follows:

$$-d(Rx)/dt = k(Rx) \tag{3.7}$$

Hydrocarbons resistant to hydrolysis include saturated and unsaturated aliphatics, aromatics, phenols, alcohols, ketones and polycyclic aromatic hydrocarbons (Harris, 1982).

Elimination. Abiotic mechanisms have been shown to be operable in the degradation of chlorinated hydrocarbons. The solvent 1,1,1-trichloroethane (TCA) is transformed abiotically to yield 1,1-dichloroethylene (1,1-DCE) by elimination and acetic acid by hydrolysis (Cline et al., 1986). Vogel and McCarty (1987a and 1987b) confirmed the process of elimination of H^+ and Cl^- from TCA producing 1,1-DCE in neutral dilute aqueous solutions at 20°C.

Abiotic homogeneous elimination of HCl from 1,1,2,2-tetrachloroethane (TeCA) produced 1,1,2-trichloroethene (TCE) in a 0.1 M phosphate-buffered distilled water solution within a pH range of 5 to 9, and temperature range of 30 to 95°C (Cooper et al., 1987). The half life of TeCA at 25°C is reportedly on the order of 102 days at pH 7 and 1.02 days at pH 9.

Dehydrogenation. Dehydrogenation can be described as the transfer of H^+ ions and electrons from a reduced substrate to a terminal H^+ and electron acceptor, A, in the absence of O_2 (Tabatabi, 1982):

$$XH_2 + A \rightarrow X + AH_2 \tag{3.8}$$

Dehydrogenation is most often mediated in soils by microbial enzymes. Enzymes are proteins produced by living cells that can catalyze chemical reactions in soils. The enzyme causes reactions to proceed at accelerated rates but do not themselves undergo permanent alteration. Extra-cellular enzymes are important in the degradation of large molecules which can not enter the microbial cell (Alexander, 1967). Enzymatic coupling of aromatic compounds has been demonstrated (Maloney et al., 1986) and it is known that toluene can strongly inhibit dehydrogenase activity in soil at elevated levels (Tabatabi, 1982). Likewise, the oxidation inhibitors added to petroleum products to retard microbial growth during storage also limit microbial degradation in soils. Dehydrogenase activities of soils exposed to leaded gasoline, diesel fuel, kerosene or motor oil are reduced compared to those of soils exposed to crude oils which do not contain oxidation inhibitors (Frankenberger & Johanson, 1982). Additional discussion of enzyme mediated transformation processes is contained in Section 11.

Redox. Redox reactions may be the most significant of the abiotic processes that lead to transformation of contaminants in soils. Redox refers to the reduction/oxidation process whereby reduction of one compound is accompanied by the oxidation of another compound. Oxidation occurs when a reactive electron-deficient substrate (oxidant) removes electrons from more electron-rich substrate. The latter substrate is oxidized (loses electrons) while the oxidant is reduced (gains electrons). The redox half reactions for the oxidation of glucose by H_2O is written (Valentine, 1987):

$$C_6H_{12}O_6 + 6H_2O \rightarrow 6CO_2 + 24H^+ + 24e^- \tag{3.9}$$

$$6O_2 + 24H^+ + 24e^- \rightarrow 12H_2O \tag{3.10}$$

$$C_6H_{12}O_6 + 6O_2 \rightarrow 6CO_2 + 6H_2O \tag{3.11}$$

Organic oxidation frequently involves the gain of oxygen and loss of hydrogen. Table 3-3 presents a general schematic of oxidation processes.

The majority of naturally occuring redox reactions in soils are ex-

Table 3–3. Oxidation Processes

H-Atom transfer

$$RO_n + H-\overset{|}{\underset{|}{C}}- \rightarrow RO_nH + \cdot\overset{|}{\underset{|}{C}}-$$

Addition to double bonds

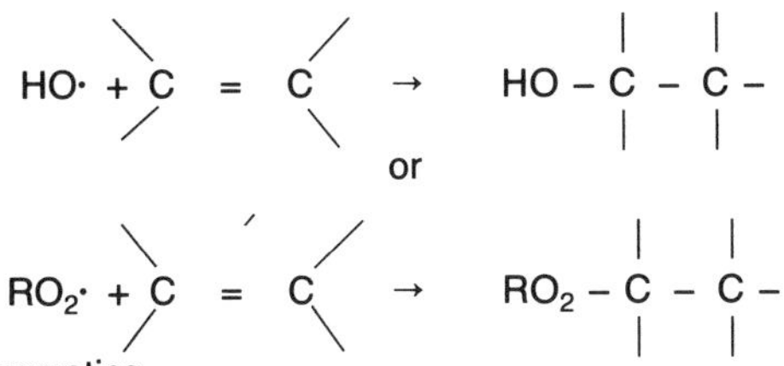

HO · addition to aromatics

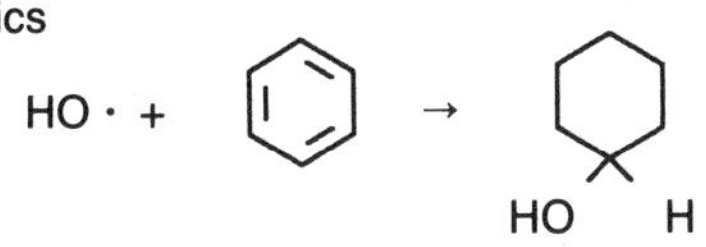

RO_2 · transfer of O-atoms to nucleophilic species

$$RO_2\cdot + NO \rightarrow RO\cdot + NO_2$$

R = alkyl or H

Source: Mill et al. (1980).

pected to be mediated by soil humus and metallic clay complexes which serve as catalysts for oxidation reactions. Oxidation of n-alkanes proceeds via terminal oxidation with the formation of a saturated primary alcohol at the expense of oxygen. Aromatic ring fission requires dehydroxylation followed by introduction of oxygen. Electron acceptors other than oxygen can mediate oxidation in soil environments. Denitrifying bacteria have been shown to be capable of degrading alkanes in the absence of O_2 if NO_3 is present to serve as an electron acceptor (Thomas and Ward, 1989).

Oxidants present in the environment include peroxy radicals (RO_2), hydroxyl radicals (OH), oxygen (O_2) and ozone (O_3) (Mill, 1980). Free radicals, metal ions or other catalysts combine with triplet oxygen to form peroxy radicals which remove hydrogen atoms from the substrate to form hydroperoxide (Larson et al., 1979).

Incorporation of oxygen into hydrocarbons increases water solubility; for instance, 2-naphthol solubility is approximately 1000 mg/L, whereas naphthalene's solubility is approximately 35 mg/L (Larson et al., 1979). Oxygenated hydrocarbon compounds were also found to be toxic or inhibitory to heterotrophic microorganisms by Larson et al. (1979).

Ozonation—Ozone is used extensively for the oxidation of organic constituents in wastewaters and drinking water supplies. Ozone acts to oxidize compounds directly by the formation of hydroxyl free radicals from decomposed ozone or by interaction of ozone with the solute to induce oxidation. Ozone occurs in the atmosphere but is not expected to be a major factor in in-situ degradation unless introduced to the soil.

Ozone attack on carbon-hydrogen bonds results in transformation of aldehydes to carboxyl acids, alcohols to carboxylic acids and/or aldehydes or ketones, ethers to alcohols and esters and hydrocarbons to alcohols and ketones (Bailey, 1972).

Ozone reacts readily with benzene, toluene, xylene, and benzopyrene, resulting in formation of products which are usually more susceptible to biodegradation and less toxic (Rice et al., 1981). Maloney et al. (1985) have determined that ozonation of dissolved organic carbon, humic substances, increases their biodegradability. However, oxidation of certain organics such as pesticides by ozone, for example, has been shown to lead to production of degradation products which are more toxic than the parent compounds (Rice et al., 1981).

Ozone decomposes very rapidly in water, yielding hydroxyl radicals that can react with aromatic compounds such as benzene (rate constant $k = 6.7 \times 10^{-5}$ l/mol.s) and cyclohexene ($k = 8.8 \times 10^{-5}$ l/mol.s) (Hoigne and Bader, 1976). However, only 0.5 mole of OH radicals per mole of ozone decomposed was shown to be effective in reacting with the organics.

The following conclusions are among those drawn by Rice et al. (1981) with respect to ozonation:

- Saturated aliphatic hydrocarbons are unreactive with ozone.
- Alcohols are slowly oxidized to aldehydes and ketones, which then slowly oxidize to acids.
- Benzene is slowly oxidized, and other aromatics are easily oxidized except when electron-withdrawing constituents are present.
- Reaction products are more polar and more readily biodegradable.
- Complete reaction to CO_2 and water rarely occurs.

Utilization of ozone for in-situ oxidation of petroleum contaminated soil may not be practical for several reasons: its reactive nature and rapid decomposition necessitates on-site production; relatively high cost; and the availability of other oxidants, such as hydrogen peroxide.

Hydrogen Peroxide—Hydrogen peroxide (H_2O_2) is fully miscible in water and commercially available in aqueous solutions over a wide range of concentrations. The oxidation of aliphatic and aromatic hydrocarbons by use of hydrogen peroxide is similar to that which occurs with ozone. However, peroxide can be more easily introduced into the subsurface environment by injection wells and is commonly used for accelerating in-situ biodegradation by increasing the available oxygen.

Polymerization. Polymerization is the joining of small molecules together to form large molecules. In soils, clay minerals are responsible for the polymerization of unsaturated organics. Benzene, toluene, phenol and quinone have been shown to polymerize spontaneously on smectite surfaces if Cu(II) or Fe(III) are present to serve as a catalyst (Bohn et al., 1979).

Humic substances are naturally occurring heterogeneous refractory organics of high molecular weight that can be categorized based on their solubility (Aiken et al., 1985):

- Humin—insoluble in water
- Humic acid—insoluble in water under acidic conditions ($pH < 2$) but becomes soluble at higher pH
- Fulvic acid—soluble in water

The characteristics of the humic substances within a soil are relatively stable as biotic inputs are balanced by mineralization. With time, humic substances are ultimately converted to kerogen, a coal-like material, which is insoluble in non-oxidizing acids, bases and most organic solvents.

The presence of naturally occurring soil humic substances will influence the rate and abiotic transformation mechanisms of hydrocarbons introduced to the environment from leaking underground storage tanks. The quantity of naturally occurring organic matter present in soils varies significantly, with soils that have developed in anaerobic environments, such as swamps, having significant accumulations of organics, while those that developed under aerobic conditions have lower organic matter contents in general. The amount of rainfall and the temperature under which the soil formed will also influence its organic matter content, with increasing organic matter generally being accumulated to a greater extent under conditions which favor biological activities. The major accumulations of organic matter are associated with the most biologically active zone of soils. This zone is generally found from the ground surface to the depth to which plant roots and soil organisms (earthworms) are active. Exceptions to these generalities are those cases where buried soil horizons or petrogenetic organic matter exists.

Organic matter inputs to soil include plant residues and microbial cells but can also include introduced hydrocarbon contaminants. Plant or-

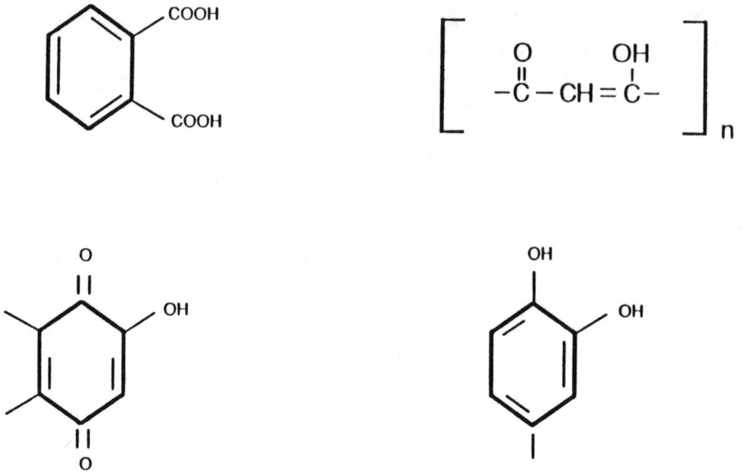

Figure 3–4. Humic substances component structures. *Source:* Stevenson, 1985.

ganic matter is comprised of cellulose, hemicellulose, lignin and various simple sugars, amino acids, aliphatic acids ethers, alcohols and proteins (Alexander, 1964). Humic substance are formed in soils as organic matter decomposes in soils. Although degradative processes lead to humic formation, polymerization also has a role in its formation.

Underground storage tanks are generally installed at a depth of 3 to 12 feet beneath the soil surface. At these depths the organic matter content of most mineral soils is less than 1 percent. The backfill and bedding materials used for UST installations is also expected to be low in organic matter content. The naturally occurring humic substances required for initiation of redox reactions with soil clay minerals may, therefore, not be present to any great extent within UST soil materials. However, the release of hydrocarbons from an UST can introduce the reducing agents required for initiation of abiotic polymerization type reactions.

Fulvic acids also play a role in the transport of insoluble organic substances, such as alkanes, in water by acting as carriers (Stevenson, 1985). Petroleum can become associated with humic materials forming non-ionic polar organic macromolecules.

Figure 3–4 presents some postulated structures for components of humic substances. Recent research using carbon magnetic resonance (CMR) and wet chemistry oxidative degradation studies have provided insight into humic structure. Although no generally accepted structure for humic substances is found in the literature, Ghosal and Chian (1985) have determined that the presence of 3 or more condensed aromatic ring systems in humic substances is almost absent, but heterocylic compounds derived from carbohydrates are present. Chen and Schnitzer (1976), in evaluating the viscosity of fulvic humic acids, determined that the materials behaved as linear polyelectrolytes, not like structures exclusively composed of condensed rings. The molecular weights of the materials they studied ranged from 1081 to 6171 with diameters of 25 to 107 A. The characteristics of the humic materials present in soil will affect the degree to which abiotic reactions take place.

Evaluation of organic matter from an Aquoll indicates that the humic acids extracted from clays had a higher content of aliphatics and lower content of aromatics than did the humic acids extracted from the larger sized soil separates, silt and sand (Catroux and Schnitzer, 1987).

Abiotic factors have been shown to play a vital role in the polymerization of phenolic compounds leading to the formation of humic substances (Wang et al., 1986). These findings are significant with respect to the discharge of hydrocarbons from an UST because mineralization of the materials may not be required to achieve remediation. Transformation of the hydrocarbons into stable humic substances would not only provide for elimination of the contaminant but impart the beneficial properties associated with organic matter to the soil.

Figure 3-5 presents the biodegradation pathway for hydrocarbons and the postulated abiotic polymerization reactions leading to formation of humic substances. Evidently, the interaction of biotic and abiotic pathways may play a major role in the transformation of hydrocarbons introduced into the soil systems.

The abiotic reaction leading to humic substance formation requires that a catalyst be present in the soil system. Catalysts that have been shown to be effective in oxidizing organic compounds include Cu (Voudrias and Reinhard, 1986), Mn, Fe and layer silicates (McBride, 1987). These catalysts are present in most soils as oxide coatings or mineral components. The presence of cations and anions on oxide surfaces can reduce the reaction rate by limiting release of Mn (II) to solution (Stone and Morgan, 1984a).

Primary minerals such as olivines, pyroxenes, and amphiboles have been shown to accelerate polymerization rates to a greater extent than micas and feldspars, whereas microclene and quartz were ineffective in catalyzing polymerization. McBride (1987) investigated the adsorption and oxidation of catechol and hydroquinone by Fe and Mn oxides. Oxidation of the organic compounds was explained by the formation of a Fe (III)-hydroquinone complex at the Fe-oxide surfaces for a sufficient period of time to allow for the election transfer. The oxidized molecule

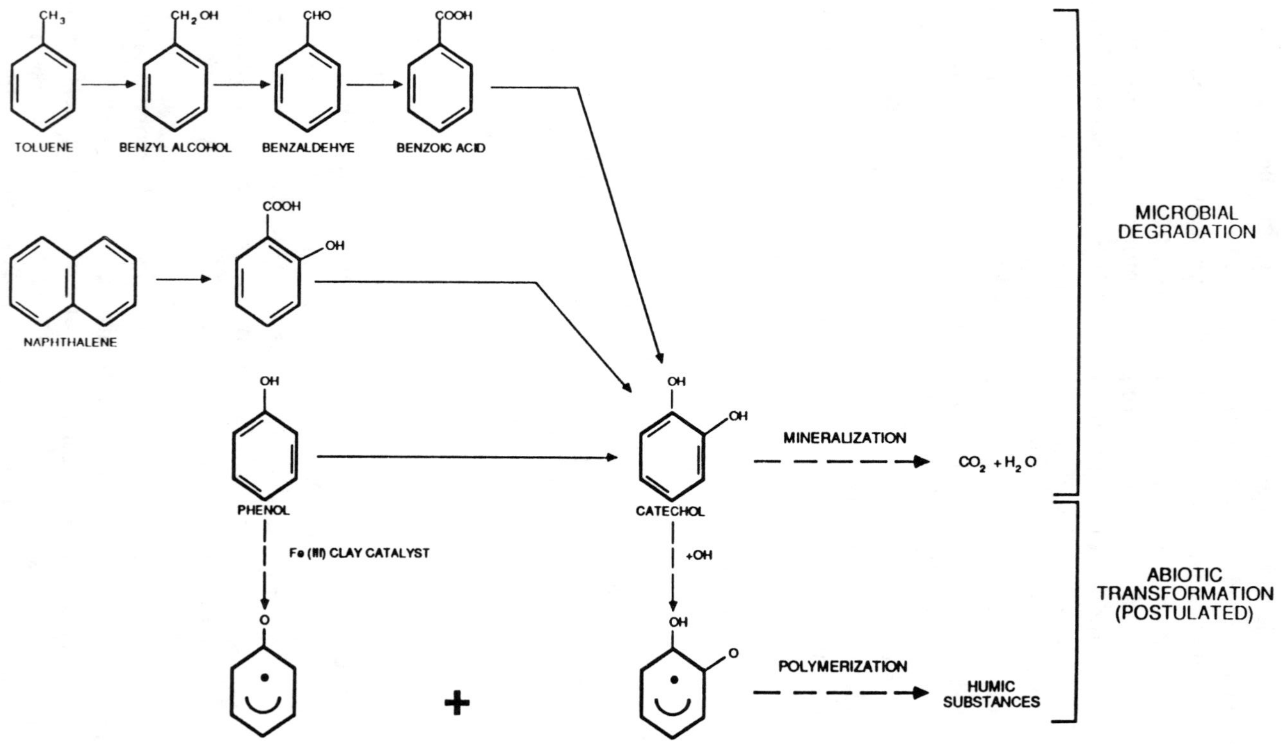

Figure 3–5. Abiotic/biotic humic substance formation pathways. *Source:* Adapted from Alexander, 1964 and Voudrias and Reinhard, 1986.

subsequently is released into solution by dissociation of the complex followed by rapid oxidation of Fe (II).

Reduction of Mn(III) and Mn(IV) by hydroquinone occurs on the oxide surface (Stone and Morgan, 1984a). Hydroquinone forms a surface complex with the oxide prior to electron transfer, and the rate of the surface chemical reaction is rate limiting, not the rate of surface complex formation.

In the oxidation of hydroquinone, Fe oxide serves as the catalyst resulting in the reduction of Fe(III). McBride (1987) postulates that the oxidized organic desorbs from the oxide or that electron transfer from the reduced Fe occurs prior to O_2 adsorption. Significant quantities of Mn(II) were released into solution upon phenol oxidation in McBride's (1987) study. He concludes that Mn oxide was the primary oxidizing agent since Mn(II) oxidation by dissolved O_2 is slow except at elevated pHs.

In a study on the dissolution of Mn oxides by aromatic and aliphatic substances, Stone and Morgan (1984b) found that saturated alcohols, aldehydes, ketones and carboxylic acid were non-reactive except for pyruvic and oxalic acid. Catechols, hydroquinones, methoxyphenols, resorcinols and ascorbate all reduced and dissolved Mn oxides. These molecules, except ascorbate, form core structure for humic substances and the researchers, therefore, conclude that humic substances should dissolve Mn oxides under natural conditions. Complexation of organics onto Fe and Mn oxides facilitates electron transfer required for abiotic transformation (Stone and Morgan, 1984a).

Longmire (1986) studied the impact of leakage of approximately 39,000 gallons of gasoline on the reduction of iron in soils. The strongly reducing conditions resulting at this UST site resulted in dissolution of iron, which was detected as uncomplexed and complexed Fe in the reduced groundwater. The dissolution of Fe in this case may be related to the depression of the soil Eh due to consumption of O_2 by microorganisms. However, dissolution may also be due to abiotic reactions with catechol or phenolics in the soil. In either case, the ferrous iron generated can serve to complex introduced organics to further the transformation processes.

Voudrias and Reinhard (1986) indicate that obtaining mechanistic information for reactions catalyzed by mineral surfaces is difficult because:

- active surface sites are difficult to characterize and quantify
- competition for reaction sites by substrates occurs
- several products may be formed from a single substrate

- recovery of products from mineral matrix is difficult
- of the presence of undefined active phase
- reactions may be mass transfer limited

Summary. Abiotic transformation of hydrocarbons in soils proceeds by a number of reaction types including elimination, dehydrogenation, redox and polymerization. These abiotic reactions are closely related to the microbial degradation pathways from which precursors to redox-polymerization reactions are derived. Determination of the rates of reactions in soils for a specific loci may be quantifiable provided estimates on the characteristics of the soil and contaminant can be defined. When mixtures of contaminants are involved, such as petroleum hydrocarbons, however, the utility of such an exercise seems questionable.

A systematic approach that allows a user to estimate the degree of abiotic transformations in soil may be a useful tool for those involved in remediation of petroleum-contaminated soil. Soil factors that play a role in abiotic transformation processes include:

- humic content and characterization
- particle size distribution
- moisture retention characteristics
- mineralogy including sesquoxide content
- redox potential, and
- pH

Values for these parameters are easily obtained for soils, and general ranges can be established for use in a guidance document, allowing an estimate of the potential (low, medium, high) for abiotic transformations.

Applicability to Locus 3. Moist soil conditions provide optimum conditions for migration of contaminants to soil sites where abiotic oxidation can be initiated.

3.3 STORAGE CAPACITY IN LOCUS

3.3.1 Introduction and Basic Equations

The mass of contaminants per unit volume of soil is given by:

$$m_v = C_w \Theta_w 10^3 \tag{3.12}$$

Table 3–4. Selected Values for Soil Bulk Density

Soil Type	Bulk Density (g/cm³)
1 Surface Soils Found in Agricultural Areas:	
A) Well-Decomposed Organic Soil	0.2–0.3
B) Cultivated Surface Mineral Soils	1.25–1.40
C) Clay, Clay Loam, Silt Loam	1.00–1.60
D) Sands and Sandy Loams	1.20–1.80
2 Materials Used in Road & Airfield Construction:	
A) Silts and Clays	1.3–2.0
B) Sand and Sandy Soils	1.6–2.2
C) Gravel and Gravelly Soils	1.8–2.3
3 Very Compact Subsoils	Up to 2.3 or 2.5

Source: Donigian et al. (1984).

where m_v = mass of contaminants per cubic meter of soil (mg/m^3)
C_w = concentration of contaminants in water film of locus no. 3 (mg/L)
Θ_w = water-filled porosity (dimensionless, $0<\Theta_w<1$)

A calculation of mass per unit weight of soil is given by:

$$m_g = C_w \, \Theta_w \, / \, \rho_b \tag{3.13}$$

where m_g = mass of contaminants per kilograms of dry soil (mg/kg)
ρ_b = dry soil bulk density (kg/L)

Selected values of Θ_w representative of soils at their field capacity and wilting point are given in Figure 12–4 (in Section 12). Selected values of soil bulk density are given in Table 3–4.

3.3.2 Guidance on Inputs for, and Calculations of, Maximum Values

3.3.2.1 Concentration of Contaminants in Water (C_w)

Use the solubility for the liquid contaminant in water. For the hydrocarbon components of a typical gasoline (i.e., excluding consideration of additives), a reasonable value for C_w at 20°C is 200 mg/L. For gasoline with oxygenated additives (e.g., ethanol or MTBE), a value of 2000 mg/L is recommended.

3.3.2.2 Soil Moisture Content: Water-Filled Porosity (Θ_w)

Use the data shown in Figure 12–4. Specifically, the line representing the lower limit of the field capacity for different soils probably represents a maximum for the amount of immobile pore water in locus no. 3. This line shows Θ_w ranging from about 0.10 for sandy soils to about 0.30 for clayey soils.

3.3.2.3 Soil Bulk Density (ρ_s)

A recommended value for ρ_b is 2.0 g/cm^3. This should be used for compacted subsoils unless site-specific information indicates a higher or lower value should be used. See Table 3–4 for selected values of ρ_b.

3.3.3 Guidance on Inputs for, and Calculations of, Average Value

3.3.3.1 Concentration of Contaminants in Water (C_w)

Use a value that is about 10 percent of the solubility limit. This is an arbitrary value but the dilution by a factor of 10 from the maximum value represents a reasonable amount of dilution with clean pore water. For a gasoline with no additives, this value would be about 20 mg/L; with an oxygenated additive it would be about 200 mg/L.

3.3.3.2 Soil Moisture Content: Water-Filled Porosity (Θ_w)

It is recommended that the line in Figure 12–4 that represents the wilting point be used. This line shows Θ_w ranging from about 0.05 for sandy soils to 0.25 for clayey soils.

3.3.3.3 Soil Bulk, Density (ρ_s)

Use same value as for "maximum" case (i.e., 2.0 g/cm^3, unless special conditions apply).

3.4 EXAMPLE CALCULATIONS

3.4.1 Storage Capacity Calculations

3.4.1.1 Maximum Value

Estimate the maximum storage capacity for an additive-free gasoline in a loam soil. We use equation 3.12:

$$m_v = C_w \Theta_w 10^3$$

with C_w = 200 mg/L (recommended in Section 3.3.2)
Θ_w = 0.20 (lower limit of field capacity according to Figure 12–4)

Thus, $m_v = (200)(0.20)(10^3) = 40{,}000 \text{ mg/m}^3$
$= 40 \text{ g/m}^3$
= 20 mg/kg (with equation 3.13 and soil bulk density of 2 kg/L)

3.4.1.2 Average Value

Estimate the average (typical) storage capacity for an additive-free gasoline in a sandy soil. We use equation 3.12:

$$m_v = C_w \Theta_w 10^3$$

with C_w = 20 mg/L (recommended in Section 3.3.3)
Θ_w = 0.05 (wilting point as shown in Figure 12–4)

Thus, $m_v = (20)(0.05)(10^3) = 1{,}000 \text{ mg/m}^3$
$= 1 \text{ g/m}^3$
= 0.5 mg/kg (with equation 3.13 and soil bulk density of 2 kg/L)

3.4.2 Transport Rate Calculations

3.4.2.1 Diffusion Flux

Estimate the diffusive flux of benzene through a 1-cm long (straight) tube of pore water given a water temperature of 2°C, and concentrations of benzene of 0 and 5 mg/L at either end of the tube.

Since a tortuosity factor is not involved with a straight tube, equation 3.1 is used:

$$J = D \, \Delta C$$

with $D = 0.58 \times 10^{-5} \text{ cm}^2\text{/s}$ @ 2°C (Table 3–2)
$\Delta C = 5 \text{ mg/L} = 6.4 \times 10^{-5} \text{ mol/L} = 6.4 \times 10^{-8} \text{ mol/cm}^3$

Thus, $J = (0.58 \times 10^{-5})(6.4 \times 10^{-8})$
$= 3.7 \times 10^{-13} \text{ mol/cm}^2\text{s} = 2.9 \times 10^{-11} \text{ g/cm}^2 \text{ s}$
$= 2.5 \times 10^{-4} \text{ mg/cm}^2 \text{ day}$

Note that the total surface area of all the pores in the soil exposed to flowing soil gas would have to be known to calculate a flux in units of mass per unit time.

3.4.2.2 Air/Water Concentrations

Estimate the vapor phase concentration of toluene for a soil gas that is in equilibrium with pore water containing 100 mg/L of toluene (mol. wt. = 92).

Equation 3.3 is used:

$$C_a = H\,C_w$$

with $H = 8 \times 10^{-3}$ atm m^3/mol @ 25°C (Figure 3–3)
$C_w = 100 \text{ mg/L} = 1.09 \times 10^{-3} \text{ mol/L} = 1.09 \text{ mol/m}^3$

Thus, $C_a = (8 \times 10^{-3})(1.09) = 8.7 \times 10^{-3} \text{ atm} = 6.6 \text{ mm Hg}$

3.5 SUMMARY OF RELATIVE IMPORTANCE OF LOCUS

3.5.1 Remediation

Contaminants held in solution in the immobile pore water of the unsaturated zone usually comprise a very small fraction of the total mass of leaked material. In spite of this low storage capacity, the locus may play an important role, first in the long term retention of some contaminants in the unsaturated zone, and second, as an impedance to remediation technologies such as vacuum extraction.

The long term retention is associated with the fact that the water is primarily held in the narrow pores away from the larger pores that may see the cleansing action of infiltrating water or flowing soil gas. Hydrocarbon contaminants of relatively low volatility and high solubility (specifically the BTEX fraction) will have enhanced concentrations and long residence times in this locus.

The phenomenon of impedance to vacuum extraction is related to the phenomenon of long term residence. In this case, in addition to having to yield up its initial contaminant load (acquired prior to vacuum extraction), the waters in this locus will act as an absorbent to all contaminant vapors, especially those of high water solubility. This ability to trap vapors mobilized by the vacuum extraction system will slow down the remediation process. As noted above, diffusion of dissolved contaminants in the water of this locus probably controls the rate of mass recovery in cases where there is no liquid contaminant (free product) present. Only when the soil is taken to dryness—not an impossible feat—will this effect be eliminated.

3.5.2 Loci Interactions

The major loci interactions are intermixing with (or advecting to) mobile pore water (locus no. 12), volatilization into soil gas (locus no. 1), and sorption to soils (locus no. 4). In all cases, diffusion of contaminants through the aqueous phase may be a rate-limiting step. Interactions with soil bacteria may be important in some cases; however, the low water content associated with this locus may be detrimental to bacterial activity and population growth.

3.5.3 Information Gaps

Perhaps the most significant data gap for this locus is in the understanding of the mobility of the aqueous phase, both in an absolute sense (e.g., micropore water velocities) and a relative sense (i.e., relative to locus no. 12). Without a knowledge of just how immobile the locus no. 3 waters are, it is impossible to predict quantitatively the rate of contaminant transport between them and other loci.

The importance of biodegradation in this locus is also unclear due—as mentioned above—to the possible hindrance of low water content on bacterial activity and growth.

3.6 LITERATURE CITED

Aiken, G.R., D.M. McKnight, R.L. Wershaw and P. MacCarthy (Eds.). 1985. An Introduction to Humic Substances in Soil, Sediment and Water. *In*: Humic Substances in Soil Sediment and Water. John Wiley and Sons, New York.

Alexander, M. 1961. Introduction to Soil/Microbiology. John Wiley and Sons, New York.

Amdurer, M., R.T. Fillman, J. Roetzer, and C. Russ. 1986. Systems to Accelerate In-Situ Stabilization of Waste Deposits. EPA/540/2-86-002.

Bailey, P.S. 1972. Organic Groupings Reactive Toward Ozone: Mechanisms in Aqueous Media. *In*: Ozone in Water and Wastewater Treatment. F.L. Evans (Ed). Ann Arbor Scientific Publishers, Ann Arbor, MI.

Bohn, H.L., B.L. McNeal and G.A. O'Connor. 1979. Soil Chemistry. Wiley-Interscience, New York.

Catroux, G. and M. Schnitzer. 1987. Chemical, Spectroscopic and Biological Characteristics of the Organic Matter in Particle Size Fractions Separated from an Aquisol. Soil Sci. Soc. America J., 51(5):1200–1206.

Chen, Y. and M. Schnitzer. 1976. Viscosity Measurements on Soil Humic Substances. Soil Sci. Soc. America J., 40: 866–872.

Cline, P.V., J.J. Delfino and W.J. Cooper. 1986. Hydrolysis of 1,1,1-Trichloroethane; Formation of 1,1-Dichloroethene. *In*: Proceedings of the NWWA/API Conference on Petroleum Hydrocarbons and Organic Chemicals in Groundwater. November 12–14. Houston, TX.

Cooper, W.J., M. Mehran, O.J. Riusech, and J.A. Joens. 1987. Abiotic Transformations of Halogenated Organics. 1. Elimination Reaction of 1,1,2,2-Tetrachloroethane and Formation of 1,1,2-Trichloroethene. Enviro. Sci. Technol. 21(11):1112–1114.

Donigian, A.S., Jr., T.Y.R. Lo and E.W. Shanahan. 1984. Groundwater Contamination and Emergency Response Guide, Part III, Noyes Publications, Park Ridge, NJ.

Frankenberger, W.T. Jr. and J.B Johanson. 1982. Influence of Crude Oil and Refined Petroleum Products on Soil Dehydrogenase Activity. J. Environ. Qual., 11(4):602–607.

Ghosal, M. and E.S.K. Chian. 1985. An Evaluation of Aromatic Fraction in Humic Substances. J. Soil Sci. Soc. Am., 49:616–618.

Gossett, J.M. and A.H. Lincoff. 1981. Solute-Gas Equilibria in Multi-Organic Systems. Report No. AFOSR-TR-81-0858 prepared for the Air Force Office of Scientific Research by Cornell University, Ithaca, NY. (Available from NTIS as AD A109 082.)

Harris, J.C. 1982. Rate of Hydrolysis. *In*: Handbook of Chemical Property Estimation Methods, W. Lyman, W. Reehl and D. Rosenblatt (Eds.) McGraw Hill, New York.

Hayduk, W. and H. Laudie. 1974. Prediction of Diffusion Coefficients for Nonelectrolytes in Dilute Aqueous Solutions. AICLE Journal, 20(3):611–15.

Hoigne, J. and H. Bader. 1976. The Role of Hydroxyl Radical Reactions in Ozonation Processes in Aqueous Solutions. Water Resources, 10:377–386.

Huang, P.M. 1987. Impact of Mineralogical Research on Soil and Environmental Sciences. *In*: Future Developments in Soil Science Research. SSSA, Madison, WI.

Larson, R.A., T.L. Batt, L.L. Hunt and K. Rogenmuser. 1979. Photo Oxidation Products of Fuel Oil and Their Antimicrobial Activity. Env. Sci. Technol., 13(8):965–969.

Leighton, D.T., Jr. and J.M. Calo. 1981. Distribution Coefficients of Chlorinated Hydrocarbons in Dilute Air-Water Systems for Groundwater Contamination Applications. J. Chem. Eng. Data, 26:382–85.

Lion, L.W. and D. Garbarini. 1983. Partitioning Equilibria of Volatile Pollutants in Three-Phase Systems. Report No. ESL-TR-83-51, Engineering and Services Laboratory, Air Force Engineering and Services Center, Tyndall AFB, Florida. (Available from NTIS as AD A137 207.)

Longmire, P. 1986. Iron Dissolution Resulting from Petroleum—Product Contamination in Soil and Groundwater. 1. Thermodynamic Considerations. Proceedings of NWWA/API Conference on Petroleum Hydrocarbons and Organic Chemicals in Groundwater. November 12–14. Houston, TX.

Mackay, D. and W.Y. Shiu. 1981. A Critical Review of Henry's Law Constants for Chemicals of Environmental Interest. J. Phys. Chem. Ref. Data, 10(4):1175–99.

Mackay, D., W.Y. Shiu, and R.P. Sutherland. 1979. Determination of Air-Water Henry's Law Constants for Hydrophobic Pollutants. Env. Sci. Technol., 13(3):333–337.

Maloney, S.W., J. Marem, J. Mallerialle, and F. Fiessinger. 1986. Transformation of Trace Organic Compounds in Drinking Water by Enzymatic Oxidative Coupling. Env. Sci. Technol., 20(3):249–253.

Maloney, S.W., I.H. Suffet, K. Bancroft, and H.M. Neukrug. 1985. Ozone-GAC Following Conventional U.S. Drinking Water Treatment. JAWWA, 8:66–73.

McBride, M.B. 1987. Adsorption and Oxidation of Phenolic compounds by Iron and Manganese Oxides. Soil Sci. Soc. Am. J. Vol. 51.

Mill, T., D.G. Hendry and H. Richardson. 1980. Free-Radical Oxidants in Natural Waters. Science, 207:886–887.

Rice, R.G., C.M. Robson, G.W. Miller and A.G. Hill. 1981. Uses of Ozone in Drinking Water Treatment. J. Amer. Water Works Assn., 73(1): 44–57.

Stevenson, F.J. 1985. Geochemistry of Soil Humic Substances. In: Humic Substances in Soil Sediment and Water. John Wiley and Sons, New York.

Stone, A.T. and J.J. Morgan. 1984a. Reduction and Dissolution of Manganese (II) and Manganese (IV) Oxides by Organics, 1: Reaction with Hydroquinone. Environ. Sci. Technol. 18:450–456.

Stone, A.T. and J.J. Morgan. 1984b. Reduction and Dissolution of Manganese (II) and Manganese (IV) Oxides by Organics. 2: Survey of the Reactivity of Organics. Environ. Sci. Technol. 18:617–624.

Tabatabi, M.A. 1982. Soil Enzymes. *In*: Methods of Soil Analysis. Agronomy Monograph No. 9, Part 2, 2nd edition, A.L. Page, R.H. Miller and D.R. Keeny (eds.), American Society of Agronomy, Madison, WI.

Thomas, R.G. 1982. Volatilization from Water. *In*: Handbook of Chemical Property Estimation Methods, W. Lyman, W. Reehl and D. Rosenblatt (eds.) McGraw-Hill, New York.

Thomas, J.M. and C.H. Ward. 1989. In Situ Biorestoration of Organic Contaminants in the Subsurface. Environ. Sci. Technol., 27(7): 760–766.

Tucker, W.A. and L.H. Nelken. 1982. Diffusion Coefficients in Air and Water. *In*: Handbook of Chemical Property Estimation Methods, W. Lyman, W. Reehl and D. Rosenblatt (eds.), McGraw-Hill, New York.

Valentine, R.L. 1987. Nonbiological Transformation. *In*: Vadose Zone Modeling of Organic Pollutants. Lewis Publishers Inc., Chelsea, MI.

Vogel, T.M. and P.L. McCarty. 1987a. Abiotic and Biotic Transformations of 1,1,1-Trichloroethane under Methanogenic Conditions. Env. Sci. Technol., 21(12):1208–1213.

Vogel, T.M. and P.L. McCarty. 1987b. Rate of Abiotic Formation of 1,1-Dichloroethylene from 1,1,1-Trichloroethane in Groundwater. J. Contam. Hydro., 1:299–308.

Voudrias, E.A. and M.R. Reinhard. 1985. Abiotic Organic Reactions at Mineral Surfaces. *In*: Geochemical Process at Mineral Surfaces. Davis and Hayes (Eds.), ACS Symposium Series No. 323.

Wang, T.S.C., P.M. Huany, C.H. Chou, and J.H. Chem. 1986. The Role of Soil Minerals in the Abiotic Polymerization of Phenolic Compounds and Formation of Humic Substances. *In*: Interactions of Soil Minerals with Natural Organics and Microbes. Huang and Schutzer (eds.), SSSA Spec. Pub. No. 17, Madison, WI.

Witherspoon, P.A. and L. Bonoli. 1969. Correlation of Diffusion Coefficients for Paraffin, Aromatic, and Cydoparaffin Hydrocarbons in Water. Ind. Eng. Chem. Fundam., 8(3):589–91.

SECTION 4

Contaminants Sorbed to "Water-Wet" Soil Particles or Rock Surface (After Migrating Through the Water) in Either the Unsaturated or Saturated Zone

CONTENTS

List of Tables 108
List of Figures 108
4.1 Locus Description 109
4.1.1 Short Definition 109
4.1.2 Expanded Definition and Comments 109
4.2 Evaluation of Criteria for Remediation 110
4.2.1 Introduction 110
4.2.2 Mobilization/Remobilization 111
4.2.2.1 Partitioning into Mobile Phase 111
4.2.2.2 Transport with Mobile Phase 121
4.2.3 Fixation 121
4.2.3.1 Partitioning onto Immobile (Stationary) Phase 121
4.2.3.2 Other Fixation Approaches 121
4.2.4 Transformation 121
4.2.4.1 Biodegradation 121
4.2.4.2 Chemical Oxidation 121
4.3 Storage Capacity in Locus 122
4.3.1 Introduction and Basic Equations 122
4.3.2 Guidance on Inputs for, and Calculations of, Maximum Value 122
4.3.3 Guidance on Inputs for, and Calculations of, Average Values 122
Comments 123
4.4 Example Calculations 123
4.4.1 Storage Capacity Calculations 123
4.4.1.1 Maximum Value Calculations 123
4.4.1.2 Average Value Calculations 124

4.4.2 Transport Rate Calculations 124
4.4.2.1 Time Required to Desorb 50, 90, and 95 Percent of Contaminant 124
4.4.2.2 Retardation of Bulk Transport 125
4.5 Summary of Relative Importance of Locus 126
4.5.1 Remediation 126
4.5.2 Loci Interaction 126
4.5.3 Information Gaps 127
4.6 Literature Cited 127

TABLES

4–1 Ranges of Values of Porosity 115
4–2 Key Parameters Affecting Sorptive Behavior 116
4–3 Estimated Soil Sorption Constants for Selected Gasoline Constituents 118
4–4 Selected Soil Characteristics and Sorption Coefficient Data 119
4–5 Loci Interactions With Hydrocarbons Sorbed onto Soil Surfaces 127

FIGURES

4–1 Schematic Cross-Sectional Diagram of Locus No. 4–Contaminants Sorbed to "Water-wet" Soil Particles or Rock Surfaces (After Migrating Through Water) in Either the Unsaturated or Saturated Zone 110
4–2 Schematic Representation of Important Transformation and Transport Processes Affecting Other Loci 111

SECTION 4

Contaminants Sorbed to "Water-Wet" Soil Particles or Rock Surface (After Migrating Through the Water) in Either the Unsaturated or Saturated Zone

4.1 LOCUS DESCRIPTION

4.1.1 Short Definition

Contaminants sorbed to "water-wet" soil particles or rock surface (after migrating through the water) in either the unsaturated or saturated zone.

4.1.2 Expanded Definition and Comments

In this locus definition, contaminants must first be dissolved in water and then sorbed to a soil or rock surface. The concentration of sorbed contaminants is, therefore, limited by the initial dissolution process. Although a partitioning process involving fairly rapid sorption and desorption is presumed to exist, the sorbed contaminants are considered immobile. The sorbed contaminants are either absorbed into a thin layer of naturally-occurring organic matter surrounding the soil particles, or adsorbed in a thin layer to exposed mineral (e.g., clay) surfaces. They are not able to form a separate liquid phase on the particle's surface. Sorbed contaminants that subsequently diffuse into the solid rock or mineral particles are covered by locus no. 10. Figures 4–1 and 4–2 present a schematic cross-sectional diagram of locus no. 4 and a schematic representation of the transformation and transport processes affecting other loci, respectively.

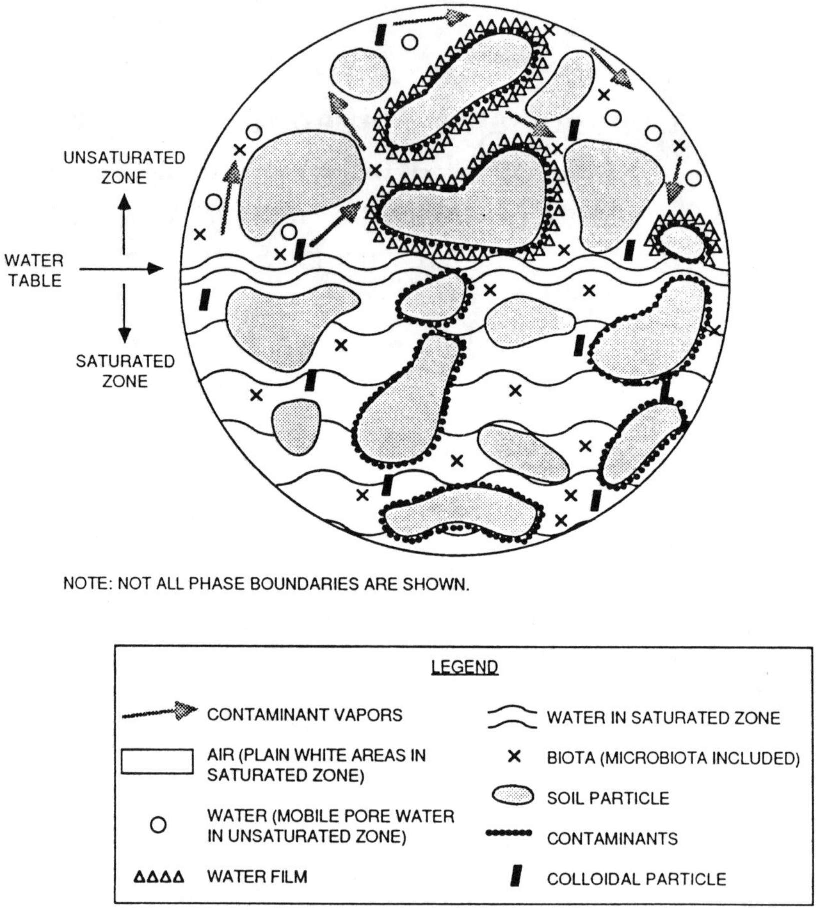

Figure 4–1. Schematic cross-sectional diagram of locus no. 4–Contaminants sorbed to "water-wet" soil particles or rock surfaces (after migrating through water) in either the unsaturated or saturated zone.

4.2 EVALUATION OF CRITERIA FOR REMEDIATION

4.2.1 Introduction

Sorption of contaminants onto soil surfaces is generally treated by assuming equilibrium conditions between the sorbed and the solution phase. At equilibrium, the relationship between the sorbed and solution phase concentration can be described by the Freundlich isotherm (described below).

There is considerable ongoing debate among researchers as to the

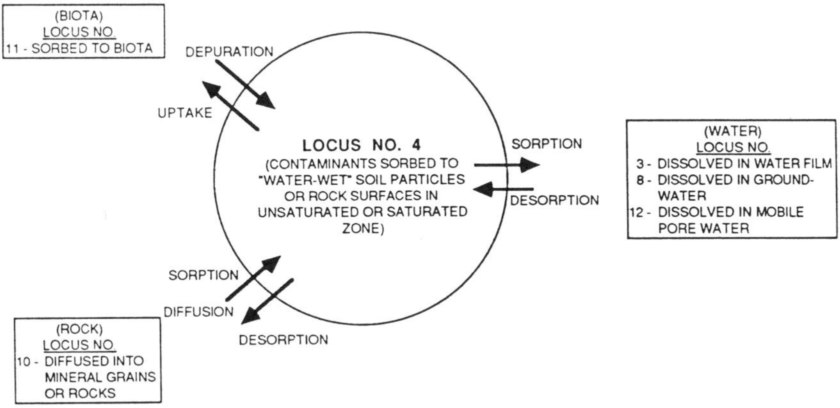

Figure 4–2. Schematic representation of important transformation and transport processes affecting other loci.

exact mechanisms involved in governing sorptive behaviors. Thermodynamically, equilibrium sorption is defined as the state in which the fugacities of contaminants in the sorbed and solution phases are equal. The general approach in defining sorptive behavior has been to identify key parameters that affect sorption and desorption. An understanding of these key parameters then facilitates the assessment of remedial actions.

Mobilization of sorbed contaminants can be accomplished by desorption into the aqueous phase in both saturated and unsaturated zones. Immobilization can be accomplished by enhancing sorption. Biodegradation and other transformation processes can, for many compounds, act as a removal mechanism. However, the rates for such processes are somewhat slow and biodegradation is not often attempted as the sole corrective action.

4.2.2 Mobilization/Remobilization

4.2.2.1 Partitioning into Mobile Phase

The only mobile phase in contact with this locus is water (loci nos. 3, 8, or 12). The contaminants initially dissolved in water are considered to have sorbed into a thin layer of naturally occurring organic matter surrounding the soil particles or adsorbed in a very thin layer to exposed clay surfaces. A partitioning process involving rapid sorption and desorption is presumed to be controlling the movement of contaminants into and out of the immobile surface region of the soil particles.

Desorption of sorbed contaminants from the soil surfaces into the mobile aqueous phase may be expressed by the Freundlich equation, which states:

$$C_s = K_d \cdot (C_w)^N \tag{4.1}$$

where C_s = concentration of contaminant in soil (mg/kg)
C_w = concentration of contaminant in water (mg/L)
K_d = sorption coefficient (L/kg)
N = measure of deviation from linearity

Rao and Davidson (1980) compiled a list of mean values of N for several pesticides and pesticide related compounds. The mean values were derived from sorption experiments conducted with a number of soils. The reported mean values range from 0.70 to 1.18 and the overall mean value is 0.87. N values from experiments conducted with gasoline constituents are not readily available in the literature.

The sorption coefficient K_d is determined by plotting the sorbed phase concentration versus the solution phase concentration. For dilute solutions such as those normally encountered in natural environments, equation 4.1 may be written as the linear equation (i.e., with $N = 1$) shown below:

$$C_s = K \cdot C_w \tag{4.2}$$

where K = sorption coefficient for linear isotherm (L/kg)

K_d or K values for the same compound can range over several orders of magnitude depending on soil characteristics. The estimation technique for K is discussed later in this section.

Karickhoff et al. (1979) observe that for hydrophobic organics (such as most gasoline constituents), if the concentration of contaminants in the aqueous phase is below 10^{-5} mole/liter, or is less than one-half of the water solubility (whichever is lower), then the sorption isotherms are linear.

Rao and Davidson (1980) indicate that for a solution phase concentration of 10 mg/L and a N value of 0.7, the amount adsorbed could be over-predicted by a factor of two by assuming linearity. For large solution concentrations, the amount adsorbed could be overestimated by an order of magnitude or more. Concentration of contaminants in water and soil is expected to be high in the region impacted by underground releases. For such systems, equation 4.1 is more appropriate for defining sorption-desorption.

Kinetically, sorption has been viewed as a two-step process involving rapid initial sorption followed by a slower sorption process. Karickhoff (1980) attributed the first step to fast adsorption onto the external soil surfaces and the second step to slow diffusion of the solute into the interstices of the soil particle (e.g., locus 10). It is apparent that the contact time between the solid and the solution phase should be long enough for reaching equilibrium conditions.

Several models exist that describe solute transfer between the immobile and mobile soil-water regions in terms of the two stage process. While the first stage is usually treated as instantaneous, the second stage is either a kinetically-controlled or a diffusion-controlled process.

Wu and Gschwend (1986) describe sorption kinetics in terms of intraparticle diffusion. They describe an effective intraparticle diffusivity as:

$$D_{eff} = \frac{D_m \Theta^2}{(1 - \Theta) \cdot K \cdot \rho_b} \tag{4.3}$$

where D_{eff} = effective intraparticle diffusivity (cm^2/s)
D_m = pore fluid diffusivity of the sorbate (cm^2/s)
Θ = porosity (dimensionless)
K = sorption coefficient (cm^3/g)
ρ_b = bulk density of the dry sorbent (g/cm^3)

D_{eff} can be calculated from this equation if values for other parameters are available. D_m for neutral organics is typically 10^{-5} cm^2/s at 25°C. A value of 2.5 g/cm^3 may be used as a rough estimate for ρ_s (Arthur D. Little, 1987). Wu and Gschwend (1986) chose a value of 0.13 for the porosity as typical for silts. K may be estimated using equation 4.9.

Wu and Gschwend (1986) showed that sorption rates were lower for larger soil aggregates and more hydrophobic compounds. The results also indicated that desorption rates can be described in terms of reversible exchange mechanisms. They further stated that if the particle size distribution is sufficiently narrow to allow estimation of an average particle size, then the analytical solution to equation 4.3 may be used to estimate the time scale of sorption and desorption. An estimate of time for 50 percent desorption or sorption can be estimated from (Arthur D. Little, 1987),

$$t_{0.5} = 0.031\, r^2/D_{eff} \tag{4.4}$$

$t_{0.5}$ = time for 50 percent of the chemical to desorb or time for 50 percent of the ultimate amount to sorb (sec)
r = average particle radius (cm)
D_{eff} = intraparticle diffusivity (cm^2/s)

Wu and Gschwend (1986) gave a value of 2×10^{-3} cm for the particle radius of river sediments. However, the range of particle sizes for natural soils and sediments can span an order of magnitude or more. For natural soils, a weighted average radius should be used. The weighted average radius can be calculated by using the equation (Arthur D. Little, 1987),

$$r = (\Sigma r_i^2 \cdot m_i/m_t)^{1/2} \tag{4.5}$$

where r_i = mean radius of the i-th size fraction (cm)
m_i = mass of particles in the i-th fraction (gm)
m_t = total mass of particles (g)

Values of r_i, m_i, and m_t can be obtained by conducting a standard particle size analysis.

Equations 4.3 through 4.5 thus can provide valuable information on the time required for immobilizing contaminants sorbed onto soil surfaces, provided equilibrium conditions exist. Arthur D. Little (1987) further presented equations for estimating time required to desorb 90 ($t_{0.9}$) and 95 ($t_{0.95}$) percent of the sorbed contaminant as follows,

$$t_{0.9} \text{ (sec)} = 0.18\ r^2/D_{eff} \tag{4.6}$$

$$t_{0.95} \text{ (sec)} = 0.25\ r^2/D_{eff} \tag{4.7}$$

Bulk transport of hydrocarbons dissolved in groundwater is discussed in detail in Section 8. A term which is relevant for discussions under this loci is the retardation factor, R. Retardation of a contaminant's movement relative to the bulk mass of groundwater can be stated as follows (Huyakorn and Pinder, 1983):

$$R = v_w/v_c = 1 + K\rho_b/\Theta \tag{4.8}$$

where v_w = average velocity of water (cm/sec)
v_c = average velocity of the contaminant (cm/sec)
K = sorption coefficient (cm^3/g)
ρ_b = soil bulk density (g/cm^3)
Θ = soil porosity (dimensionless)

Table 4–1. Range of Values of Porosity

Unconsolidated Deposits	Porosity, %
Gravel	25–40
Sand	25–50
Silt	35–50
Clay	40–70
Rocks	
Fractured basalt	5–50
Karstlimestone	5–50
Sandstone	5–30
Limestone, dolomite	0–20
Shale	0–10
Fractured crystalline rock	0–10
Dense crystalline rock	0–5

Source: Freeze and Cherry (1979).

The range of porosity values for natural soils is listed in Table 4-1.

In general, sorption and desorption are often assumed as having the same partition coefficient. Several researchers have stated that such assumptions are not valid. Irreversibility and hysteresis have been observed indicating that the adsorption and desorption mechanisms might be different. Jaffe and Ferrara (1983) stated that for hydrophobic substances with high carbon content, if the product of sorption coefficient and soil density is larger than one, then desorption rate is slower than the rate of sorption. Several other researchers, however, have attributed this phenomenon to experimental conditions.

The key parameters that influence sorptive behavior are listed in Table 4-2. Many studies have established relationships between a sorption property, K_{oc} (sorption coefficient normalized for organic carbon) with a solution property (solubility or octanol/water partition coefficient, K_{ow}). For a complete list of such relationships, the reader is referred to A.D. Little, Inc. (1987).

The K_{oc} values so obtained can then be used to estimate the sorption coefficient by using the relationship

$$K = K_{oc} \times f_{oc} \tag{4.9}$$

Where f_{oc} is the weight fraction of organic carbon in the soil. Soil carbon content can span over two orders of magnitude. For example, in surface soils of Oklahoma and Pennsylvania, organic carbon ranged from 0.5 to over 6 percent (Banerjee et al., 1985a; USDA, 1960). The range is probably much higher spanning greater areas. In the saturated zone, most soils are expected to have a carbon content less than 0.1

Table 4–2. Key Parameters Affecting Sorptive Behavior

Soil Properties	Solute Properties	Solution Properties
Organic matter content	Water solubility	Dissolved organic carbon (both true solutions and colloids)
Clay content	Octanol/water partition coefficient	Temperature
Cation exchange capacity	Polarity	Salinity
Surface area	Chemical class	pH
		Presence of cosolvents
		Presence of other solutes

percent (Newsom, 1985). Banerjee et al. (1985a) and Banerjee et al. (1985b) measured organic carbon contents of two subsurface core samples collected from surface down to same depth in the subsurface. In one set of 20 contiguous sections of soil, organic carbon content ranged from 0.08 to 0.89 percent. In another set of six samples collected from surface down to the saturated zone at 90 cm intervals, the organic carbon content of the surface soil was reported at 1.33 percent and the other five samples, including two from the saturated zone, had carbon content ranging from 0.03 to 0.05 percent. Although organic carbon is generally associated with the finer sized particles, the organic carbon content can not be estimated from other soil properties.

Organic carbon content can be measured in the laboratory, and it is strongly recommended that the fraction organic carbon content to be used in equation 4.9 be a measured value.

For hydrocarbons such as gasoline dissolved in water, the contaminants should be viewed as the individual constituents of gasoline. For a few constituents, measured K_{oc} values are available in the literature. One convenient source of estimated values is the Superfund Public Health Evaluation Manual (U.S. EPA, 1986).

K_{oc} values can also be estimated from equations presented in Arthur D. Little, Inc. (1987) as follows:

$$\log K_{oc} = 0.779 \log K_{ow} + 0.46 \quad (4.10)$$

$$\log K_{oc} = -0.602 \log S + 0.656 \quad (4.11)$$

where S = water solubility (moles/L)
K_{ow} = octanol/water partition coefficient (dimensionless)

Experimentally measured K_{oc} values for gasoline constituents are not readily available in the literature.

It is generally believed that for hydrophobic compounds, the relationships based on K_{ow} are superior to those based on water solubility. However, for gasoline constituents with low K_{ow} values, solubility-based relationships are probably superior to those based on K_{ow} (W. Lyman, pers. comm.) The sources for K_{oc} values should be the Superfund Public Health Evaluation Manual, equation 4.10, or equation 4.11.

The chemical species that are most likely to desorb from the soil surfaces are also the ones that are least likely to sorb. These include the compounds with high water solubility and low K_{ow} values. Water solubility, log K_{ow} and estimated K_{oc} values for gasoline constituents are listed in Table 4-3.

The estimation techniques described in the preceding paragraphs are applicable for most soils in the unsaturated zone where the carbon content is expected to be greater than 0.1 percent. Several researchers (Schwarzenbach and Westall, 1981; Karickhoff, 1984) have shown that the K_{oc} relationships may not be valid for soils with carbon content less than 0.1 percent, which is the type of soil expected to be present in the saturated zone. McCarty et al. (1981) state that sorption studies conducted with pure minerals indicate that hydrophobic solutes are sorbed by inorganic surfaces. They defined a critical level of organic carbon (f_{oc}*) in inorganic matrices below which the inorganic phase contributes to the total sorption. f_{oc}* may be expressed as,

$$f_{oc}^* = \frac{A_s}{200} \quad \frac{1}{K_{ow}0.84} \tag{4.12}$$

where A_s = silica-specific surface area (m^2/g) (McCarty et al., 1981, used a value of 13 m^2/g)
K_{ow} = octanol/water partition coefficient

For benzene, $f_{oc}^* = 0.001$, which translates to an organic carbon content of 0.1 percent. For compounds with higher K_{ow} values, f_{oc}* would be lower, indicating that carbon-based sorption is more important for those compounds.

In such low carbon soils, the sorption coefficient may be expressed as (McCarty et al., 1981):

$$K = K_{oc} \times f_{oc} + K_{oi} \times f_{oi} \tag{4.13}$$

where K = sorption coefficient (L/kg)
f_{oi} = inorganic fraction

Table 4–3. Estimated Soil Sorption Constants for Selected Gasoline Constituents

Gasoline Component	Log K_{ow}[a]	Water[b] Solubility at 25°C (mg/L)	Soil Sorption Constant, Koc (L/kg)[d] Estimated From: K_{ow}	S
n-Butane	2.89	61.4	490	240
Isobutane	2.76	48.9	420	270
n-Pentane	3.39	41.2	910	320
Isopentane	3.37	48.5	880	300
1-Pentene	2.84	148	460	180
n-Hexene	4.00	12.5	1,900	600
1-Hexene	3.39	50	910	320
2-Methylpentane	3.80	14.2	1,500	560
Cyclohexane	3.44	59.7	960	290
Benzene	2.13	1780	190	62
n-Heptane	4.66	2.68	4,300	1,300
2-Methylhexane	4.41	2.54	3,200	1,300
Methylcyclohexane	3.97	15	1,800	580
Toluene	2.69	537	380	110
n-Octane	5.18	0.66	8,200	2,600
2,4-Dimethylhexane	4.82	1.5[c]	5,200	1,800
Ethylbenzene	3.15	167	680	200
m-Xylene	3.20	162	720	210
2,2,4-Trimethylhexane	5.23	0.8[c]	8,700	2,500
1,3,5-Trimethylhexane	3.42	72.6	940	320
2,2,5,5-Tetramethylhexane	5.64	0.13[c]	14,000	5,900
1,4-Diethylbenzene	4.35	15[c]	2,900	670
Dodecane	7.12	0.005	88,000	28,000

No Additives Considered.
a. From Leo (1983).
b. From Mackay and Shiu (1981), except as noted.
c. Estimated by Lyman (personal communication, 1987).
d. Based upon limited experimental data, the values estimated from solubility are considered more reliable.

K_{oi} = partition coefficient in the inorganic phase (L/kg)

The inorganic fraction or clay content of natural soils may range from less than 1 to greater than 99 percent.

At present, relationships are not available for estimating K_{oi} based on solute properties. However, qualitative and semi-quantitative estimates of the contributions to sorption due to the inorganic fraction of soils is possible. Karickhoff (1984) states that if the ratio of the total clay content to the carbon content is greater than 60, then the contribution due to mineral components are significant to the total sorption of neutral organics. Another study (Banerjee et al., 1985b) was conducted to determine the magnitude of the non-carbon based sorption of benzene, a gasoline constituent, on low carbon soils. The extent of carbon-based

Table 4–4. Selected Soil Characteristics and Sorption Coefficient Data

Soil	f_{oc} (%)	Clay (%)	K (L/kg)
J-10	0.26	13.5	0.12
B-1	0.031	11	0.038
C-1	0.031	14	0.026
N-6	0.028	23	0.035

Source: Banerjee et al. (1985b).

sorption and K_{oc} can be established using the soil sorption coefficient determined using the soil with 0.26 percent carbon content. The organic carbon content, clay content, and sorption coefficients for benzene on these soils are listed in Table 4–4. The K_{oc} value so determined was 46 (i.e., $K_{oc} = 0.12/0.0026 = 46$ since sorption to inorganic fraction was not significant). This K_{oc} value is preferred over those listed in Table 4–4 since all other sorption coefficient values were determined under the same experimental conditions. The partition coefficient, K_{oi}, can thus be calculated by substituting appropriate values into equation 4.13. For example, for soil B-1,

$$0.038 = 46 \times \frac{0.031}{100} + K_{oi} \times \frac{11}{100}$$

Therefore, $K_{oi} = 0.22$.

K_{oi} values for C-1 and N-6 are 0.12 and 0.1, respectively. The results indicate that 45 to 63 percent of the total sorption could be attributed to mineral based sorption for these low carbon soils.

It should be noted that carbon-based sorption estimates are not generally valid for non-neutral organics either. However, for gasoline constituents listed in Table 4–3, it is safe to assume that K_{oc} will provide a good estimate of the sorption/desorption potential.

Sorption potential is expected to decrease if the aqueous phase contains dissolved organic matter or colloidal particles. The effect of the presence of colloidal particles are discussed in detail in Section 9. The presence of organic co-solvents such as methanol, MTBE, or acetone will also increase the desorption potential. The relative sorption coefficient may be determined by using (Nkedi-Kizza et al., 1985):

$$\ln(K_m/K) = -abf \tag{4.14}$$

where K = sorption coefficient in water (L/kg)
K_m = sorption coefficient in mixed solvent system (L/kg)
a = empirical constant (dimensionless)

f = fraction of organic co-solvent (dimensionless)

and, b = a parameter related to the solute's surface area and interfacial free energy (dimensionless)

Nkedi-Kizza et al. (1985) found that the parameter b is unique to each sorbate for a specified amount and type of co-solvent.

The parameter, a, can be calculated from the relationship:

$$\ln K_m = -a (\ln x_m) + d \tag{4.15}$$

where X_m = mole-fraction solubility of the contaminant in mixed solvent

d = constant related to the contaminants melting point

Nkedi-Kizza et al. (1985) used a value of 0.83 for a. An example calculation, showing the effect of organic co-solvent on contaminant solubility is presented in Section 8.

The effect of temperature on sorption coefficients of neutral organics for soils in the subsurface does not appear to be significant. Ambient temperature in the subsurface environment is not expected to vary beyond a factor of two. Most researchers (Wu and Gschwend, 1986; Wauchope et al., 1983) found little or no effect for temperature variations over a factor of two.

For solutions containing significant amount of inorganic salts, such as groundwater affected by salt water intrusion, sorption potential is expected to increase.

Several researchers have indicated that competitive sorption may occur for organic cations, chlorinated phenols and inorganic ions. Sorptive competition between immobile adsorbates and mobile colloidal adsorbates may also occur. However, results from several studies indicate that these phenomena may not be important for neutral organics. Chiou et al. (1983) showed that nonionic organics sorb on soil independently from mixtures. However, Abdul and Gibson (1986) noted competitive sorption in experiments conducted with fluorene and naphthalene. They observed that K values for naphthalene and fluorene were greater by a factor of 1.1 and 1.3, respectively, in single compound experiments compared to experiments having a mixture. Considering the imprecision associated with determining sorption coefficients, even under similar experimental conditions, the extent of decreased sorption coefficients observed for multiple solutes does not appear to be significant.

4.2.2.2 Transport with Mobile Phase

Transport of contaminants dissolved in water is discussed in Sections 8 and 12.

4.2.3 Fixation

4.2.3.1 Partitioning onto Immobile (Stationary) Phase

Contaminants adhering to the soil surface are already considered to be immobile. Sorption, however, is not an effective mechanism for fixation since desorption into the mobile phase could continue for an extended period of time.

4.2.3.2 Other Fixation Approaches

None of the available processes can significantly affect fixation.

4.2.4 Transformation

4.2.4.1 Biodegradation (Details in Section 11)

Since sorbed contaminants are in close contact with a dissolved aqueous phase, contaminants in locus no. 4 are very likely subjected to biodegradation. The concentrations in the dissolved and sorbed phases are related by a partition coefficient, as explained earlier. Therefore, as microbes remove contaminants from the aqueous phase, more contaminant molecules desorb into aqueous phase to allow continued biodegradation.

It is important to realize, however, that transport of necessary substrates to the microbes, including nitrogen, phosphorous, and particularly oxygen, may limit the rate of biodegradation. This is especially true for oxygen transport below the water table.

4.2.4.2 Chemical Oxidation (Details in Section 3.2.4.2)

Under wet soil conditions, anaerobic conditions can develop resulting in a shift in pH. Such conditions can increase the occurrence of reduced clay-metal complexes which can catalyze abiotic transformations and lead to polymerization.

4.3 STORAGE CAPACITY IN LOCUS

4.3.1 Introduction and Basic Equations

The maximum amount of gasoline dissolved in water that can be held at the soil surfaces can be stated as,

$$C_s = K\, C_w$$

where C_s = concentration of contaminants in soil (mg/kg)
K = sorption coefficient (L/kg)
C_w = concentration in aqueous phase (mg/L)

4.3.2 Guidance on Input for, and Calculations of, Maximum Value

- The sorption coefficient can be estimated from equation 4.9, which states

$$K = K_{oc} \times f_{oc}$$

- It is recommended that fraction organic carbon be a measured value. For soils with carbon content of less than 0.1 percent, assume 0.1 percent carbon to maximize sorption.
- For retention of gasoline, assume C_w = water solubility of gasoline. Also, estimate K_{oc} from equation 4.11.
- K_{oc} values for gasoline constituents are listed in Table 4–3. For constituents not listed in Table 4–3, obtain literature value or use equations 4.10 or 4.11.
- For maximum value calculation, assume water solubility of gasoline as 200 mg/L (source of this value is discussed in Section 8.3). This value applies to non-oxygenated gasolines.
- For calculating storage capacity for individual gasoline constituents, estimate solubility in water by using

$$C_{iw} = X_{ig}\, S_i \qquad (4.17)$$

where S_i = solubility of constituent i in water (mg/L)
X_{ig} = mole fraction of constituent i in gasoline
C_{iw} = concentration of constituent i in water (mg/L)

4.3.3 Guidance on Inputs for, and Calculations of, Average Value

Use average value of 20 mg for concentration of gasoline. The source of this value is discussed in detail in Section 8.4.

Comments:

- It is assumed that concentration of contaminants in the aqueous phase is linearly related to concentration in soil.
- Effects of salinity, co-solutes, co-solvents, and non-carbon-based sorptions are ignored.

4.4 EXAMPLE CALCULATION

Problem: Calculate mass of gasoline and benzene for release of non-oxygenated gasoline at a site where the soil organic carbon content is 1 percent.

4.4.1 Storage Capacity Calculations

4.4.1.1 Maximum Value Calculations

- Storage capacity for gasoline:

$$S = 200 \text{ mg/L} = 0.002 \text{ molar (assuming molecular weight of gasoline as 100)}$$

$$f_{oc} = 1/100 = .01$$

Use equation 4.11 to estimate K_{oc},

$$\begin{aligned} \log K_{oc} &= -0.602 \times \log(0.002) + 0.656 \\ &= 2.28 \\ K_{oc} &= 191 \end{aligned}$$

Then $K = (191)(0.01) = 1.91 \text{ L/kg}$

and $C_s = 1.91\text{L/kg} \times 200 \text{ mg/L} = 3.82 \times 10^2 \text{ mg/kg}$

- Storage capacity for benzene:

Estimated concentration of benzene in gasoline saturated water

$$C_w = 64.8 \text{ mg/L} \quad \text{(from Table 8-3)}$$

$$K_{oc} = 62 \quad \text{(from Table 4-3)}$$

Therefore, $K = 0.62$

and $C_s = 0.62 \times 64.8 \text{ mg/kg}$

$= 40.2 \text{ mg/kg}$

4.4.1.2 Average Value Calculations

For calculating average retention of gasoline, the concentration of gasoline in water is assumed to be 20 mg/L.

K_{oc} = 191 (from above)

Then, K = 1.91 L/kg

and C_s = 1.91 L/kg × 20 mg/L = 38 mg/kg

All other parameter values remain the same.

4.4.2 Transport Rate Calculations

In this section, we evaluate the transport and fate of gasoline constituents initially dissolved in water which has sorbed onto soil surfaces. Two questions are specifically addressed here: 1) What would be the time scale for desorbing the contaminants and 2) what would be the effect of sorption on bulk transport of gasoline dissolved in water.

4.4.2.1 Time Required to Desorb 50, 90, and 95 percent of Contaminant

The guiding equations are:

$$D_{eff} = \frac{D_m \Theta^2}{(1 - \Theta)\rho_s K} \tag{4.3}$$

$$t_{0.5} = 0.031\ r^2/D_{eff} \tag{4.4}$$

$$t_{0.9} = 0.018\ r^2/D_{eff} \tag{4.6}$$

$$t_{0.95} = 0.25\ r^2/D_{eff} \tag{4.7}$$

Input values for benzene:

$D_m = 10^{-5}$ cm²/s(assumed)
Θ = 0.13 (from Wu and Gschwend, 1986)
ρ_s = 2.5 g/cm³
$K = K_{oc} \times f_{oc}$ = 62 (from Table 4–3) × 0.005 (assuming an f_{oc} content of 0.5 percent)

$$\text{Then}\quad D_{eff} = \frac{10^{-5} \times (0.13)^2}{(1 - 0.13) \times 2.5 \times 0.31}$$

$$= 2.5 \times 10^{-7}\ \text{cm}^2/\text{sec}$$

Then, assuming a particle radius of 2×10^{-3} cm, the time required to desorb 50 percent of the sorbed benzene,

$$t_{0.5} = \frac{0.031 \times (2 \times 10^{-3})^2}{2.5 \times 10^{-7}} \text{ sec}$$

$$= 0.5 \text{ second}$$

The time required to desorb 90 and 95 percent of the sorbed benzene are:

$$t_{0.9} = \frac{0.18 \times (2 \times 10^{-3})^2}{2.5 \times 10^{-7}} \text{ sec} = 2.9 \text{ sec}$$

and

$$t_{0.95} = \frac{0.25 \times (2 \times 10^{-3})^2}{2.5 \times 10^{-7}} \text{ sec} = 4 \text{ sec}$$

For a compound such as dodecane, which has a very high K_{oc} value, the time taken to desorb a certain fraction of the sorbed contaminate would be much higher. The time required to desorb 90 percent of dodecane is calculated below:

$$K_{oc} \text{ of dodecane} = 88{,}000$$

$$\text{Then, } K = 88{,}000 \times 0.005 = 440$$

$$\text{Then, } D_{eff} = 1.76 \times 10^{-10} \text{ cm}^2/\text{s}$$
$$\text{and } t_{0.9} = 4091 \text{ sec} = 1.14 \text{ hours}$$

4.4.2.2 Retardation of Bulk Transport

The governing equation for calculating the retardation of a contaminant relative to the bulk movement of water is,

$$R = 1 + \frac{K \rho_b}{\Theta}$$

Input values for benzene:

$K\rho_b$: same as above
Θ: 0.4 for silt (from Table 4–1)

$$R = 1 + \frac{0.31 \times 2.5}{0.4}$$

$$= 2.9$$

Due to the uncertainties associated with predicting K_{oc} from K_{ow} and solubility, a retardation factor of 2 or below is not significant. In the saturated zone, where f_{oc} can be below 0.001, the sorption coefficient will be much lower than 0.31 and consequently, R will be less than 2. For such environments, benzene should be considered as moving with the bulk flow of water.

The retardation factor for dodecane in an environment similar to that presented for benzene will be

$$R = 1 + \frac{88{,}000 \times 0.005 \times 2.5}{0.4}$$

$$= 2750$$

4.5 SUMMARY OF RELATIVE IMPORTANCE OF LOCUS

4.5.1 Remediation

Sorbed contaminants on soil surfaces in the unsaturated and the saturated zone can act as a source of contamination for an extended period of time. Natural flushing would continue to "wash" contaminants away from the soil surfaces and into the groundwater, acting as a ready source of contamination. Therefore, sorption onto soil surfaces is not a desirable process for long term remediation. Flushing of the system and removal combined with treatment, surface capping and subsurface barriers are probably the most effective remediation techniques. Biodegradation, although a slow process, can potentially be an effective remedial alternative.

4.5.2 Loci Interactions

Table 4–5 lists all the loci that interact with hydrocarbons sorbed onto soil surfaces from the aqueous phase. Bulk transport is influenced by sorption-desorption reactions which influences exchange of contaminants (loci nos. 3, 8, 9, 11, 12). Diffusion in and out of the soil surfaces depends on particle size distribution and geometry of soil particles (loci nos. 3, 8, 9, 11, 12). Volatilization into the soil gas from these loci is probably not significant.

Table 4–5. Loci Interactions with Hydrocarbons Sorbed onto Soil Surfaces

Process	Phases in Contact	Interacting Loci	Relative Importance
Mobility			
Diffusion	Wet soil	3, 8, 9, 11, 12	Modest
Desorption	Wet Soil	3, 8, 9, 11, 12	High
Immobility			
Sorption	Rock	10, 13	Low

4.5.3 Information Gaps

In order to improve predictive capabilities for transport and fate of liquid hydrocarbons sorbed onto soil surfaces, the following information is needed:

1. Sorption isotherm data on gasoline constituents that include low carbon soils.
2. Influence of additives on sorption of gasoline constituents.
3. Data showing errors associated with assuming linear isotherms.
4. Development of models relating soil characteristics to sorption in low carbon soil.

4.6 LITERATURE CITED

Abdul, A.S. and T.L. Gibson. 1986. Equilibrium Batch Experiments with Six Polycyclic Aromatic Hydrocarbons and Two Aquifer Materials. Hazardous Waste & Hazardous Materials, 3(2):125–137.

Arthur D. Little, Inc. 1987. Prediction of Soil and Sediment Sorption for Organic Compounds. Office of Water Regulations and Standards, U.S. EPA, Washington, D.C.

Banerjee, P., M.D. Piwoni, and K. Ebeid. 1985a. Sorption of Organic Contaminants to a Low Carbon Subsurface Core. Chemosphere, 14(8):1057–1067.

Banerjee, P., M.D. Piwoni, and K. Ebeid. 1985b. Sorption of Organic Solvents to Subsurface Soil. Presented at the annual SETAC Symposium. St. Louis, MO.

Chiou, C.T., P.E. Porter, and D.W. Schmedding. 1983. Partition Equilibria of Nonionic Organic Compounds Between Soil Organic Matter and Water. Environ. Sci. Technol., 17:227–231.

Freeze, R.A. and J.A. Cherry. 1979. Groundwater, Prentice-Hall, Inc., Englewood Cliffs, NJ.

Huyakorn, P.S. and G.F. Pinder. 1983. Computation Methods in Subsurface Flow, Academic Press.

Jaffe, P.R. and R.A. Ferrara. 1983. Desorption Kinetics in Modeling of Toxic Chemicals. Journal of Environmental Engineering, 109(40):859–867.

Karickhoff, S.W., D.S. Brown, and T.A. Scott. 1979. Sorption of Hydrophobic Pollutants on Natural Sediments. Water Research, 13:241–248.

Karickhoff, S.W. 1980. Sorption Kinetics of Hydrophobic Pollutants in Natural Sediments. *In*: Contaminants and Sediments, Volume 2 (R.A. Baker, ed.). Ann Arbor Science Publishers, Inc., Ann Arbor, MI.

Karickhoff, S.W. 1984. Organic Pollutant Sorption in Aquatic Systems. Journal of Hydraulic Engineering, 110:707–735.

Leo, A.L. 1983. Log P and Parameter Database. Obtained from Pomona College, Claremont, CA.

McCarty, P.L., M. Reinhard, and B.E. Rittman 1981. Trace Organics in Groundwater. Environ. Sci Technol., 15(1):40–51.

Mackay, D. and W. Shiu. 1981. A Critical Review of Henry's Law Constants for Chemicals of Environmental Interests. J. Phys. Chem. Ref. Data, 10:1175–1199.

Newsom, J. 1985. Transport of Organic Compounds Dissolved in Groundwater. Groundwater Monitoring Review, Spring.

Nkedi-Kizza, P., P.S.C. Rao, and A.G. Hornsby. 1985. Influence of Organic Cosolvents on Sorption of Hydrophobic Organic Chemicals by Soils. Environ. Sci. Technol., 19(10):975–979.

Rao, P.S.C and J.M. Davidson. 1980. Estimation of Pesticide Retention and Transformation Parameters Required in Nonpoint Source Pollution Models. In: Environmental Impact of Nonpoint Source Pollution (M.R. Overcash and J.M. Davidson, eds.). Ann Arbor Science Publishers, Inc., Ann Arbor, MI.

Schwarzenbach, R.P. and J. Westall. 1981. Transport of Nonpolar Organic Compounds from Surface Water to Groundwater: Laboratory Sorption Studies. Environ. Sci Technol., 15:1360–1367.

U.S. Department of Agriculture (USDA). 1960. Soil Survey: Erie County, PA.

U.S. 1986. EPA Superfund Public Health Evaluation Manual, Office of Emergency and Remedial Response, Washington, D.C.

Wauchope, R.D., K.E. Savage, and W.C. Koskinen. 1983. Adsorption-Desorption Equilibriums of Herbicides in Soil: Naphthalene as a

Model Compound for Entropy-Enthalpy Effects. Weed Science, 31:744–751.

Wu, S. and P.M. Gschwend. 1986. Sorption Kinetics of Hydrophobic Organic Compounds to Natural Sediments and Soils. Environ. Sci. Technol., 20(7):717–725.

SECTION 5

Liquid Contaminants in the Pore Spaces Between Soil Particles in the Saturated Zone

CONTENTS

List of Tables 132
List of Figures 132
5.1 Locus Description 133
5.1.1 Short Definition 133
5.1.2 Expanded Definition and Comments 133
5.2 Evaluation of Criteria for Remediation 133
5.2.1 Introduction 133
5.2.2 Mobilization/Remobilization 138
5.2.2.1 Partitioning into Mobile Phase 138
5.2.2.2 Transport with Mobile Phase 138
Physical Dislodgement/Displacement ... 138
Hydraulic Removal 140
Mobilization Schemes 143
5.2.3 Fixation 144
5.2.3.1 Partitioning onto Immobile (Stationary) Phase 144
5.2.3.2 Other Fixation Approaches 144
5.2.4 Transformation 144
5.2.4.1 Biodegradation 144
5.2.4.2 Chemical Oxidation 144
5.3 Storage Capacity in Locus 145
5.3.1 Introduction and Basic Equations 145
5.3.2 Guidance on Inputs for, and Calculations of, Maximum Value 145
5.3.3 Guidance on Inputs for, and Calculations of, Average Values 145
Comments 146
5.4 Example Calculations 146
5.4.1 Storage Capacity Calculations 146
5.4.1.1 Maximum Storage Capacity 146
5.4.1.2 Average Storage Capacity 146

5.4.1.3 Maximum Mass 147
5.4.1.4 Average Mass 147
5.4.2 Transport Rate Calculations 147
5.4.2.1 Dissolution Rates 147
5.4.2.2 Transport Rate of Residual Liquid Contaminant 147
5.5 Summary of Relative Importance of Locus 148
5.5.1 Remediation 148
5.5.2 Loci Interaction 149
5.5.3 Information Gaps 150
5.6 Literature Cited 150

TABLES

5-1 Typical Fluid Properties of Hydrocarbons 142
5-2 Water-Hydrocarbon Experimental Results 145
5-3 Loci Interactions with Residual Liquid Contaminant in Groundwater 149

FIGURES

5-1 Schematic Cross-Sectional Diagram of Locus No. 5—Liquid Contaminants in the Pore Spaces Between Soil Particles or Rock Surfaces in the Saturated Zone 135
5-2 Schematic Representation of Important Transformation and Transport Processes Affecting Other Loci 136
5-3 Sketches of Low Capillary Number Trapping Mechanisms Using the Pore Doublet Model: Snap-off at Top, By-passing Below 136
5-4 Effect of Aspect Ratio on Hydrocarbon Trapping in a Non-uniform Table 137
5-5 Residual Hydrocarbon Saturation Ratio, Relating Final Residual Saturation (S_{or}) to Initial Residual Saturation (S^*_{or}) as a Function of Capillary Number (N_c) 140
5-6 Hydraulic Gradient Necessary to Initiate Blob Mobilization (at N^*_c) in Soils of Various Permeabilities for Hydrocarbons of Various Interfacial Tension 140
5-7 Hydraulic Gradient Necessary to (a) Initiate Blob Mobilization for Various Hydrocarbons and (b) Obtain Reported Percentage Recovery of Residual Hydrocarbon for Interfacial Tension = 10 dyne/cm .. 142
5-8 Percentage of Total Volume of a Heterogeneous Sand in which Residual Hydrocarbon has been Reduced by the Amount Shown on the Curves 143

SECTION 5

Liquid Contaminants in the Pore Spaces Between Soil Particles in the Saturated Zone

5.1 LOCUS DESCRIPTION

5.1.2 Short Definition

Liquid contaminants in the pore spaces between soil particles in the saturated zone.

5.1.2 Expanded Definition and Comments

The liquid contaminant in this locus may be in either the "water-wet" configuration (i.e. particle surfaces wet by water) or the "oil-wet" configuration (i.e. particle surfaces wet by oil); no air is present. The liquid contaminants are likely to be present as a discontinuous phase, derived from a fluctuating water table level. A continuous phase is more likely for a liquid contaminant such as tetrachloroethylene or PCBs which are denser than water. The liquid contaminants in locus no. 5 are subject to relatively weak buoyancy forces due to density differences and entrainment forces due to moving groundwater. Thus the liquid material is considered relatively immobile and dissolution is the principal loss mechanism.

This locus differs from locus no. 13 which considers liquid contaminants in fractured bedrock, not in porous media. Figures 5-1 and 5-2 present a schematic cross-sectional diagram of locus no. 5 and a schematic representation of the transformation and transport processes affecting other loci, respectively.

5.2 EVALUATION OF CRITERIA FOR REMEDIATION

5.2.1 Introduction

Liquid contaminant in pore spaces between particles in the saturated zone is most likely to be present as a discontinuous phase in pore volume otherwise filled with water, and is derived from displacement of the main body of liquid contaminant by a fluctuating groundwater table (described in locus no. 7). Once the main liquid contaminant body is displaced, a portion is retained in the porous media because of capillary forces, becoming immobile at lower residual saturations. Residual saturation is the volume of discontinuous immobile liquid contaminant per unit void volume.

Residual liquid contaminant is retained in pore spaces in the saturated zone by two mechanisms: by-passing and snap-off (Mohanty, et al., 1980; Chatzis, et al., 1983) illustrated in Figure 5–3. The doublet model of pore structure (Chatzis, et al., 1983) consists of a tube split into two pores, one generally narrower than the other, which rejoin. Liquid contaminant is trapped in the upper tube by snap-off, which strongly depends on pore shape and wettability (Figure 5–3). For pores with a high aspect ratio (i.e. where the pore throats are much smaller than the pore bodies), snap-off is common. For pores with a low aspect ratio, liquid contaminant is easily displaced (Figure 5–4a). The remaining liquid contaminant in the upper pore is displaced downstream. If the downstream junction is unable to form a stable meniscus (boundary) in the downstream tube, two menisci between water and liquid contaminant develop: one that is displaced downstream from the upper pore and another in the lower pore. Liquid contaminant in the lower pore becomes trapped by by-passing (Figure 5–3). As the aspect ratio increases, the proportion of liquid contaminant trapped by snap-off increases. Heterogeneity in soil or rock promotes trapping through by-passing.

Naturally occurring hydraulic gradients in groundwater are too small to displace the highly-curved menisci between liquid contaminant and water in the pore space, and cannot "squeeze" the liquid contaminant through the pore throats. Thus, the menisci block pores and interfere with water flow. For relatively low liquid contaminant saturations, these capillary forces cause disconnections of the liquid contaminant phase, resulting in oil blobs. This residual liquid contaminant in the saturated zone can occupy from 15 to 40 percent or more of the pore volume (Felsenthal, 1979; Morrow and Chatzis, 1982; Chatzis, et al., 1984). Porous media structure and particle size have much less influence on residual saturation in the saturated zone than in the unsaturated zone.

When considering the problem of remediation of residual liquid contaminant in the saturated zone, some fundamental questions are:

(1) What is the amount of residual liquid contaminant in the porous media interstices?
(2) Which constituents can be remobilized from, or immobilized in, the interstices?

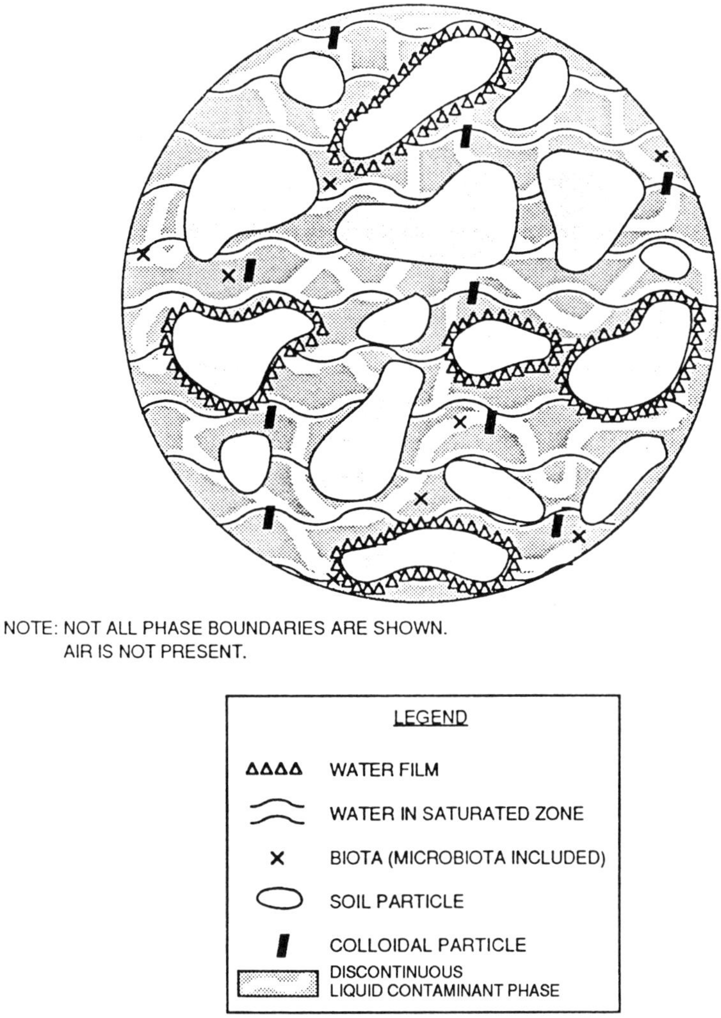

Figure 5–1. Schematic cross-sectional diagram of locus no. 5–liquid contaminants in the pore spaces between soil particles or rock surfaces in the saturated zone.

The influence of capillary forces in this locus is an important factor for remediation. In order for corrective action mechanisms to mitigate residual liquid contaminant, these capillary forces must be either surmounted (or enhanced) for mobilization (or immobilization) to occur.

Mobilization of residual liquid contaminant can be accomplished either: (1) by dissolving the liquid contaminant into groundwater; or (2)

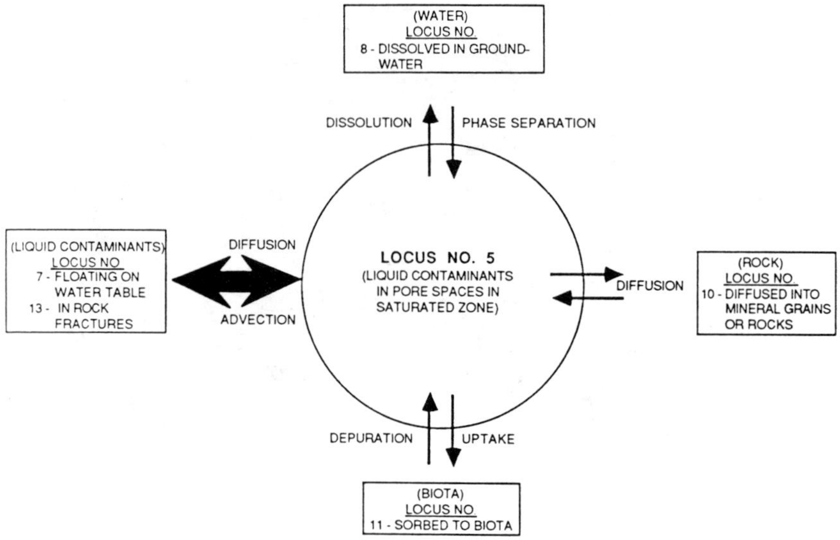

Figure 5–2. Schematic representation of important transformation and transport processes affecting other loci.

by physically dislodging the droplets or blobs from soil pore volume, entraining the droplets in the groundwater flow. Dissolution of residual hydrocarbon is not as effective overall as physical removal, and acts as a continual source of contamination. Physical removal of residual liquid contaminant can be relatively effective and is accomplished either by hydraulic sweeping (i.e., increasing hydraulic gradient) and/or by lowering of interfacial tensions between the residual liquid and water with surfactants.

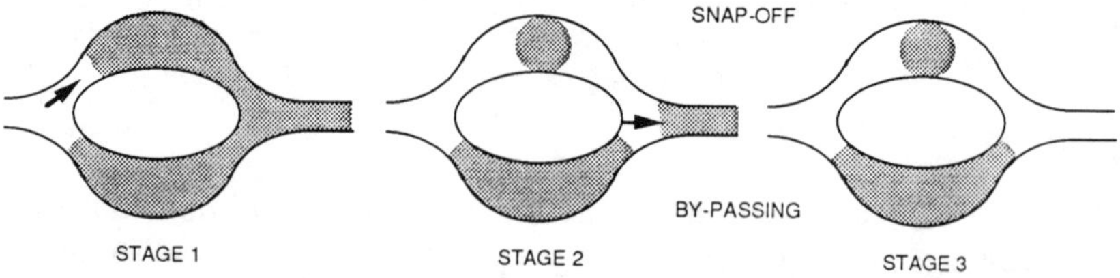

Figure 5–3. Sketches of low capillary number trapping mechanisms using the pore doublet model: snap-off at top, by-passing below. *Source:* Chatzis et al., 1983. (Copyright Society of Petroleum Engineers, 1983. Reprinted with permission.)

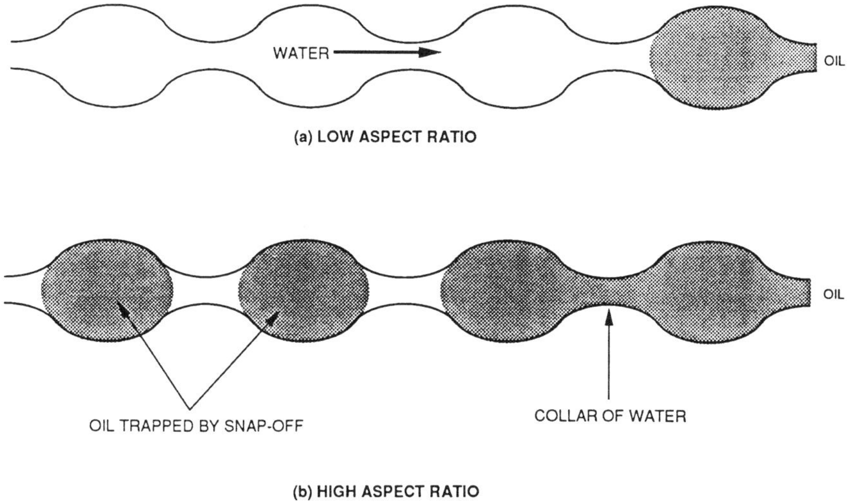

Figure 5–4. Effect of aspect ratio on hydrocarbon trapping in a non-uniform table. *Source:* Chatzis et al., 1983. (Copyright Society of Petroleum Engineers. 1983. Reprinted with permission.)

Immobilization of residual liquid contaminant can be accomplished by enhancing capillary trapping via lowering the hydraulic gradient.

Transformation of residual liquid contaminant by biodegradation (locus no. 11) is seldom attempted as the only corrective action for removal of residual liquid contaminant; the quantity of liquid contaminant at many of these sites is simply too large. Biodegradation, in conjunction with mobilization mechanisms, can result in effective removal of residual liquid contaminant, leading to ultimate restoration of the aquifer.

In locus no. 5 two fluid phases are present: water and liquid contaminant. In such a two-fluid phase system, the liquid contaminant is usually considered to be the non-wetting fluid (i.e. water preferentially wets the particle surfaces rather than the liquid contaminant). The case where the liquid contaminant is considered the wetting fluid is discussed in loci nos. 2 and 7. Only the "water-wet" configuration will be considered in this discussion.

5.2.2 Mobilization/Remobilization

5.2.2.1 Partitioning into Mobile Phase

Partitioning of residual liquid contaminant (e.g., gasoline) in pore water depends on: (1) the surface area of the liquid contaminant in contact with groundwater (locus no. 8); and (2) the solubilities of the constituents of a liquid contaminant in water and their concentrations in that blend (see locus no. 7 for a more detailed discussion of dissolution). Equilibrium dissolution is approximated by:

$$C_i = X_i S_i$$

$$i = 1, 2, 3 \ldots n \text{ constituents} \tag{5.1}$$

where C_i = concentration of constituent i in the water phase (mg/L)
X_i = mole fraction of constituent i in liquid contaminant
S_i = solubility of constituent i in pure water (mg/L)

Calculations are more difficult if significant amounts of oxygenated additives are present. The difficulty is compounded by the lack of data for these additives (e.g., amount in gasoline blend, solubility, fluid properties, etc.). The gasoline constituents with the highest solubilities are the light aromatics: benzene, toluene, xylenes, and ethylbenzene. Many gasoline additives, such as MTBE, ethanol, and methanol, have much higher solubilities. Solubilities of chemicals found typically in gasolines are presented in Tables 4–3 and 7–1.

5.2.2.2 Transport with Mobile Phase

Once residual liquid contaminant in the saturated phase has become mobilized either by dissolution or dislodgment and entrainment in the groundwater, transport occurs with groundwater flow. For dissolved constituents, the advective-dispersive equation (8.1) describes bulk transport and is presented in locus no. 8. Residual liquid contaminant can become dislodged and entrained in the groundwater as a result of an artificially increased hydraulic gradient. Because residual liquid contaminant exists as a discontinuous phase, continuum flow equations may not describe adequately the transport of entrained liquid contaminant.

Physical Dislodgement/Displacement. The capillary forces which resist mobilization can be overcome by viscous forces associated with hydraulic gradient or by gravity buoyancy forces. In this locus, residual liquid contaminant (e.g., gasoline) is subjected to relatively weak buoyancy forces due to density differences. Capillary and viscous forces are more significant than buoyancy forces with regard to mobilization of residual liquid contaminant in pore spaces in the saturated zone.

The dimensionless ratio of capillary forces to viscous forces is known as the capillary number (N_c):

$$N_c = \frac{k \cdot \rho_w \cdot g \cdot J}{\partial} \tag{5.2}$$

where k = intrinsic permeability of porous medium (cm^2)
ρ_w = the density of water (g/cm^3)
∂ = interfacial tension (dyne/cm)
g = gravitational acceleration (cm/sec^2)
J = hydraulic gradient (cm/cm)

Experimental studies by Wilson and Conrad (1984) established a strong correlation between displacement of residual liquid contaminant and capillary number, when the hydraulic gradient was greater than the critical value needed to initiate motion of some of the residual liquid contaminant blobs or droplets. This critical value of the capillary number depends on the residual saturation.

Liquid contaminant saturation is the volume of liquid contaminant per unit void volume:

$$S_o = V_o/V_v \tag{5.3}$$

where S_o = organic or liquid contaminant saturation (cm^3/cm^3)
V_o = volume of hydrocarbons (cm^3)
V_v = volume of voids (or pore space)(cm^3)

As the liquid contaminant is displaced, its saturation is reduced to a residual:

$$S_{or} = V_{do}/V_v \tag{5.4}$$

where S_{or} = residual liquid contaminant saturation (cm^3/cm^3)
V_{do} = volume of discontinuous liquid contaminant (cm^3)
V_v = volume of voids (cm^3)

The relationship between the capillary number and residual saturation is shown in Figure 5–5; residual saturation is normalized by its initial value (i.e., ratio of final residual saturation, S_{or} to initial saturation, S_{or}*). N_c* represents the critical capillary number at which motion is initiated. N_c** denotes the capillary number necessary to displace all of

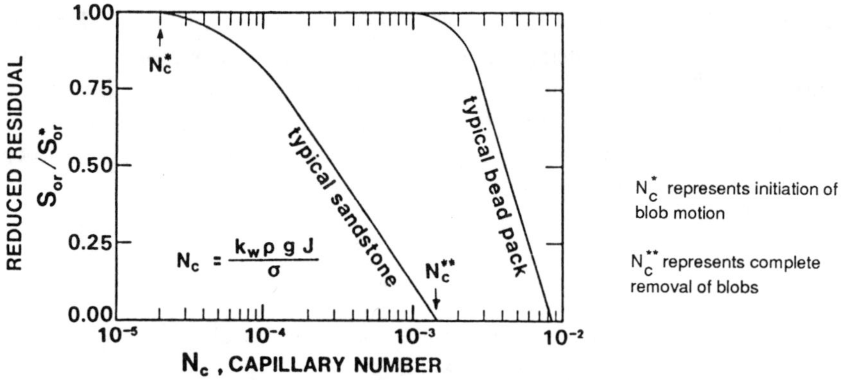

Figure 5–5. Residual hydrocarbon saturation ratio relating final residual saturation (S_{or}) to initial residual saturation (S^*_{or}) as a function of capillary number (N_c). *Source:* Wilson and Conrad, 1984. (Copyright Water Well Journal Publishing Co., 1984. Reprinted with permission.)

the residual hydrocarbon. For sandstone, $N_c^* = 2 \times 10^{-5}$ and $N_c^{**} = 2 \times 10^{-3}$. Experiments on carbonate rocks produce curves lying to the left of the sandstone curve in Figure 5–5 (Morrow, 1984).

Hydraulic Removal. Some of the liquid contaminant will be removed when $N_c > N_c^*$ and all of it will be removed when $N_c \geq N_c^{**}$. Figures 5–6 and 5–7a are plots of hydraulic gradient of the water (J) versus intrinsic permeability of the porous medium (k) using the capillary number of equation 5.2 for the critical value N_c^* and for various liquid contaminant water interfacial tensions, Table 5–1. These plots describe hydraulic gradients necessary to initiate blob mobilization.

For example, n-octane with a low solubility (S = 0.66 mg/L) at 20°C will tend not to dissolve in the water phase; it has an interfacial tension of 50 dyne/cm. As evident from Figure 5–6, in a coarse sand with $k = 10^{-5}$ cm^2, a gradient of J = 0.1 is necessary to initiate blob mobility. This is a steep gradient, but not unreasonable. The upper right curve of Figure 5–6 represents the gradient necessary for complete removal of all hydrocarbon for $\partial = 10$ dyne/cm.

Figure 5–7b is a plot of the percentage of residual hydrocarbon recovered as a function of gradient and soil permeability for a specific interfacial tension, $\sigma = 10$ dyne/cm. The top curve labeled N_c^{**} in Figure 5–6 is the 100 percent recovery curve for the same hydrocarbon. Removal of other liquid contaminants can be examined by multiplying these results by $\sigma/10$. For the coarse sand example, which required a gradient of 0.1 to mobilize n-octane blobs, a gradient of 1.0 is required to completely

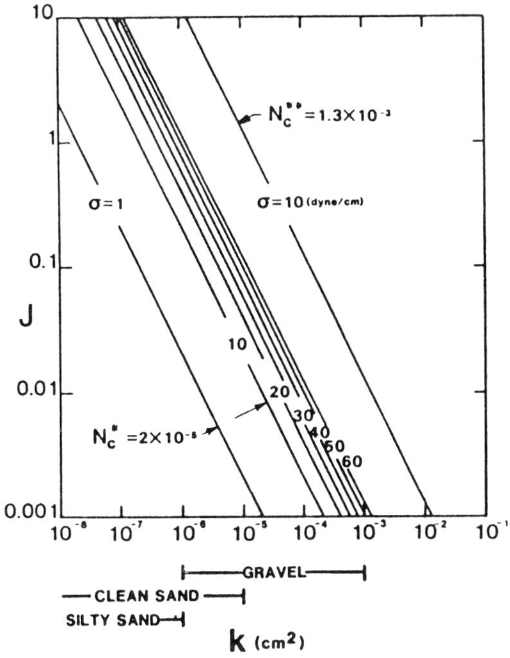

Figure 5–6. Hydraulic gradient necessary to initiate blob mobilization (at N^*_c) in soils of various permeabilities for hydrocarbons of various interfacial tension. *Source:* Wilson and Conrad, 1984. (Copyright Water Well Journal Publishing Co., 1984. Reprinted with permission.)

remove a liquid contaminant with σ = 10 dyne/cm. For n-octane, the necessary gradient is (50/10) = 5 times as large, of J = 5; this is an impossibly large gradient.

The residual saturation curves in Figures 5–5 through 5–7 were derived for homogeneous media. Heterogeneity, however, plays a significant role when considering the total volume of residual hydrocarbon recovered. A spatially variable medium-to-coarse sand was statistically evaluated by Smith (1981) and used by Wilson and Conrad (1984). The results are shown in Figure 5–8. The curves depict the percentage of residual hydrocarbon swept from each portion of the soil volume. Thus, for a gradient of 10 and σ = 10 dyne/cm, only 7 percent of the pore space has been swept of 75 percent or more of the residual hydrocarbon, while over 80 percent of the pore space has been swept of 50 percent or more. This plot is crude because it neglects spatial correlation, but it demonstrates that heterogeneity significantly affects the removal/mobilization of residual liquid contaminant.

In summary, residual liquid contaminant is easier to mobilize for

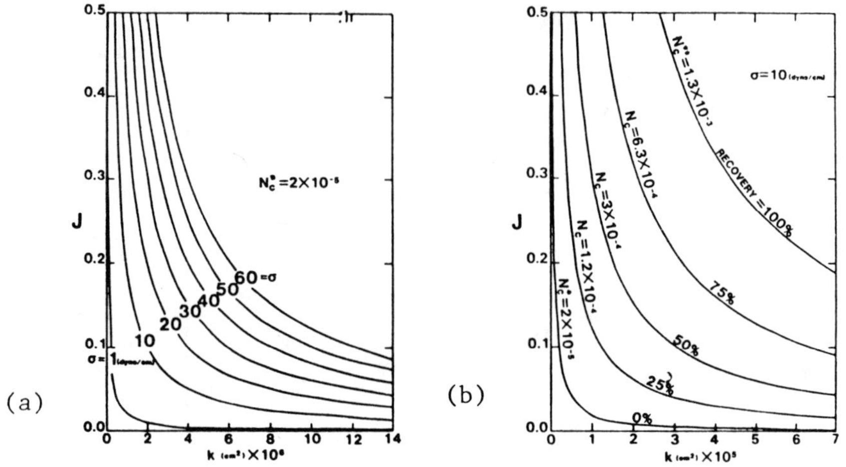

Figure 5–7. Hydraulic gradient necessary to (a) initiate blob mobilization for various hydrocarbons and (b) obtain reported percentage recovery of residual hydrocarbon for interfacial tension = 10 dyne/cm. *Source:* Wilson and Conrad, 1984. (Copyright Water Well Journal Publishing Co., 1984. Reprinted with permission.)

contaminants with lower interfacial tensions with water, for more permeable porous media, and in the laboratory where extremely high hydraulic gradients can be produced. Liquid contaminant is more difficult to mobilize in the field where there are severe practical constraints on the hydraulic gradient; it may be possible to mobilize some of the residual liquid contaminants, but it is difficult to mobilize it all.

Table 5–1. Typical Fluid Properties of Hydrocarbons

Fluid	Interfacial Tension (dyne/cm)	Density (g/cm^3)	Viscosity ($\times 10^{-2}$ g/cm.sec)	Temperature (°C)
Water	0.0	0.9982	1.002	20
n-Butanol	1.6	0.8098	2.948	20
n-Pentanol	4.4	0.8144[a]		25
n-Hexanol	6.8	0.8136[a]		25
Benzene	35.0	0.87865	0.652	20
n-Heptane	50.2	0.68376	0.409	20
n-Octane	50.8	0.7025	0.542	20
n-Hexane	51.0	0.6603	0.326	20

a. 20°C.

Source: Wilson and Conrad, 1984. (Copyright Water Well Journal Publishing Co., 1984. Reprinted with permission.)

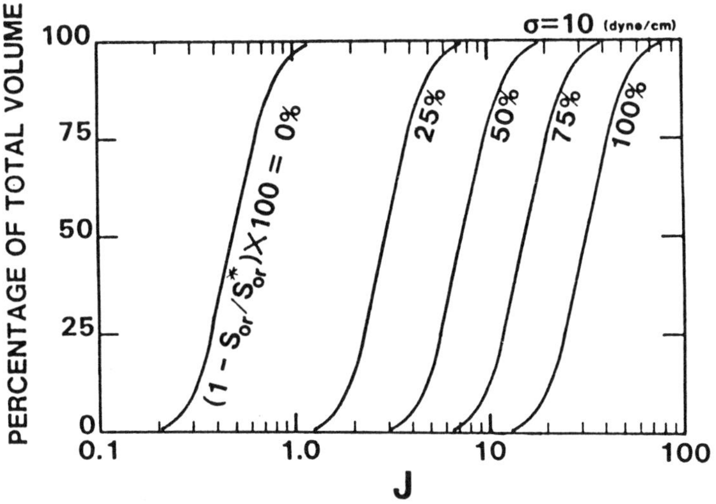

E(log k) = -6.4; $\sigma_{\log k}$ = 0.17; and N^*_c = 0.00002

Figure 5–8. Percentage of total volume of a heterogeneous sand in which residual hydrocarbon has been reduced by the amount shown on the curves. *Source:* Wilson and Conrad, 1984. (Copyright Water Well Journal Publishing Co., 1984. Reprinted with permission.)

Mobilization Schemes. Mobilization of residual liquid contaminants is possible by a number of different schemes involving an increased hydraulic gradient or lowering of interfacial tensions. These schemes are in different stages of development and understanding, and usually do not result in mobilization of all the residual liquid contaminants. Hydraulic gradient can be maximized in the field by employing closely-spaced vertical sweeps and/or hydraulic fracturing of aquifer (Wilson and Conrad, 1984). A scheme for mobilization which can be used in conjunction with increased gradients, is the reduction of interfacial tensions (Melrose and Brander, 1974; Taber, 1981) by surfactants and alkaline flooding. There does not appear, however, to be a safe means of implementation without further contamination of the groundwater. Thermal or steam floods may be effective in lowering interfacial tensions, but also increases the dissolution of certain constituents. These schemes have been utilized to some degree by the oil industry for enhanced oil recovery (EOR). EOR methods have not been applied to groundwater pollution problems, but may prove to be feasible.

5.2.3 Fixation

5.2.3.1 Partitioning onto Immobile (Stationary) Phase

The residual liquid contaminant liquid phase is already considered to be relatively immobile. In the "water-wet" configuration, liquid contaminant constituents could partition onto the porous medium particles after migrating through a water film surrounding a water-wetted porous medium particle or rock surface (see locus no. 4 for discussion). Residual liquid contaminant could be made more immobile if the wetting conditions changed to an "oil-wet" configuration, perhaps brought about by lowering the water table (see locus no. 7 for discussion).

5.2.3.2 Other Fixation Approaches

Residual liquid contaminant in pore spaces has already undergone capillary trapping and remains relatively immobile. Immobilization can be enhanced by decreasing the hydraulic gradient, thus lowering dissolution and the capillary number (see Figure 5–5). Even with static groundwater flow, however, dissolution occurs; it is not possible to completely fix or isolate liquid contaminant in pore spaces between the particles of the porous medium in the saturated zone.

5.2.4 Transformation

5.2.4.1 Biodegradation (Details in Section 1)

Below the water table, the supply of oxygen is limited, reducing the rate of aerobic degradation. The rise and fall of the water table might increase the dissolved oxygen level if the layer of liquid contaminant upon the water table (locus no. 7), is very thin or absent.

5.2.4.2 Chemical Oxidation (Details in Section 3.2.4.2)

Reducing conditions encountered in the saturated zone can increase the likelihood of contaminant catalyst interaction and thereby lead to abiotic transformations. Polymerization reactions can also proceed under anaerobic conditions.

Table 5–2. Water-Hydrocarbon Experimental Results

Media	S_{or} (%)	R (L/m^3)	Θ
Glass Beads	14	57	0.38
Sandstones	27–40	54–80	0.20

Source: Wilson and Conrad (1984).

5.3 STORAGE CAPACITY IN LOCUS

5.3.1 Introduction and Basic Equations

The maximum amount of residual liquid contaminants that can be stored in the pore space between the particles of the porous medium in the saturated zone is:

$$R = S_{or} \cdot \Theta \cdot 10^3 \qquad (5.5)$$

where R = retention (liters of residual liquid contaminant/cubic meters of soil, L/m^3)
S_{or} = residual saturation ($0 \leq S_{or} \leq 1$)
Θ = soil porosity ($0 \leq \Theta \leq 1$)

- R and S_{or} values have largely been determined experimentally on natural sandstone cores or on glass beads (see Table 5–2).
- The mass of residual liquid contaminant can be calculated by multiplying (R)(ρ_g), where ρ_g is the density of the liquid contaminant.

5.3.2 Guidance on Inputs for, and Calculations of, Maximum Value

Residual hydrocarbon saturation (S_{or}) as determined by experiments run at low capillary number ranged between 27 percent and 43 percent (Chatzis and Morrow, 1981).

- The upper value of 43 percent is recommended for calculating the maximum value.
- For appropriate porosity values, see Table 12–2 or Figure 12–4.

5.3.3 Guidance on Inputs for, and Calculations of, Average Value

- Experimentally determined residual hydrocarbon saturation (S_{or}) run at various capillary numbers were found to have a mean of 28 percent (Felsenthal, 1979). S_{or} = 28 percent is recommended for calculating the average value.

Comments

- Retention values neglect subtraction or addition of liquid contaminant by partitioning processes.
- The retention values will likely decrease with time in response to loss of constituents to biodegradation and dissolution into groundwater.
- Retention is not significantly affected by modest changes in temperature.
- Soil or rock heterogeneity and porosity are the most significant factors affecting retention.
- Retention in the saturated zone is significantly larger than in the unsaturated zone and is essentially independent of particle size.
- Most hydrocarbon retention values reported in the literature refer to vadose zone experiments.

5.4 EXAMPLE CALCULATIONS

5.4.1 Storage Capacity Calculation

5.4.1.1 Maximum Storage Capacity

If the total volume of voids for a sandstone is 4 cubic meters ($V_v = 4m^3$), and the volume of residual hydrocarbon is 1.7 cubic meters ($V_{or} = 1.7m^3$), then the residual saturation can be determined using:

$$S_{or} = V_{or}/V_v$$
$$S_{or} = (1.7)/(4)$$
$$S_{or} = 43\%$$

Using the maximum residual saturation value ($S_{or} = 43\%$), and a porosity value for a very fine-grained sandstone ($\Theta = 0.19$), the maximum amount of retention (R) from equation 5.5 is:

$$R = (0.43.)(0.19.)10^3$$

$$R = 81.7 \text{ liters of residual hydrocarbon/m}^3 \text{ of sandstone}$$

5.4.1.2 Average Storage Capacity

Using an average residual saturation value ($S = 28\%$), for the same porosity ($\Theta = 0.19$), the average amount of retention is:

$$R = (0.28.)(0.19.)10^3$$

$$R = 53.2 \text{ liters of residual hydrocarbon/m}^3 \text{ of sandstone}$$

The mass of residual gasoline per unit volume of porous medium can be determined if the density of the liquid contaminant (ρ_g) is known:

$$m_r = \rho_g \cdot (R/10^3)$$

Where: m_r = residual mass of gasoline/unit volume of porous medium (kg/m^3)
ρ_g = 710 kg/m^3 (from Table 7-1)

5.4.1.3 Maximum Mass

$$m_r = (710)\,(82/10^3) = 58.2 \text{ kg/m}^3$$

5.4.1.4 Average Mass

$$m_r = (710)\,(53/10^3) = 37.6 \text{ kg/m}^3$$

5.4.2 Transport Rate Calculations

5.4.2.1 Dissolution Rates (Details in Section 7.4.2)

The rate of dissolution from residual liquid contaminant in the pore space in the saturated zone depends strongly on the partition coefficient and mixing within each phase (e.g., greater mixing with an increased groundwater flow). Dissolution rates are discussed in detail in locus no. 7.

5.4.2.2 Transport Rate of Residual Liquid Contaminant

Since residual liquid contaminant in porous medium interstices in the saturated zone exists as a discontinuous phase, continuum flow models do not describe adequately the transport of entrained liquid contaminant. At present, there is not adequate understanding of how discontinuous blobs or droplets are transported in groundwater. These droplets may move as single blobs and may or may not coalesce. Bigger blobs may breakup into single blobs which have a greater propensity for retrapping. If surfactants are present, the likelihood of coalescence is diminished. Overall, the heterogeneity of the porous media will undoubtedly have a significant influence on blob distribution and transport.

A useful calculation might be to determine the hydraulic gradient necessary to initiate blob mobilization. For example, in a typical sandstone the critical capillary number required to initiate blob mobilization, $N_c^* = 2 \times 10^{-5}$ (Chatzis and Morrow, 1981), can be used to determine the hydraulic gradient.

Using equation 5.2:

$$N_c = \frac{k \cdot \rho \cdot g \cdot J}{\partial}$$

and rearranging to solve for J:

$$J = \frac{N^*_c\, \partial}{k \cdot \rho \cdot g}$$

$$\begin{aligned} N_c^* &= 2 \times 10^{-5} \text{ (given in example)} \\ \partial &= 50 \text{ dyne/cm (from Table 5–1)} \\ k &= 10^{-5}\ \text{cm}^2 \text{ (for a coarse sandstone)} \\ g &= 980.7\ \text{cm/sec}^2 \\ \rho_w &= 1.0\ \text{gm/cm}^3 \\ J &= \frac{(2 \times 10^{-5})(50)}{(10^{-5})\,(1.0)\,(980.7)} \\ J &= 0.1 \end{aligned}$$

For a coarse sandstone, n-octane would require a steep gradient of 0.1 to initiate blob motion. A hydraulic gradient of 10 would be required to completely remove all residual n-octane from the coarse sandstone.

5.5 SUMMARY OF RELATIVE IMPORTANCE OF LOCUS

5.5.1 Remediation

The largest amount of liquid contaminant stored in the saturated zone is most likely the liquid contaminant in porous media interstices. This residual liquid contaminant can be present in amounts of one or two orders of magnitude greater than that dissolved in the groundwater or sorbed onto particulate matter of the porous media. Residual liquid contaminant in pore space in the saturated zone provides a continual source of groundwater contamination by dissolution of the more soluble constituents. Because of dissolution, it is not particularly feasible to promote retention of this residual liquid contaminant. Removal of the residual appears to be more pragmatic and could possibly result in more effective aquifer restoration. This removal could be accomplished by a number of mobilization and transformation mechanisms. Dislodgment of some of the residual liquid contaminant by increased hydraulic gradi-

Table 5–3. Loci Interactions with Residual Liquid Contaminant in Groundwater

Process	Phases in Direct Contact[a]	Interacting Loci	Relative Importance
Mobility			
Dissolution (Phase separation)	water	8	high
Bulk Transport (Displacement, Entrainment)	water	8	high
	wet soil	4	moderate
	rock	10	very low
	rock	13	high
	liquid hydrocarbon	7	high
Immobility			
Sorption	wet soil	4	moderate
	rock	13	low
Wetting Conditions	water	7, 8	moderate-high

a. Biota (locus no. 11) are potentially in direct contact with all phases.

ent and/or decreased interfacial tensions could reduce the quantity of liquid contaminant to a level appropriate for biodegradation. The efficacy of these corrective action mechanisms (or combination of mechanisms) is not currently known. Specific studies of each mechanism however, indicates a substantial potential for the remediation of residual liquid contaminant in groundwater.

5.5.2 Loci Interactions

The loci interacting with residual liquid contaminant in groundwater during partitioning and mass transport process are summarized in Figure 5–2 and Table 5–3. For residual liquid contaminant, the principal partitioning process is dissolution into the groundwater (locus no. 8) by mixing due to advection and diffusion. Bulk transport into the water-filled pore space in the saturated zone is the result of a fluctuating water table (locus no. 7). Bulk transport out of the locus is influenced by the hydraulic gradient of the groundwater (locus no. 8) and the permeability of the media (loci nos. 5, 10, and 13). The degree to which the residual liquid contaminant is retained is related to the sorptive properties of the porous media (loci nos. 4, 10, and 13) and wetting conditions (loci nos. 7, 8). Residual liquid contaminant is considered to be relatively immobile and dissolution is the chief loss mechanism.

5.5.3 Information Gaps

For a better understanding and definition of the behavior of residual liquid contaminant in pore spaces in the saturated zone, and the influence and utility of partitioning process on this residual liquid contaminant, the following information is needed:

1. fundamental data on solubility and dissolution rates as well as the composition and physicochemical properties of liquid contaminant;
2. specific concentration of additives in different liquid contaminants and their behavior in an aquifer;
3. sorption data (k_d,k_{oc}) that represent natural soil systems of interest as well as nonequilibrium effects (DOC, OC, time, temperature, etc.);
4. data on the influence of surfactants on mobility of residual liquid contaminant;
5. fundamental understanding of transport of a discontinuous fluid-phase in a porous media;
6. further understanding of trapping mechanisms in relation to media heterogeneity; and,
7. aerobic and anaerobic biodegradation in groundwater. Studies to date have focused on surface and ocean water.

5.6 LITERATURE CITED

Chatzis, I. and N. Morrow. 1981. Correlation of Capillary Number Relationships for Sandstones. S.P.E. Annual Technical Conference and Exhibit, San Antonio, TX.

Chatzis, I., N.R. Morrow, and H.T. Lim. 1983. Magnitude and Detailed Structure of Residual Oil Saturation. Soc. Petroleum Engineering Journal, Vol. 23, no. 2.

Chatzis, I., M.S. Kuntamukklua, and N.R. Morrow. 1984. Paper S.P.E. 13213. 1984 S.P.E. Annual Technical Conference and Exhibit, Houston, TX.

Felsenthal, M. 1979. A Statistical Study of Core Waterflood Parameters. Journal of Petroleum Technology, 31:1303–1304.

Melrose, J. C. and C. F. Brander. 1974. Role of Capillary Forces in Determining Displacement Efficiency for Oil Recovery by Waterflooding. Journal of Canadian Petroleum Technology, Vol. 13.

Mohanty, K.K., H.T. Davis, and L.D. Scriven. 1980. Physics of Oil Entrapment in Water-Wet Rock. Paper S.P.E. 9406, 1980 S.P.E. Annual Technical Conference and Exhibit, Dallas, TX.

Morrow, N.R. (ed). 1984. Measurement and Correlation of Conditions for Entrapment and Mobilization of Residual Oil. Report No.

NMERDI 2-70-3304, New Mexico Energy Research and Development Institute, Santa Fe, NM

Morrow, N.R. and I. Chatzis. 1982. Measurement and Correlation of Conditions for Entrapment and Mobilization of Residual Oil. DOE/BC/10310-20, Department of Energy, Washington, D.C.

Smith, L. 1981. Spatial Variability of Flow Parameters in a Stratified Sand. Math. Geol. Vol. 13.

Taber, J.J. 1981. Research on Enhanced Oil Recovery, Past, Present, and Future. *In*: Surface Phenomena in Enhanced Oil Recovery 1981, D.O. Shah (Ed.) Plenum Publishers.

Wilson, J.L. and S. H. Conrad. 1984. Is Physical Displacement of Residual Hydrocarbons a Realistic Possibility in Aquifer Restoration? *In*: Proceedings of Petroleum Hydrocarbons and Organic Chemicals in Ground Water Prevention, Detection, and Restoration, Houston, TX, pp. 274–298.

SECTION 6

Liquid Contaminants in the Pore Spaces Between Soil Particles in the Unsaturated Zone

CONTENTS

List of Figures 154
6.1 Locus Description 155
  6.1.1 Short Definition 155
  6.1.2 Expanded Definition and Comments 155
6.2 Evaluation of Criteria for Remediation 155
  6.2.1 Introduction 155
  6.2.2 Mobilization/Remobilization 156
    6.2.2.1 Partitioning into/onto Mobile Phases ... 156
      Partitioning into Air 156
      Partitioning into Water 156
    6.2.2.2 Transport of/with Mobile Phase 157
      Introduction 157
      Intrinsic Permeability and Hydraulic Conductivity 158
      Physical Property Temperature-Dependence 160
      Hydraulic and Fluid Conductivity 160
      Relative Permeability 161
      Enhancement of Transport 162
      Darcy's Law 163
      Capillary Tension 165
      Mobility Enhancement 167
  6.2.3 Fixation 167
  6.2.4 Transformation 169
    6.2.4.1 Biodegradation 169
    6.2.4.2 Chemical Oxidation 169
6.3 Storage Capacity in Locus 169
  6.3.1 Introduction and Basic Equations 169
  6.3.2 Guidance on Inputs for, and Calculation of, Maximum Value 169

6.3.1.1 Basic Equations 170
6.3.1.2 Guidance on Parameter Values 170
6.3.3 Guidance on Inputs for, and Calculation of, Average Value 170
6.3.3.1 Basic Equations 170
6.3.3.2 Guidance on Parameter Values 172
6.4 Example Calculations 174
6.4.1 Storage Capacity Calculations 174
6.4.1.1 Maximum Quantity 174
6.4.1.2 Average Quantity 174
6.4.2 Transport Rate Calculations 176
6.5 Summary of Relative Importance of Locus 176
6.5.1 Remediation 176
6.5.2 Loci Interaction 177
6.5.3 Information Gaps 177
6.6 Literature Cited 177
6.7 Additional Reading 178

FIGURES

6–1 Schematic Cross-Sectional Diagram of Locus No. 6—Liquid Contaminants in the Pore Spaces Between Solid Particles or Rock Surfaces in the Unsaturated Zone 157
6–2 Schematic Representation of Important Transformation and Transport Properties Affecting Other Loci 158
6–3 Flow Response of Soils to Hydraulic Gradient 164
6–4 Hydraulic Conductivity Versus Capillary Suction for Different Soils 167

SECTION 6

Liquid Contaminants in the Pore Spaces Between Soil Particles in the Unsaturated Zone

6.1 LOCUS DEFINITION

6.1.1 Short Definition

Liquid contaminants in the pore spaces between soil particles in the unsaturated zone.

6.1.2 Expanded Definition and Comments

The liquid contaminants in locus 6 are in the "water-wet" configuration, (i.e, no substantial contact between liquid contaminant and the surfaces of the soil particles). Soil air is present and liquid contaminant-air interfaces exist. When a spill front from a large liquid release passes through the unsaturated zone, the liquid contaminants may form a continuous phase, filling a large fraction of the void volume. After the spill front has passed, a discontinuous phase is more likely. The mobility of liquid contaminants in locus 6 depends on the volume of the spill, its physicochemical properties, and the hydraulic properties of the porous medium. The liquid contaminants move in response to gravity and capillarity. Figures 6–1 and 6–2 present a schematic cross-sectional diagram of locus no. 6 and a schematic representation of the transformation and transport processes affecting other loci, respectively.

6.2 EVALUATION OF CRITERIA FOR REMEDIATION

6.2.1 Introduction

The evaluation of criteria for remediation of liquid contaminant in locus no. 6 is based on the physical and chemical processes which control the movement of the liquid contaminant through the unsaturated zone. These physical and chemical factors (e.g., hydraulic conductivity, kinematic viscosity, and capillary tension) are discussed in Sections 6.2.2 and 6.2.3.

6.2.2 Mobilization/Remobilization

6.2.2.1 Partitioning into/onto Mobile Phase

Partitioning into Air (Details in Section 1.2.2). Since the liquid contaminant enters locus no. 6 as a mobile phase, discussion of partitioning of the liquid contaminant should focus on how the liquid contaminant leaves the mobile phase. A detailed discussion of the movement of the liquid contaminant through locus no. 6 is presented in Section 6.2.2.2 and applies to movement of liquid water in locus no. 12 as well.

Depending on whether air and water are present and in contact with the liquid contaminant in locus no. 6, the contaminant may volatilize into the air or dissolve into the water. One of the factors controlling the volatilization of a liquid contaminant is its pure chemical vapor pressure. A liquid contaminant with a high pure chemical vapor pressure (e.g., isobutane) may readily volatilize into the air. If the liquid contaminant has a relatively low vapor pressure (e.g., benzene), it may not evaporate as readily. The pure chemical vapor pressure of a liquid increases exponentially with temperature. The variation of temperature in a natural environment is discussed in some detail in Section 6.2.2.2.

Another factor affecting the volatilization of the liquid contaminant is the volume of pure air into which the liquid contaminant volatilizes. Once the saturation vapor pressure in the air above the liquid contaminant is reached, the air and liquid contaminant phases are in equilibrium; no further net exchange of mass between the two phases occurs. If an ambient or artificial pressure gradient exists, air containing the volatilized contaminant is replaced with fresh air into which more liquid contaminant volatilizes.

Diffusion may limit volatilization if it is the predominant transport mechanism for moving the volatilized contaminants away from the liquid-vapor interface. The factors which influence the extent of volatilization and subsequent vapor phase transport are discussed in detail in locus 1.

Partitioning into Water (Details in Section 7.2.2). Although most organic contaminants are sparingly soluble in water, some liquid contaminant may dissolve. Aromatic hydrocarbons (e.g., benzene, toluene, xylene) dissolve more readily into water than chemicals such as isobu-

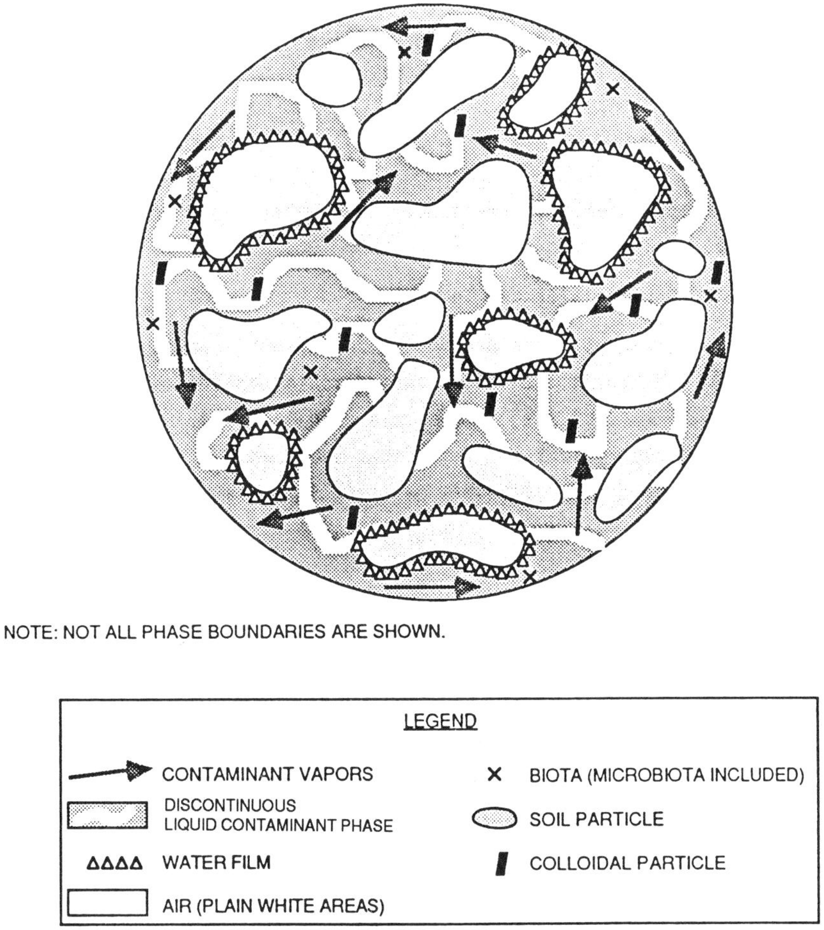

Figure 6–1. Schematic cross-sectional diagram of locus no. 6–liquid contaminants in the pore spaces between solid particles or rock surfaces in the unsaturated zone.

tane. The factors influencing dissolution of liquid contaminant into water are discussed in detail in locus no. 7. Solubilities of compounds typically found in gasoline are presented in Table 7-1.

6.2.2.2 Transport of/with Mobile Phase

Introduction. The mobilization/remobilization of a separate immiscible contaminant phase within the pore spaces of the unsaturated zone is controlled by physical properties and conditions of the porous medium,

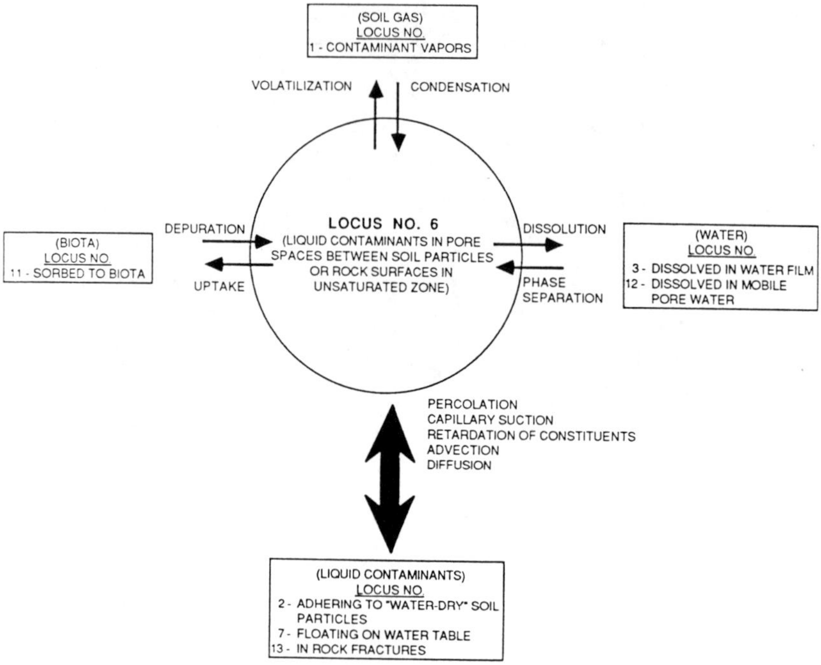

Figure 6–2. Schematic representation of important transformation and transport properties affecting other loci.

contaminant phase, and other immiscible phases present in locus no. 6. Three immiscible fluids—liquid contaminant, liquid water, and air—may be in direct physical contact with one another. In the sections below, the influence of the properties of the porous medium and various phases on the mobilization and subsequent transport of the liquid contaminant through the unsaturated zone are discussed.

The physical properties of the porous medium and of the liquid contaminant (e.g., intrinsic permeability, liquid density, and liquid viscosity) responsible for transport of the liquid contaminant through locus no. 6 are also responsible for initiating the movement of the liquid contaminant in locus no. 6. These same properties which allow movement may change spatially and temporally within a natural environment, arresting the movement of the contaminant liquid within locus no. 6.

Intrinsic Permeability and Hydraulic Conductivity. One of the key parameters governing the initiation of movement of liquid contaminant is the intrinsic permeability, k, of the porous medium. Intrinsic permeability is solely a function of the grain-size distribution of the porous medium and generally decreases from sands to loams to silts to clay soils. Liquid contaminant may move through the unsaturated zone easily, with little resistance to flow through a porous medium with a high intrinsic permeability.

Intrinsic permeability is a measure of the resistance of the unsaturated zone to the flow of a liquid and is independent of the physical properties of the liquid. A more useful flow parameter, however, is the fluid conductivity, $K_{sat,l}$ which is proportional to the intrinsic permeability of the porous medium and inversely proportional to the kinematic viscosity of the liquid, ν_l. For water, the term hydraulic conductivity is adopted; fluid conductivity describes the ease with which a porous medium transmits any fluid.

$$K_{sat,w} = k\ g/\nu_w = k\ \rho_w\ g/\mu_w \tag{6.1}$$

where $K_{sat,w}$ = saturated hydraulic conductivity (cm/sec)
k = intrinsic permeability (cm^2)
ν_w = kinematic viscosity of water (cm^2/sec)
ρ_w = liquid density of water (g/cm^3)
g = gravity constant (g = 980.7 cm/sec^2)
μ_w = dynamic viscosity of water (g/cm sec)

The kinematic viscosity of a liquid is the ratio of its dynamic viscosity to its density. A high density liquid with a low dynamic viscosity has a low kinematic viscosity; it moves more quickly through a porous medium than a liquid with a higher kinematic viscosity. Values for dynamic viscosity, density, and kinematic viscosity of water and several selected organic contaminant liquids at 20°C are tabulated below (Lyman et al., 1982):

Compound	Water	Benzene	Isopentane
liquid density, (g/cm^3)	1.0	0.89	0.62
liquid dynamic viscosity, $\times 10^{-2}$, (g/cm · sec)	1.0	0.64	0.52
liquid kinematic viscosity, $\times 10^{-2}$, (cm^2/sec)	1.0	0.72	0.84

The lower kinematic viscosity values indicate that both liquid benzene and isopentane will flow faster than water through the same porous medium. A more extensive table of kinematic viscosity values is presented in Table 7-1.

Physical Property Temperature-Dependence. Both liquid density and dynamic viscosity are temperature-dependent, increasing as temperature decreases. Dynamic viscosity is, however, more strongly dependent on temperature than is density. As temperature decreases, the kinematic viscosity of the liquid increases; colder liquids have a higher kinematic viscosity and move more slowly through the same porous medium than a warmer liquid.

Because temperature varies in the subsurface on a daily and seasonal basis, the kinematic viscosity of any liquids (e.g., water or liquid contaminant) may vary similarly. Temperature varies in the subsurface with a diurnal (i.e., daily) cycle, warming to a peak by mid-day and cooling during the night (Rosenberg et al., 1983). Soil temperature also fluctuates seasonally, freezing the top several meters of the soil in some latitudes. Beyond two meters, however, temperature of the soil remains fairly constant daily and seasonally.

Soil type and moisture content also influence the response of a soil to temperature changes. Generally, sandy soils transmit heat more efficiently than loamy and clayey soils. The same increase in surface temperature produces a larger temperature change in the sandy soil than in the clayey soil for the same depth. The presence of water in the soil ameliorates temperature extremes because of the high heat of vaporization of water. Soil type, moisture content, latitude, and season are important factors governing the fluctuations of temperature in the subsurface, which in turn affects the kinematic viscosity of the water or liquid contaminant in locus no. 6 and its mobility within the locus.

Hydraulic and Fluid Conductivity. Hydraulic conductivity is measured in the field or in more accurate laboratory experiments on extracted soil samples. The intrinsic permeability may be calculated from the experimentally-determined hydraulic conductivity of the soil materials to a given liquid. The calculated intrinsic permeability may then be used to estimate the fluid conductivity of the same soil material for another liquid with a different kinematic viscosity. A sample calculation is given below based on properties of a synthetic gasoline blend in Table 7–1.

i. measured $K_{sat,w}$ for soil with water, $K = 10^{-2}$ cm/sec
density for water, $\rho_w = 1.0$ g/cm^3
dynamic viscosity for water, $\mu_w = 10^{-2}$ g/cm sec

ii. calculated k for soil, $k = 10^{-7}$ cm^2

iii. density for gasoline, $\rho_l = 0.7$ g/cm^3
dynamic viscosity for gasoline, $\mu_l = 4.92 \times 10^{-3}$ g/cm sec

iv. estimated K for soil with gasoline, $K_{sat,g} = 1.44 \times 10^{-2}$ cm/sec

The sample calculation indicates that the same porous medium is able to conduct the synthetic gasoline about forty four percent faster than water because of the difference in liquid kinematic viscosity.

Relative Permeability. Variation in saturation of a fluid in locus no. 6 affects the fluid conductivity of the partially-saturated porous medium to that fluid. If there are several immiscible fluids present (i.e., air, water, and liquid contaminant) the increase in saturation of one immiscible liquid reduces the fluid conductivity of the porous medium relative the other immiscible fluids. When the porous medium is relatively unsaturated with the liquid contaminant, the fluid conductivity of the porous medium for the liquid contaminant is at a minimum and the liquid contaminant may be relatively immobile.

For an immiscible fluid in a porous medium, the variation in fluid conductivity with the degree of saturation is known as relative permeability. A commonly-used expression for estimating the relative permeability for two or more immiscible fluids is the Brooks-Corey relationship (Personal communications with D.K. Sunada, Civil Engineering Dept., Colorado State University, Ft. Collins, 1987). The Brooks-Corey relationship estimates relative permeability for an immiscible liquid from the saturation of the porous medium relative to that fluid and the empirical grain-size distribution parameter:

wetting liquid

$$k_{rw} = [(S-S_r)/(1-S_r)]^{(2+3\,b)/b} \tag{6.2}$$

non-wetting liquid

$$k_{nrw} = [1-(S-S_r)/(1-S_r)]^2 \cdot [1-(S-S_r)/(1-S_r)^{(2+b)/b}] \tag{6.3}$$

where k_{rw} = wetting liquid relative permeability (dim.)
k_{nrw} = non-wetting liquid relative permeability (dim.)
S = ratio of volume of water to pore volume (cm^3/cm^3)
S_r = ratio of irreducible water volume to pore volume (cm^3/cm^3)
b = grain-size distribution parameter (dim.)

and the range in values of relative permeability for the wetting and non-wetting liquids are:

$$0 \leq k_{rw} \leq 1$$

$$0 \leq k_{nrw} \leq 1$$

- Sample calculation of wetting and non-wetting liquid relative permeability at b = 2.8 and S_r = 0.05.

Wetting Saturation (S)	k_{rw}	k_{nrw}
0.0	0.0	1.0
0.25	0.003	0.580
0.50	0.062	0.200
0.75	0.322	0.028
1.0	1.0	0.0

This example shows the importance of saturation on the relative permeability of a fluid in the presence of another immiscible fluid; the saturation must be high for the fluid to flow readily.

The increase in fluid conductivity with saturation is due to the decrease in frictive resistance to flow per unit cross-sectional area which decreases with increasing pore size. Small pores exert a higher resistance to flow per unit cross-sectional area than larger pores and have, therefore, lower fluid conductivity than larger pores for the same liquid.

At low saturation levels, the smallest pores of the medium are filled with a liquid because of high capillary tension exerted by the smaller pores. Because the liquid exists within the smaller pores at higher capillary tension, the fluid conductivity of the medium to that liquid is small. As saturation increases, larger pores begin to fill with liquid, and the fluid conductivity of the porous medium to the liquid increases to a maximum at complete saturation.

Enhancement of Transport. The preceding discussion described how fluid conductivity affects mobilization and transport of the liquid contaminant within locus no. 6. If mobilization and transport of the liquid contaminant is desired, some enhancement of the fluid conductivity of the porous medium to the liquid contaminant is necessary. Such an enhancement of the fluid conductivity might be accomplished by changing one or more of the parameters (i.e., intrinsic permeability, liquid density, dynamic viscosity, and saturation) upon which the fluid conductivity depends.

As stated above, the fluid conductivity is directly proportional to the intrinsic permeability of the porous medium. It is possible, albeit perhaps impractical, to enhance the intrinsic permeability of the porous medium by removing some of the fine-grained material of the porous medium. Such a removal of fine-grained material often occurs when a newly-installed water well in unconsolidated porous materials first pumps water.

The removal of fine-grained material from the porous medium by the newly-installed well usually is limited to the immediate volume of porous medium surrounding the well and usually ceases shortly after initiation of pumping, if the water well has been properly designed. The removal of fine-grained material from the porous medium may enhance locally the intrinsic permeability, but would be expensive to implement on a large scale. Furthermore, substantial removal of porous medium materials may result in subsidence of the land surface with an attendant decrease in the fluid conductivity because of compaction. In summary, the intrinsic permeability probably cannot be significantly enhanced to induce mobilization and transport.

It is difficult to imagine how the liquid density of the liquid contaminant could be enhanced to increase the fluid conductivity of the porous medium to the liquid contaminant; small increases in density would come with a lowering of the temperature. The dynamic viscosity of the liquid contaminant possibly could be altered with the addition of some emulsifying compound, altering the liquid contaminant to a more fluid state. The reduced dynamic viscosity of the emulsified liquid contaminant would allow the liquid contaminant to flow more readily through the unsaturated porous medium of locus no. 6.

The relative permeability of the porous medium of locus no. 6 to the liquid contaminant could be increased by increasing the saturation of liquid contaminant in locus no. 6. But a healthier approach toward increasing the potential for mobilization would be to reduce the water saturation, the other principal immiscible phase in locus no. 6. The reduction of water content in locus no. 6 would reduce the relative permeability of the porous medium to water and increase it for the liquid contaminant; the liquid contaminant would then be more mobile.

Darcy's Law. The role of hydraulic conductivity or fluid conductivity can be visualized as the slope of the line produced by plotting the flow per unit area against the hydraulic gradient which produces that flow. Figure 6–3 depicts the flow response of a sandy soil and a clayey soil to changes in hydraulic gradient. The steeper slope associated with the sandy soil reflects its higher hydraulic conductivity relative to the clayey soil.

This simple linear relationship is known as Darcy's law. Darcy's law states that the rate of flow of a volume of the liquid contaminant through the porous medium of locus no. 6 is proportional to the stress

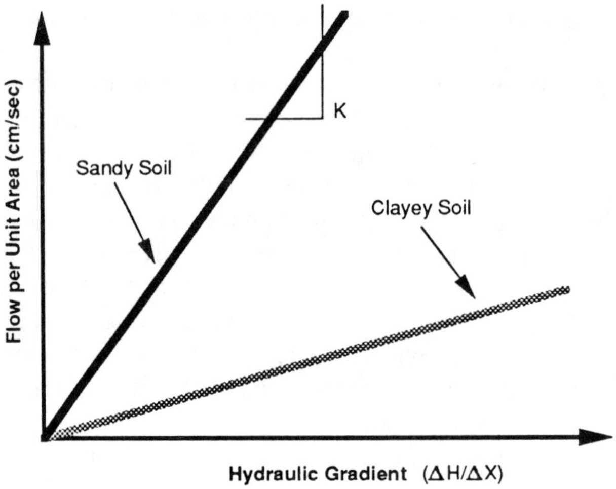

Figure 6–3. Flow response of soils to hydraulic gradient. *Source:* Hillel, 1980. (Copyright Academic Press, Inc., 1980. Reprinted with permission.)

applied to the liquid; the higher the stress, the more liquid flows. The stress is applied by ambient and artificial pressure gradients and from the downward force of gravity. A one-dimensional, vertical expression for Darcy's law is:

$$q = (k\, k_r/\mu)\,(dP/dz + \rho\, g\, dh/dz) \tag{6.4}$$

where
μ = dynamic liquid viscosity (g/cm sec)
k = intrinsic permeability (cm^2)
q = discharge of liquid per unit cross-sectional flow area (cm^3/sec cm^2)
k_r = relative permeability (dim.)
dP/dz = vertical ambient or artificial pressure gradient (g cm sec^{-2}/cm^2 cm)
dh/dz = gravity gradient (cm/cm)
z = depth (cm)
ρ = liquid density (g/cm^3)
g = gravity constant, $g = 980.7\ cm/sec^2$

Generally, in most natural settings of a liquid contaminant in the unsaturated zone, it is difficult to apply an external artificial pressure stress to the unsaturated zone having anything more than a limited areal effect (Personal communication with D.L. Marrin, Consultant, La-Jolla, CA, 1987). By the same token, ambient pressure gradients, while certainly affecting the air phase of locus no. 6, generally are of small magnitude and have a small effect on the movement of liquid contaminant. By neglecting the effects of artificial and ambient pressure gradients on the liquid contaminant in locus no. 6, the remaining force which transports the liquid contaminant is gravity. Assuming a unit vertical gravity gradient and no external pressure gradients:

$$dh/dz = 1$$

and

$$dP/dz = 0$$

Darcy's Law becomes:

$$q = k\ k_r\ \rho\ g/\mu \tag{6.5}$$

Capillary Tension. Gravity produces a downward movement of the liquid contaminant locus no. 6. Acting against the downward-driving force of gravity is the dissipative force of capillary tension. Capillary tension is the force arising in any partially saturated porous medium (i.e., soils, sponges, paper towels) as a result of the surface tension of the liquid and its contact angle with the solid grains of the porous medium. The result is a suction or tension exerted on the liquid by the pores of the porous medium. Capillary tension also varies with saturation of the porous medium, being strongest at the lowest saturation levels. An empirical relationship between capillary tension and saturation was reported by Clapp and Hornberger (1978):

$$\psi = \psi_s \cdot (\Theta/\Theta_s)^{-b} \tag{6.6}$$

where
- ψ = capillary tension at some moisture content (cm H_2O)
- ψ_s = capillary tension at complete moisture saturation (cm H_2O)
- Θ = moisture content (cm^3/cm^3)
- Θ_s = moisture content at saturation (cm^3/cm^3)
- b = empirical grain-size distribution parameter

Representative values for the parameters b, ψ_s and Θ_s are given in Table 12–2 (Section 12).

Generally, the smaller the pore, the more tightly held the liquid. The range in capillary tension among the pores is from the atmospheric pressure (i.e., 1 atm) in the largest pores to large negative pressures (e.g., suction at -15 atm) in the smallest pores. Since the unsaturated medium of locus no. 6 is open to the atmosphere, there will be a capillary pressure gradient between a liquid at atmospheric pressure and a liquid held at less-than-atmospheric pressure by capillary tension. Below are example calculations with parameter values taken from Clapp and Hornberger (1978):

		Sand	Loam
b	:	4.05	5.39
Θ_s(cm^3/cm^3)	:	0.395	0.451
ψ_s(cm H_2O)	:	12.1	47.8

Same moisture contents, different materials.

	(cm$^{3\Theta}$/cm^3)	(cm$^\psi$ H_2O)
Sand	0.20	190
Loam	0.20	3827

Same materials, different moisture content.

	(cm$^{3\Theta}$/cm^3)	(cm$^\psi$ H_2O)
$Sand_1$	0.20	190
$Sand_2$	0.35	20

The sample calculations indicate that the loam holds the same amount of water as the sand at a much higher capillary tension. The calculations also show that, for the same sand, a smaller water volume is held at a higher capillary tension than a larger water volume.

The capillary pressure gradient acts on a liquid entering the unsaturated medium of locus no. 6, causing the movement of liquid in the direction of decreasing capillary pressure (i.e., from high to low pressure). The movement of liquid in locus no. 6 from high to low pressure areas typically causes the liquid contaminant to spread laterally as it moves downward.

The dependence of hydraulic conductivity and capillary tension on moisture content implies that hydraulic conductivity is also dependent on capillary tension. As the soil is dewatered, the capillary tension increases, and the hydraulic conductivity decreases. The effect of increasing capillary tension on hydraulic conductivity varies with soil texture and can be seen in Figure 6–4.

Figure 6–4 indicates that hydraulic conductivity for the sandy and clayey soils decreases with increasing capillary tension. Another important point to be drawn from Figure 6–4 is that when the capillary tension reaches a certain point, the hydraulic conductivity of the clayey soil

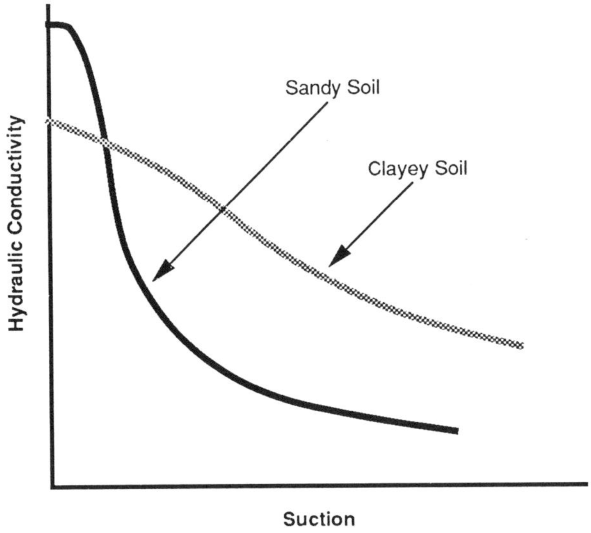

Figure 6–4. Hydraulic conductivity versus capillary suction for different soils. *Source:* Hillel, 1980. (Copyright Academic Press, Inc., 1980. Reprinted with permission.)

actually exceeds that of the sandy soil. This means that, in spite of the fact that the sandy soil may conduct fluids more readily at saturation, moisture is held more strongly by the finer pores of the clayey soil resulting in a larger hydraulic conductivity of the clayey soil at high capillary tension.

Mobility Enhancement. The mobilization of the contaminant liquid in locus no. 6 could be enhanced by reducing the effect of capillary tension on the liquid. An obvious but unfeasible method for doing this would be to add more liquid contaminant, increasing the saturation—not a desirable situation. A more appropriate alternative might be the addition of emulsifiers to the liquid contaminant which would lower the surface tension of the liquid, making the emulsified liquid contaminant more mobile.

6.2.3 Fixation

As stated above, many of the hitherto described transport parameters that allow the initiation of movement and translocation of a liquid contaminant in locus no. 6 also can retard and arrest movement of the liquid contaminant. The fixation of the liquid contaminant in locus no.

6 thus arises from natural or artificial changes induced in the transport parameters which may be both spatial and temporal.

The intrinsic permeability of the porous medium may reduce with time as the result of compaction of the ground by application of external weight or by groundwater removal. The result would compress the porous medium, reducing its ability to transmit any liquid or fluid. Such a compaction of the porous medium could be effected prior to emplacement of a new UST, giving the site added protection against migration of the liquid contaminant.

Variability in sediment sizes and distributions is often large in the natural environment and intrinsic permeability of a porous medium is spatially variable. For example, the upper portion of locus no. 6 might have an intrinsic permeability high enough to initiate movement of the liquid contaminant with translocation. Moving downward, it is conceivable that the sediments of locus no. 6 become finer, reducing the intrinsic permeability of the porous medium. At some depth, it is possible that the movement of the liquid contaminant would become arrested by a low intrinsic permeability. A new UST placed in such a natural environment with a vertical decrease in grain size and intrinsic permeability would have a natural protection against widespread vertical and lateral liquid contaminant migration.

The relative permeability of the porous medium of locus no. 6 to the contaminant liquid could be reduced sufficiently to arrest the liquid. This reduction could be accomplished by lowering the saturation of the liquid contaminant by pumping. This may be a costly and inefficient remedial alternative if the liquid contaminant exists in locus no. 6 as a discontinuous phase. Alternatively, the relative permeability of the liquid contaminant may be reduced by the increase in saturation of water. The problem with this alternative, apart from the increase in contaminated water, is that the added water might force the liquid contaminant ahead of it, displacing it instead of immobilizing it.

The density and dynamic viscosity of the contaminant liquid changes with time because of changes in the composition of the liquid contaminant. If the air phase is in contact with the liquid contaminant, volatile components of the liquid contaminant may evaporate. For example, in the case of gasoline, the lighter-molecular weight alkane chemicals (e.g., isobutane, isopentane, n-butane) volatilize readily. As the contaminant liquid contacts pure water in the unsaturated zone, the more soluble components of the contaminant liquid (e.g., simple aromatic hydrocarbons) dissolve into the water. The aged contaminant liquid left as residuum by the processes of volatilization and dissolution will be a more viscous immobile liquid compared to fresh contaminant liquid.

6.2.4 Transformation

6.2.4.1 Biodegradation (Details in Section 11)

Biodegradation is most likely to take place in an aerated aqueous environment in the unsaturated zone. Therefore, liquid contaminant in locus no. 6 is likely to be degraded only after passing into the water film which wets the soil particles around which they flow. Some microbes can live while suspended in pure hydrocarbon but these are rare and metabolize oil slowly. Therefore, locus no. 6 contaminants would likely diffuse into other loci before they could be biodegraded in place.

6.2.4.2 Chemical Oxidation (Details in Section 3.2.4.2)

Liquid contaminants in soil which do not come in contact with mineral surfaces may not undergo catalytic transformations. In order for chemical oxidation to occur in this locus, an oxidant such as ozone or peroxide must be introduced.

6.3 STORAGE CAPACITY IN LOCUS

6.3.1 Introduction and Basic Equations

Determination of the amount of liquid contaminant held in locus no. 6 is based on equations and guidance on selection of appropriate parameter values given below.

6.3.2 Guidance on Inputs for, and Calculations of, Maximum Value

The maximum quantity of liquid contaminant held in locus no. 6 occurs in the presence of a liquid contaminant spill front. Under this condition, it is assumed that:

- the migrating contaminant liquid fills the air-filled porosity;
- the liquid contaminant and the soil water are immiscible; and
- the moisture-filled porosity is assumed to be minimal so that it does not impede the vertical migration of the contaminant front nor occupies a significant fraction of the total porosity.

It is also assumed that the quantity of liquid contaminant is very much larger than any quantity of contaminant mass lost to volatilization into the soil air, dissolution into the vadose zone water, microbial degradation and uptake, and attenuation to soil particles.

6.3.1.1 Basic Equations

Mass of liquid contamination per unit volume of soil held in locus no. 6.

$$m_g = \rho_l \, \Theta_a \tag{6.7}$$

where m_g = mass of liquid contaminant per unit volume of contaminant (kg/m^3)
Θ_a = air-filled void space filled with contaminant liquid (m^3/m^3)
ρ_l = density of liquid contaminant (kg/m^3)

Air-filled porosity in locus no. 6.

$$\Theta_a = \Theta_t - \Theta_w \tag{6.8}$$

where Θ_t = total porosity (m^3/m^3)
Θ_w = water-filled porosity (m^3/m^3)

6.3.1.2 Guidance on Parameter Values

Total Porosity, Θ_t.

- Theoretical maximum porosity for a sandy soil is 47.6%.
- Refer to Table 12–1 for representative total porosity values for a variety of consolidated and unconsolidated sediments.

Water-Filled Porosity, Θ_w.

- Water-filled porosity ranges from 0 to Θ_t.
- The air-filled porosity is maximized by choosing a moisture content at wilting point, which is available for various soils from Figure 12–4.

6.3.3 Guidance on Inputs for, and Calculation of, Average Value

6.3.3.1 Basic Equations

The average quantity of liquid contaminant in locus no. 6 is defined as the quantity of liquid remaining in the locus after the passage of the contaminant spill front. The available porosity within which the volume of residual liquid contaminant may be held in locus no. 6 under these conditions is affected by the water-filled porosity:

$$\Theta_{ac} = \Theta_t - \Theta_w \qquad (6.9)$$

where Θ_{ac} = air-filled and contaminant-filled porosity (m^3/m^3)

The air-filled porosity, Θ_a, is calculated from:

$$\Theta_a = \Theta_{ac} - \Theta_c \qquad (6.10)$$

where Θ_c = contaminant-filled porosity (m^3/m^3)

With the passage of the contaminant front, some quantity of liquid contaminant remains trapped by capillary forces or by other trapping mechanisms. At this point, the gravity forces causing drainage are balanced by capillary forces and the liquid contaminant volume is at the field capacity of the soil. Clapp and Hornberger (1978) estimated the contaminant-filled porosity from:

$$\Theta_c = \Theta_s \cdot (\psi_{fc}/\psi_s)^{-1/b} \qquad (6.11)$$

where Θ_s = saturation content of liquid contaminant (m^3 of liquid contaminant per 1 m^3 of soil volume)
ψ_{fc} = capillary tension at field capacity (cm H_2O)
ψ_s = capillary tension at saturation (cm H_2O)
b = grain-size distribution parameter (dim.)

The contaminant-filled porosity can be converted to a mass of residual liquid contaminant per unit volume of soil:

$$m_{res} = \Theta_c \cdot \rho_l \qquad (6.12)$$

where m_{res} = mass or residual liquid contaminant per unit soil volume (kg/m^3)
ρ_l = density of liquid contaminant (kg/m^3)

Since not all the porosity is filled by a continuous supply of liquid contaminant as described in Section 6.3.2, liquid contaminant can be lost to dissolution to vadose zone water, volatilization into vadose zone air, microbial degradation, and chemical attenuation to soils. When these losses are subtracted from the residual mass of contaminant liquid, the mass of liquid contaminant per unit volume of soil held in locus no. 6 is:

$$m_g = m_{res} - m_w - m_v - m_b - m_s \tag{6.13}$$

where m_g = mass of liquid contaminant per unit volume held in locus no. 6 after contaminant front has passed (kg/m^3)
m_{res} = mass of liquid contaminant per unit volume held as residual quantity in locus no. 6 (kg/m^3)
m_w = mass of liquid contaminant per unit volume lost to dissolution into vadose zone water (kg/m^3)
m_v = mass of liquid contaminant per unit volume lost to volatilization into vadose zone air (kg/m^3)
m_b = mass of liquid contaminant per unit volume lost to microbial degradation (kg/m^3)
m_s = mass of liquid contaminant per unit volume lost to attenuation to soils (kg/m^3)

Losses to microbial degradation and attenuation will be assumed to be negligible for the present analysis, $m_b = m_s = 0$. The mass of liquid contaminant lost to dissolution into the vadose zone water is calculated from:

$$m_w = S \cdot \Theta_w / 10^3 \tag{6.14}$$

where S = solubility of liquid contaminant in water (mg/L)

The mass of liquid contaminant lost to volatilization into the vadose zone air is calculated from:

$$m_v = \Theta_a \cdot \rho_{sg} \tag{6.15}$$

where ρ_{sg} = vapor density of volatilized liquid contaminant-air mixture (kg/m^3)

6.3.3.2 Guidance on Parameter Values

Grain-Size Distribution Parameter, b.

- Empirical factor related to soil grain size and distribution, increasing with decreasing grain size. Refer to Table 12–2 for representative values of b for a variety of soils.

Capillary Tension at Field Capacity, ψ_{fc}.

- Use whatever capillary tension desired for field capacity. Recommend using 0.3 bar tension (i.e., 306 cm H_2O).

Capillary Tension at Saturation, ψ_s.

- Use values reported in Table 12–2 for various soils.

Saturation Content, Θ_s.

- Use values reported in Table 12–2 for various soils.

Solubility, S.

- User should supply solubility of liquid contaminant of interest.
- For gasoline containing no oxygenated additives, use 200 mg/L for solubility. See locus 3 for detailed information of solubilities of organic chemicals commonly found in gasolines.

Vapor Density, ρ_{sg}.

- For gasoline vapors, Table 1–2 indicates a range of 0.51 kg/m^3 at 0°C to 1.04 kg/m^3 at 20°C.
- Vapor density may be calculated for any temperature and gasoline from:

$$\rho_{sg} = \bar{M}_{ca}\, P_a/T\, R \tag{6.16}$$

where $\bar{M}_{ca}$ = average molecular weight of volatilized liquid contaminant-air mixture (g/mol)
P_a = atmospheric pressure, P_a = 760 mm Hg
T = temperature (K)
R = universal gas constant,
R = 0.06236 mm Hg m^3/mol K

The average molecular weight of the volatilized liquid contaminant air mixture is calculated from:

$$\bar{M}_{ca} = M_c\,(P_t/P_a) + M_a\,(1 - P_t/P_a) \tag{6.17}$$

where: M_c = molecular weight of volatilized liquid contaminant (g/mol)
M_a = molecular weight of air = 28.97 g/mol
P_t = equilibrium vapor pressure of liquid contaminant (mm Hg)

6.4 EXAMPLE CALCULATIONS

6.4.1 Storage Capacity Calculations

6.4.1.1 Maximum Quantity

For this sample calculation, a sandy soil is assumed. The available quantity of contaminant liquid is maximized by minimizing the water content in the soil. The wilting point is a convenient measuring point for this minimum water content. For sandy soils, the water content at the wilting point is about 5 percent (See Figure 12–4). The total porosity of the sandy soil is assumed to be 39.5% (Clapp and Hornberger, 1978). The air-filled porosity in which contaminant liquid may be stored is:

$$\Theta_a = \Theta_t - \Theta_w$$

$$\Theta_t = 0.395$$

$$\Theta_w = 0.05$$

$$\Theta_a = 0.345$$

Assuming a liquid contaminant density similar to the gasoline reported in Table 7–2:

$$\rho_g = 710 \text{ kg/m}^3$$

The maximum mass of liquid contaminant per unit volume of soil held in locus no. 6 is:

$$m_g = \rho_g \, \Theta_a$$

$$m_g = (710 \text{ kg/m}^3)\,(0.345)$$

$$m_g = 245 \text{ kg liquid contaminant per 1 m}^3 \text{ of soil volume}$$

The maximum of liquid contaminant mass per unit volume held in locus no. 6 is 245 kg/m^3 or 345 L/m^3.

6.4.1.2 Average Quantity

A sandy soil will also be assumed for the calculation of the average quantity of contaminant liquid held in locus no. 6. It is assumed that the water-filled porosity is 5% and the total porosity is 39.5%. As stated in

Section 6.4.1.1, the maximum available porosity for contaminant liquid is 34.5%. Using equation 6.11, the porosity occupied by contaminant liquid at field capacity is:

$$\Theta_c = \Theta_s \cdot (\psi_{fc}/\psi_s)^{-1/b}$$

Values for the grain-size distribution parameter and capillary tension at field capacity are taken from Clapp and Hornberger (1978):

$$\psi_{fc} = 306 \text{ cm } H_2O$$

$$\psi_s = 12 \text{ cm } H_2O$$

$$b = 4.05$$

$$\Theta_c = 0.155$$

The air-filled porosity is calculated:

$$\Theta_a = \Theta_{ac} - \Theta_c$$

$$0.19 = 0.345 - 0.155$$

Using the values for air-, water-, and contaminant-filled porosity, the residual mass and the masses lost to volatilization and dissolution may be calculated:

Residual Mass:

$$m_{res} = \Theta_c \, \rho_g$$

$$m_{res} = (0.155)\,(710 \text{ kg/m}^3)$$

$$m_{res} = 110 \text{ kg/m}^3 \text{ or } 155 \text{ L/m}^3$$

Mass lost to Volatilization:

$$m_v = \Theta_a \, \rho_{sg}$$

$$m_v = (0.19)\,(1.04 \text{ kg/m}^3)$$

$$m_v = 0.198 \text{ kg/m}^3$$

Mass lost to Dissolution:

$$m_w = S \, \Theta_w/10^3$$

$$m_w = (130 \text{ mg/L})\,(0.05)/10^3$$

$$m_w = 6.5 \times 10^{-3} \text{ kg/m}^3 \text{ or } 9.2 \times 10^{-3} \text{ L/m}^3$$

Mass held in locus no. 6 under average conditions:

$$m_g = m_{res} - m_v - m_w$$

$$m_g = 110 \text{ kg/m}^3 - 0.198 \text{ kg/m}^3 - 6.5 \times 10^{-3} \text{ kg/m}^3$$

$$m_g = 109.8 \text{ kg/m}^3 \text{ or } 154.6 \text{ L/m}^3$$

The mass of liquid contaminant lost to biodegradation and alteration to soil is assumed to be negligible; $m_b = m_s = 0$.

6.4.2 Transport Rate Calculations

Sample calculations are provided in Section 6.2 for selected equations.

6.5 SUMMARY OR RELATIVE IMPORTANCE OF LOCUS

6.5.1 Remediation

The presence of liquid contaminant in the unsaturated zone is a source for contamination of the soil air (locus no. 1) by volatilization. It also contaminates unsaturated zone water (loci nos. 3 and 12) by dissolution. The extent to which the liquid contaminant in locus no. 6 contaminates locus no. 1 or loci nos. 3 and 12 depends on the vapor pressure, solubility, and the kinematic viscosity of the liquid contaminant.

The amount of mass per unit volume held in locus no. 6 depends on the density of the liquid and the volume of pore space. More mass per unit volume of a denser liquid may be stored in the same volume. Sample calculations in section 6.4.1 indicate that a dry sandy soil holds a large fraction of liquid contaminant. In fact, locus no. 6 can hold as much mass per unit volume as loci nos. 5 and 7, and certainly more than loci nos. 1–4 and 8–13.

If the liquid contaminant in locus no. 6 occupies a significant fraction of the unsaturated zone porosity, if the liquid has a low kinematic viscosity, and if the intrinsic permeability is sufficient, then the contaminant liquid may be mobile. If this is the case, the pumping of liquid contaminant may be a viable remedial option. As the liquid contaminant is removed by pumping, however, some fraction remains trapped in the pores of the unsaturated zone. As described in Section 6.2.2.2, as the saturation level of the liquid contaminant drops, it will become progressively harder to withdraw more liquid contaminant. Depending on the volatility of the liquid contaminant, however, vacuum extraction may remove much of the remaining liquid contaminant.

6.5.2 Loci Interactions

Volatilization from locus no. 6 represents the primary mechanism for partitioning to locus no. 1. Liquid contaminant may dissolve into the unsaturated zone water (loci nos. 3 and 12) and attenuate to soil particles (loci nos. 2 and 4). If the liquid contaminant is of sufficient quantity or is mobile (i.e., low kinematic viscosity), it may transfer to loci nos. 7 and 8.

6.5.3 Information Gaps

Information in the following areas is necessary for improved understanding of immiscible liquid contaminant transport in the unsaturated zone:

1. Chemical composition and physicochemical properties of the liquid contaminant;
2. Information on effects of trapping mechanism and capillary tension on liquid contaminant.
3. Understanding the behavior of three-phase flow in the unsaturated zone.

6.6 LITERATURE CITED

Clapp, R.B. and G.M. Hornberger. 1978. Empirical Equations for Some Soil Hydraulic Properties. Water Resources Research, 14(4):601–604.

Hillel, D. 1980. Fundamentals of Soil Physics, Academic Press.

Lyman, W.J., W.F. Reehl, and D.H. Rosenblatt. 1982. Handbook of Chemical Property Estimation Methods, McGraw-Hill Book Co., New York.

Rosenberg, N.J., B.L. Blad, and S.B. Verma. 1983. Microclimates: The Biological Environment, 2nd ed., Wiley-Interscience.

6.7 ADDITIONAL READING

Aleman, M.A. and J.C. Slattery. 1988. Estimation of Three-phase Relative Permeabilities. Transport in Porous Media, 3:111–131.

Baehr, A.L., G.E. Hoag, and M.C. Marley. 1989. Removing Volatile Contaminants from the Unsaturated Zone by Inducing Advective Air-phase Transport. Journal of Contaminant Hydrology, 4:1–26.

Conrad, S.H., J.L. Wilson, W. Mason, and W. Peplinski. 1988. Observing the Transport and Fate of Petroleum Hydrocarbons in Soil and in Groundwater Using Flow Visualization Techniques. Proc. of American Association of Petroleum Geologists Symposium, May 10.

Cyr, T.J., V. de la Cruz, and T.J.T. Spanos. 1988. An Analysis of the Viability of Polymer Flooding as an Enhanced Oil Recovery Technology. Transport in Porous Media, 3:591–618.

Darcos, T. 1978. Theoretical Considerations and Practical Implications on the Infiltration of Hydrocarbons in Aquifers, Int. Symp. on Ground Water Pollution by Oil Hydrocarbons, Prague, pp. 127–137.

Delshad, M. and G.A. Pope. No date. Comparison of the Three-phase Oil Relative Permeability Models, submitted to Transport in Porous Media.

Faust, C.R., J.H. Guswa, and J.W. Mercer. 1988. Simulation of Three-dimensional Flow of Immiscible Fluids within and below the Unsaturated Zone, submitted to Water Resources Research.

Kuppusamy, T., J. Sheng, J.C. Parker, and R.J. Lenhard. 1987. Finite-element Analysis of Multiphase Immiscible Flow Through Soils. Water Resources Research, 23(4):625–631.

Reible, D.D., T.H. Illangasekare, D.V. Doshi, and M.E. Malhiet. 1988. Infiltration of Immiscible Contaminants in the Unsaturated Zone, submitted to Ground Water.

Schramm, M., A.W. Warrick, and W.H. Fuller. 1986. Permeability of Soils to Four Organic Liquids, and Water. Hazardous Waste and Hazardous Materials, 3(1):21.

Touma, J. and M. Vauclin. 1986. Experimental and Numerical Analysis of Two-phase Infiltration in a Partially Saturated Soil. Transport in Porous Media, 1:27–55.

Tyler, S.W., M.R. Whitbeck, M.W. Kirk, J.W. Hess, L.G. Everett, D.K. Kreamer, and B.H. Wilson. 1987. Processes Affecting Subsurface Transport of Leaking Underground Storage Tank Fluids, EPA/600/6-87/005, EPA Environmental Monitoring Systems Laboratory, Las Vegas, NV.

Van der Waarden, M., A.L.A.M. Birdie, and W.M. Groenewoud. 1971. Transport of Mineral Oil Components to Groundwater, I. Model Experiments on the Transfer of Hydrocarbons from a Residual Oil Zone to Trickling Water. Water Research, 5:213–226.

Van der Waarden, M., A.L.A.M. Birdie, and W.M. Groenewoud. 1977.

Transport of Mineral Oil Components to Groundwater, II. Influence of Lime, Clay, and Organic Soil Components on the Rate of Transport. Water Research, 11:359–365.

SECTION 7

Liquid Contaminants Floating Upon the Water Tables

CONTENTS

List of Tables 182
List of Figures 182
7.1 Locus Description 183
7.1.1 Short Definition 183
7.1.2 Expanded Definition and Comments 183
7.2 Evaluation of Criteria for Remediation 184
7.2.1 Introduction 184
7.2.2 Mobilization/Remobilization 184
7.2.2.1 Partitioning onto Mobile Phases 184
Volatilization 184
Dissolution 185
7.2.2.2 Transport of/with Mobile Phase 187
7.2.3 Fixation 194
7.2.3.1 Partitioning onto Immobile (Stationary) Phase 195
Abandonment 195
Adsorption 196
Dissolution 196
Diffusion 196
7.2.3.2 Other Fixation Approaches 196
7.2.4 Transformation 196
7.2.4.1 Biodegradation 196
7.2.4.2 Chemical Oxidations 197
7.3 Storage Capacity in Locus 197
7.3.1 Introduction and Basic Equations 197
7.3.2 Guidance on Inputs for, and Calculations of, Maximum Value 198
7.3.3 Guidance on Inputs for, and Calculations of, Average Values 199
7.4 Example Calculations 199
7.4.1 Storage Capacity Calculations 199

7.4.1.1 Maximum Storage Capacity 199
7.4.1.2 Average Storage Capacity 199
7.4.2 Transport Rate Calculations 200
7.5 Summary of Relative Importance of Locus 200
7.5.1 Remediation 200
7.5.2 Loci Interaction 201
7.5.3 Information Gaps 202
7.6 Literature Cited 202
7.7 Other References 203

TABLES

7-1 Physicochemical Properties of Selected Chemical Constituents of a Synthetic Gasoline at 20°C 188
7-2 Physicochemical Properties of Selected Chemical Constituents of an "Aged" Synthetic Gasoline at 20°C .. 190
7-3 Average Grain Size and Funicular Zone Thickness for Various Geologic Materials 192
7-4 Examples of Product Recovery After Tank Leaks 193
7-5 Total Porosities for Common Consolidated and Unconsolidated Sediments 197
7-6 Loci Interactions with Liquid Contaminant Floating on Groundwater 201

FIGURES

7-1 Schematic Cross-Sectional Diagram of Locus No. 7—Liquid Contaminants Floating on the Groundwater Table 185
7-2 Schematic Representation of Important Transformation and Transport Processes Affecting Other Loci 186
7-3 Schematic of Immiscible Liquid Contaminant Accumulating Upon a Deflected Water Table 187
7-4 Schematic of the Capillary Zone With a Typical Water Saturation Curve 193
7-5 Mobilization of Contaminants in Locus 7 194

SECTION 7

Liquid Contaminants Floating Upon the Water Tables

7.1 LOCUS DEFINITION

7.1.1 Short Definition

Liquid contaminants floating upon the water tables.

7.1.2 Expanded Definition and Comments

The leaked liquid contaminant in this locus floats upon the water table and is generally less dense than water; in some circumstances, liquids which are denser than water may also float upon the water table. If the liquid contaminant has been released in sufficient quantity, it infiltrates downward through the vadose zone (unsaturated zone), forming a continuous pool in the vicinity of the leak. The weight of the accumulating liquid contaminant deflects the water table downward. At the fringes of the pool, the liquid contaminant may form viscous stringers as it migrates into zones of low interfacial tension or high hydraulic conductivity. The liquid contaminant spreads laterally under its own weight away from the leak source, or it may move downgradient atop a sloping water table.

While aspects of this locus are obviously close to those of loci 5 and 6, the main feature is the existence of a floating layer of relatively immiscible liquid upon the water table. This locus also includes liquid contaminant floating upon perched water tables. Figure 7-1 is a schematic of locus 7 and Figure 7-2 represents the transformation and transport processes affecting other loci.

7.2 EVALUATION OF CRITERIA FOR REMEDIATION

7.2.1 Introduction

Liquid contaminant is present in locus 7 because of a leak which occurred in the subsurface or upon the land surface. The infiltrating liquid contaminant loses mass by volatilization into soil air and by dissolution into soil water. Liquid contaminant is also lost to retention in the voids of the vadose zone by capillarity. If the leak has released a quantity of liquid contaminant larger than losses to volatilization, dissolution, and capillary retention, the liquid contaminant accumulates upon the water table. For a continuous leak in a homogeneous, isotropic, and unconfined porous medium, a mound of liquid contaminant may form, which deflects the water table under its weight (see Figure 7–3).

The remainder of this section describes in some detail the factors affecting the mobilization, fixation, and transformation of liquid contaminants in locus 7. This is followed by an assessment of the liquid contaminant storage capacity of locus 7, guidance for the assessment, and sample calculations.

7.2.2 Mobilization/Remobilization

7.2.2.1 Partitioning into Mobile Phase

Locus 7 consists of free (mobile) liquid contaminants floating upon the water table. Contaminants in the free phase of this locus are present because of their relative immiscibility with water and high mobility in the subsurface. Historically, remedial efforts (e.g., free-product pumping) exploited these properties of the contaminants; alternative corrective action may involve removal of contaminants via other loci. Figure 7–2 presents these loci which may be affected by the liquid contaminants in locus 7 and the pathways or partitioning processes whereby inter-loci transfer occurs. A discussion of the factors affecting the movement and transfer of liquid contaminants in locus 7 follows.

Volatilization. Volatilization is the primary direct mechanism by which partitioning of liquid contaminant from locus 7 to locus 1 occurs. Once contaminants have moved into the soil vapor phase of locus 1, the volatilized liquid contaminants move through the vadose zone by diffusion and advection (see Section 1.2.2) where they may be treated or removed from the subsurface by various corrective actions such as vacuum extraction.

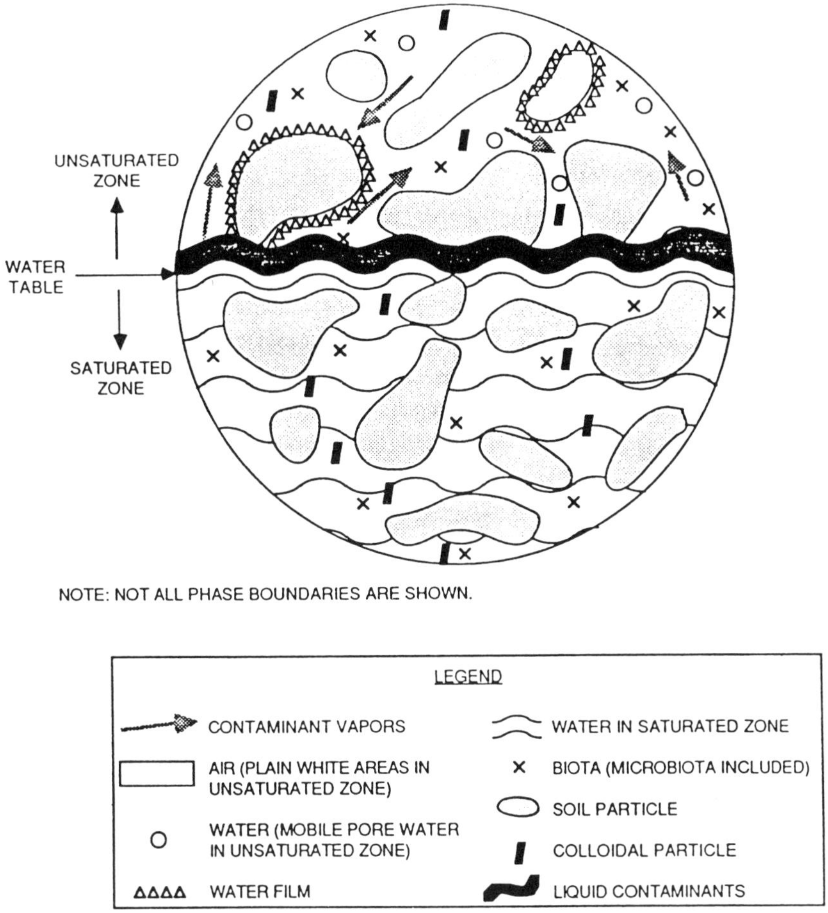

Figure 7–1. Schematic cross-sectional diagram of locus no. 7–liquid contaminants floating on the groundwater table.

Dissolution. Dissolution is the primary direct mechanism by which partitioning of liquid contaminant from locus 7 to loci 3, 8, and 12 occurs. The extent to which contaminant transfer occurs by the dissolution of liquid contaminant is limited, in part, by the aqueous solubility of the liquid contaminant.

Solubility of additive-free gasoline ranges between about 130 and 250 mg/L in water at 20°C. Solubility values for the chemical constituents of a hypothetical, 23-component gasoline are reported in Table 7–1; the mole-fraction-weighted average solubility of the synthetic gasoline is about 130 mg/L.

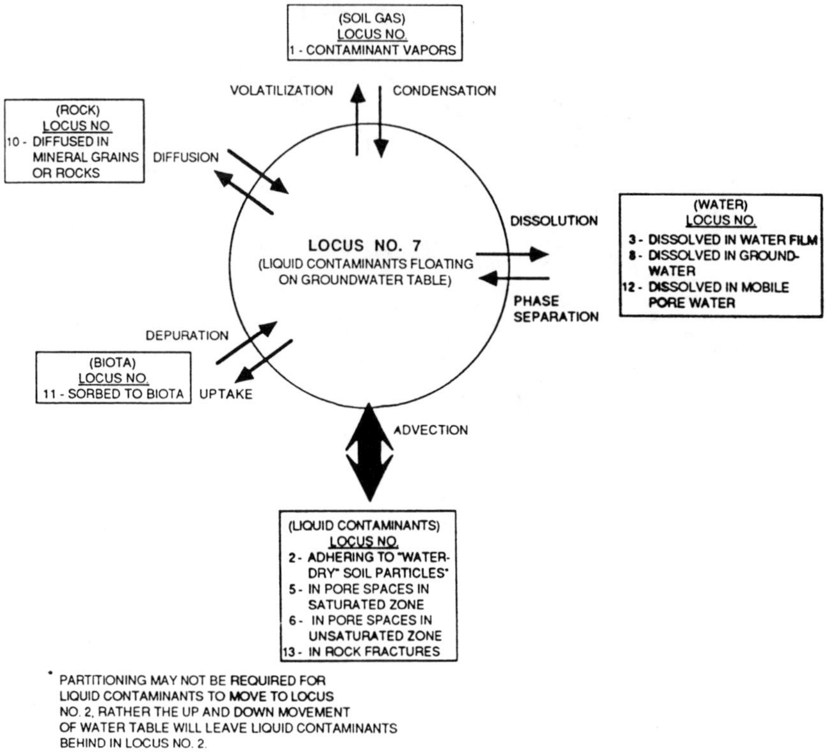

Figure 7–2. Schematic representation of important transformation and transport processes affecting other loci.

The addition of hydrophilic compounds such as methanol, ethanol, and methyl-tertiary-butylether in quantities of as much as 10% by volume may result in gasoline solubilities of greater than 2000 mg/L (Lyman, 1987). Site-specific factors such as ambient temperature and pressure, pH, salinity, particulate matter content, and concentrations of other compounds in the aqueous phase also affect dissolution.

Dissolution reduces the mass of liquid contaminant in locus 7. The rate of dissolution is affected by the concentration gradient across the liquid contaminant-water phase interface: the higher the gradient, the greater the diffusive flux across the interface. Section 1.2.2.2 describes the process and mathematics of diffusion in greater detail. The advection and diffusion/dispersion of the dissolved contaminants in the aqueous phase of locus 7 affects this concentration gradient and is discussed in Section 8.2.2.2.

Dissolution of the liquid contaminant is facilitated by contact with

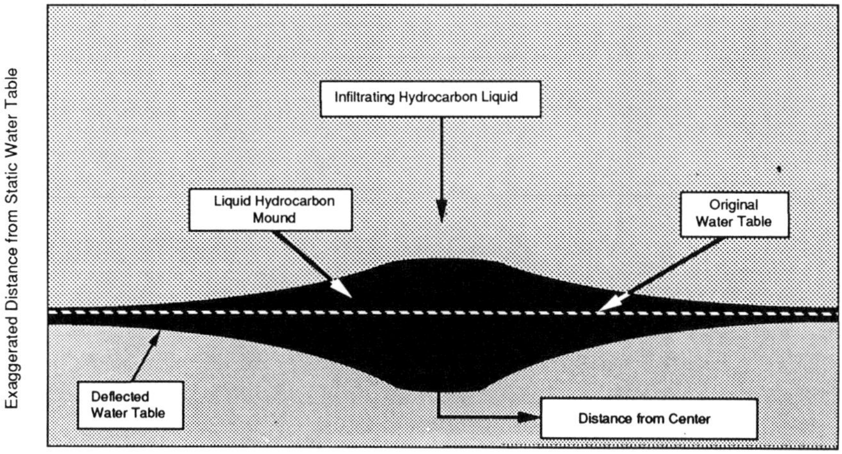

Figure 7–3. Schematic of immiscible liquid contaminant accumulating upon a deflected water table.

infiltrating rainwater and the attendant fluctuations in the water table surface. The rise of the water table covered with liquid contaminants will probably increase the dissolution of liquid contaminant into the water of the vadose zone, the fate of which is described in Sections 3.2.2.2 and 12.2.2.2.

7.2.2.2 Transport of/with Mobile Phase

Contaminants present in locus 7 are, by definition, mobile. The transport of liquid contaminants within this locus is governed by Darcy's law (see Section 6.2.2.2) with the fluid conductivity of the porous medium appropriately scaled to reflect the kinematic viscosity of the liquid contaminant.

The saturated hydraulic conductivity of a porous medium for water may be scaled by the ratio of kinematic viscosity of the water to that of the liquid contaminant to obtain a fluid conductivity of the porous medium for the liquid contaminant:

$$K_o = K_w (\nu_w/\nu_o) \tag{7.1}$$

where: K_o = fluid conductivity of a porous medium to a liquid contaminant (cm/sec)

K_w = saturated hydraulic conductivity of a porous medium to water (cm/sec)

ν_o = kinematic viscosity of liquid contaminant (cm^2/sec)

ν_w = kinematic viscosity of water (cm^2/sec)

Table 7–1. Physicochemical Properties of Selected Chemical Constituents of a Synthetic Gasoline at 20°C

Representative Chemical	Mole Weight (g/mol)	Concen-tration (% w/w)	Mole Fraction	Water Solubility[a] (mg/L)	Liquid Density (g/cc)	Liquid Dynamic Viscosity (g/cm/sec)	Liquid Kinematic Viscosity (cm^2/sec)	Pure Vapor Pressure (mm Hg)
n-Butane[c]	58.12	1.00	0.0161	61.40	0.58	1.78E−03	3.07E−03	1560
Isobutane[c]	58.12	2.00	0.0322	48.90	0.56	1.80E−03	3.24E−03	2250
n-Pentane	72.15	3.00	0.0389	41.20	0.63	2.25E−03	3.60E−03	424
Isopentane	72.15	14.00	0.1813	48.50	0.62	2.17E−03	3.50E−03	575
n-Hexane	86.18	9.00	0.0976	12.50	0.66	3.07E−03	4.66E−03	121
2-Methylpentane	86.18	8.00	0.0867	14.20	0.65	2.93E−03	4.49E−03	172
Cyclohexane	84.16	3.00	0.0333	59.70	0.78	9.63E−03	1.24E−03	77.6
Benzene	78.11	3.00	0.0359	1780.00	0.89	6.40E−03	7.23E−03	75.2
n-Heptane	100.20	1.50	0.0140	2.68	0.68	4.10E−03	6.00E−03	35.6
2-Methylhexane	100.20	5.00	0.0466	2.54	0.68	3.72E−03	5.48E−03	51.9
Methylcyclohexane	98.19	1.00	0.0095	15.00	0.77[b]	7.21E−03	9.31E−03	36.2
Toluene	92.14	5.00	0.0507	537.00	0.87	5.81E−03	6.70E−03	21.8
n-Octane	114.23	1.00	0.0082	0.66	0.70	5.43E−03	7.73E−03	10.5
2,4-Dimethylhexane	114.23	8.00	0.0654	1.50	0.70	5.00E−03	7.14E−03	23.3
Ethylbenzene	106.17	2.00	0.0176	167.00	0.87	6.68E−03	7.70E−03	7.08
m-Xylene	106.17	7.00	0.0616	162.00	0.86	6.09E−03	7.05E−03	6.16
1-Pentene	70.14	1.50	0.0200	148.00	0.64	1.97E−03	3.08E−03	531
1-Heptene	84.16	1.50	0.0167	50.00	0.67	2.59E−03	3.85E−03	150
2,2,4-Trimethylhexane	128.26	2.00	0.0146	0.80	0.72[b]	5.00E−03	6.94E−03	11.3
1,3,5-Trimethylbenzene	120.20	5.00	0.0389	72.60	0.87	7.29E−03	8.42E−03	1.73
2,2,5,5-Tetramethylhexane	142.29	1.50	0.0099	0.13	NA	6.40E−03	NC	6.47
1,4-Diethylbenzene	134.22	5.00	0.0348	15.00	0.86	7.00E−03	8.12E−03	0.697
(C_{12}-Aliphatic)	134.22	10.00	0.0696	0.01	0.86	1.49E−02	1.74E−02	0.0753
Mole-fraction-weighted averages		—	—	128.98	0.71	4.92E−03	6.36E−03	
Summations		100.00	1.0000	—	—	—	—	

a = Solubility values reported at 25°C.
b = Density reported at 16°C.
c = Liquid properties are for the hypothetical superheated liquid.
NA = Not Available.
NC = Not Calculable.

Source: CDM, Inc. (Concentrations were selected after review of several analyses of gasoline. Chemical property data are from numerous sources and include several estimates.)

Values of saturated hydraulic conductivity for soils reported in Table 12–2, and for aquifer materials in Table 8–1, may be thus scaled.

Kinematic viscosity, the ratio of dynamic viscosity of a liquid to its liquid density, affects the rate at which the liquid contaminant may flow through a porous medium. The higher the kinematic viscosity of a liquid, the more resistance to flow, resulting in slower flow. Because of its definition, low-density fluids of high dynamic viscosity have a high kinematic viscosity and flow less readily through the porous medium. A discussion of the effects of dynamic viscosity and liquid density on contaminant flow and Darcy's law appears in Section 6.2.2.2.

Kinematic viscosity values for selected constituents of a synthetic gasoline are reported in Table 7–1. As a basis for comparison, the kinematic viscosity of water at 20°C is 0.01 cm^2/sec; benzene has a kinematic viscosity of 0.00721 cm^2/sec. Pure liquid benzene flows about 40% faster than water through the same porous medium, all other factors remaining equal.

In fact, most of the chemicals reported in Table 7–1 possess kinematic viscosities less than that of water, indicating that they flow more readily in a pure liquid state than water. The mole-fraction-weighted average kinematic viscosity of the synthetic gasoline is 6.36×10^{-3} cm^2/sec.

With time, the liquid contaminant loses its more volatile constituents to the air of the vadose zone and more soluble constituents to the water of the vadose and the saturated zone. The remaining liquid contaminant becomes more dense and viscous. This "aging" of the liquid contaminant by losses to volatilization and dissolution increases the kinematic viscosity of the liquid contaminant. As a result, the "aged" liquid contaminant flows less readily through the porous medium. To estimate the effects of volatilization and dissolution upon the kinematic viscosity of the synthetic gasoline, the more volatile (vapor pressure > 20 mm Hg) and soluble (solubility > 10 mg/L) chemical constituents of the synthetic gasoline in Table 7–1 were removed (see Table 7–2); the resultant kinematic viscosity is 1.34×10^{-2} cm^2/sec, indicating that the "aged" gasoline would flow about half as fast as the unweathered synthetic gasoline.

In the vadose zone above the water table, gravity and capillarity cause liquid contaminant to migrate downward and away from the leak source. As the liquid contaminant reaches the water table and accumulates, it begins to spread out laterally until gravitational and capillary forces balance.

Hantush (1967) presents an analytical solution that predicts the rise and decay of the groundwater table in response to deep percolation. This technique may also describe the growth of a mound of liquid

Table 7–2. Physicochemical Properties of Selected Chemical Constituents of an "Aged" Synthetic Gasoline at 20°C

Representative Chemical	Mole Weight (g/mol)	Concentration (% w/w)	Mole Fraction	Water Solubility[a] (mg/L)	Liquid Density (g/cc)	Liquid Dynamic Viscosity (g/cm/sec)	Liquid Kinematic Viscosity (cm^2/sec)	Pure Vapor Pressure (mm Hg)
n-Octane	114.23	6.90	0.0800	0.66	0.70	5.43E–03	7.73E–03	10.5
2,2,4-Trimethylhexane	128.26	13.79	0.1425	0.80	0.72[b]	5.00E–03	6.94E–03	11.3
2,2,5,5-Tetramethylhexane	142.29	10.34	0.0964	0.13	NA	6.40E–03	NC	6.47
(C_{12}-Aliphatic)	134.22	68.97	0.6811	0.01	0.86	1.49E–02	1.74E–02	0.0753
Mole fraction-weighted averages:		—	—	0.18	0.74	1.19E–02	1.34E–02	
Summations:		100.00	1.0000	—	—	—	—	

a = Solubility values reported at 25°C.
b = Density reported at 16°C.
NA = Not Available.
NC = Not Calculable.

Source: CDM, Inc. (Concentrations were selected after review of several analyses of gasoline. Chemical property data are from numerous sources and include several estimates.)

contaminant upon the water table for a continuous leak of liquid contaminant:

$$h(r) \simeq [(Q/(2 \cdot \pi \cdot K)) \cdot W(r^2 \cdot S/(4 \cdot K \cdot b \cdot t))]^{1/2} \quad (7.2)$$

for distances larger than the radius of the leak, where:

$h(r)$ = liquid contaminant thickness upon the water table (m)
Q = leak rate of liquid contaminant from source (m^3/day)
π = pi (3.14159 . . .)
K = fluid conductivity of porous medium to a liquid contaminant (m/day)
r = radial distance from leak (m)
S = specific yield of porous medium ($m^3/m^2/m$)
t = leak duration (days)
b = average thickness of liquid contaminant (m)
$W(\,)$ = exponential integral function

The use of equation 7.2 requires an initial guess at the value of the average thickness of liquid contaminant, b, which is not known *a priori*. Once a guess for b is made, a range of values of liquid contaminant thickness, h(r), can be calculated, from which b may be re-estimated. Values of h(r) are re-calculated in an iterative procedure until the difference in average thickness between any two successive iterations is negligible. This technique may be used to estimate the areal extent and distribution of liquid contaminant upon the water table.

The thickness of liquid contaminant accumulating upon the water table depends on the rate at which the liquid contaminant leaks from the source, the intrinsic permeability of the porous medium, and the kinematic viscosity of the liquid contaminant. Capillary forces may cause liquid contaminant to form a layer upon the water table which varies in thickness depending on the geologic material. Fine-grained materials exhibit more pronounced capillarity than coarse-grained materials. A zone of capillary tension and partial saturation forms above the capillary fringe which has been termed the funicular zone (see Figure 7–4).

Within the funicular zone, liquid contaminant may be trapped by capillarity; the stronger the capillary force, the thicker the entrapped layer. Dietz (1971) states that oil spreads over a water table in a layer roughly equivalent to the thickness of the funicular zone. Table 7–3 lists average funicular zone thicknesses for various geologic materials.

Figure 7–5 depicts the effects of various remedial pumping schemes on liquid contaminant upon the water table in locus no. 7 (see Figure 7–5a). Initially, liquid contaminant is withdrawn from an extraction well

Table 7–3. Average Grain Size and Funicular Zone Thickness for Various Geologic Materials

Material	Average Grain Size (mm)	Funicular Zone Thickness (cm)
Very coarse sand	0.5–2.0	1.8–9.0
Moderately coarse sand	0.2–0.5	9.0–22.4
Moderately fine sand	0.05–0.2	22.4–28.1
Very fine sand	0.015–0.05	28.1–45
Silt	0.004–0.015*	45–90
Clay	<0.004*	120–240

Source: Driscoll, 1986.

screened in the liquid contaminant pool, which is assumed to be oil (see Figure 7–5b). Pumping of the oil causes a cone of depression to form in the soil at the extraction well; the underlying water also responds to the pumping, up-coning at the extraction well. If the extraction well is pumped at a sufficient rate, more and more water is withdrawn from the well instead of oil. Eventually, the cone of depression in the oil and the up-coning water may intersect; because of the low relative permeability of oil in the presence of water, no further oil is removed. Apart from the difficulty of treating contaminated water and oil simultaneously, the objective of removing as much oil as possible from locus 7 by pumping is not met.

A similar situation occurs in Figure 7–5c where water is initially withdrawn and becomes gradually replaced with oil. The cone of depression caused by pumping water causes the oil to accumulate in a greater thickness at the extraction well. Sufficient pumping of water may cause oil to intrude in the screened interval, reducing water withdrawal. The scenario depicted in Figure 7–5d of two extraction wells simultaneously pumping water and oil causes a cone of depression to form in the water, collecting oil in greater thickness which is then more efficiently extracted by the second well.

Numerous remediation attempts have been conducted, some with considerable success. Remediation of many leaks at service stations with gasoline USTs has been performed with free product recovery. These leaks are usually limited to a few gallons to tens of thousands of gallons of leaked liquid contaminant. Larger leaks have occurred at industrial facilities and airports. Table 7–4 presents recovery histories of some larger remediation efforts.

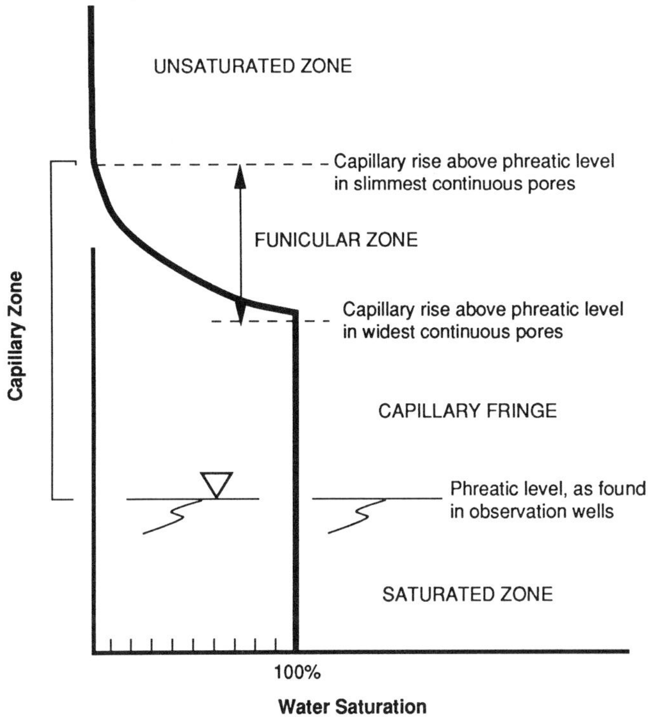

Figure 7–4. Schematic of the capillary zone with a typical water saturation curve. *Source:* Adapted from Deitz, 1971.

Table 7–4. Examples of Product Recovery After Tank Leaks

Location	Type of Facility	Ongoing Recoveries	Product Type
West Coast, USA	Abandoned bulk storage	500–1,000 gal/wk	Mixed fuels
South Korea	Fuel loading rack	1,000–1,500 gal/wk	Jet fuel/Avgas
East Coast, USA	UST/load rack	1,000–1,500 gal/wk	Jet fuel
North Coast, Venezuela	Pipeline/bulk storage	28,000 gal/day	Gasoline
Eastern USA	Bulk storage	500–700 gal/day	Fuel oil
Southwest USA	Refinery/bulk storage	3,000–5,000 gal/day	Mixed fuels
New Zealand	Refinery/bulk storage	1,500–2,000 gal/day	Gasoline/ gas oil
Midwestern USA	Retail fuel facility	700 gal/month	Gasoline

Source: Yaniga, 1984a and 1984b.

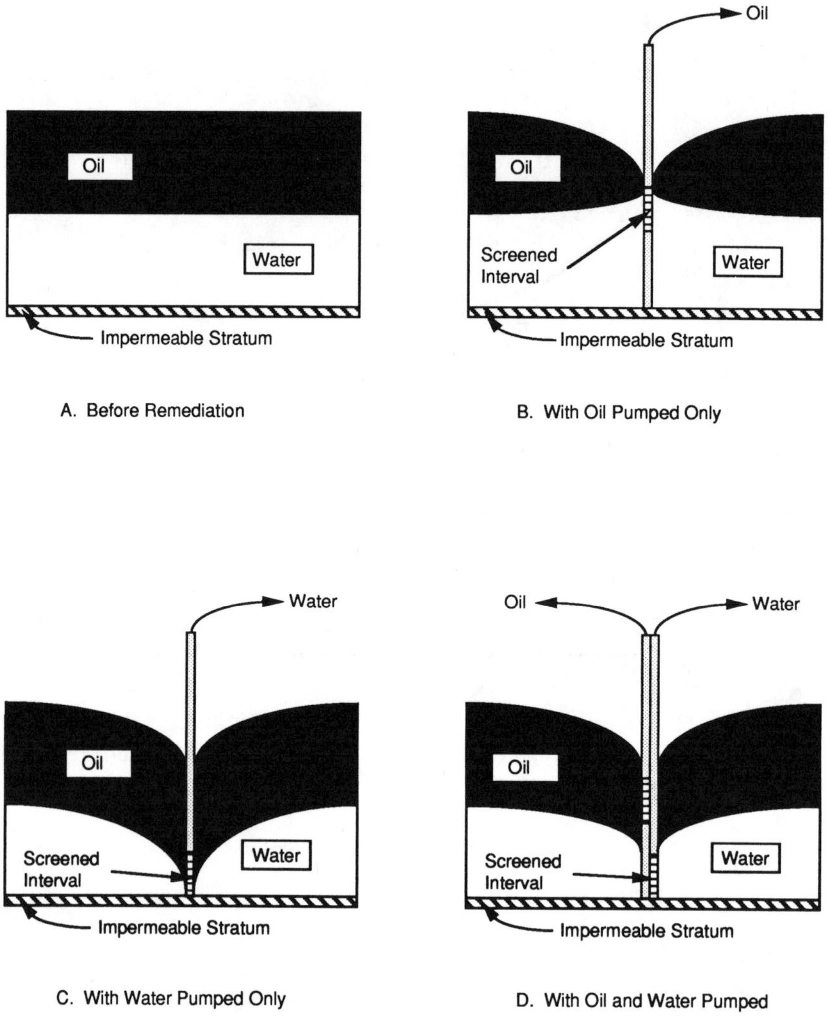

Figure 7–5. Mobilization of contaminants in locus 7.

7.2.3 Fixation

As noted in Section 7.2.2, contaminants in locus 7 are in a relatively high mobile liquid state. Contaminants present in locus 7 could be rendered less threatening if fixed into a less mobile phase. The considerations addressed in the following subsections are directed toward: 1) enhancing phase exchange and bulk transport out of the locus and into one of the loci that, presumably, hold the contaminants in a less mobile state; and, 2) fixing the locus as a whole, for possible future remediation or as a permanent corrective action.

7.2.3.1 Partitioning onto Immobile (Stationary) Phase

The processes of abandonment, dissolution, and adsorption may cause contaminants to move from locus 7 to comparatively less mobile phases.

Abandonment. The ultimate direction of partitioning between loci 2 and 7 is site-dependent. A falling water table may leave a contaminant film over water-dry soil particles at residual saturation. The contaminant film coating a water-dry soil particle is relatively immobile. Under these conditions, however, the contaminant film may transform to the more mobile vapor phase (locus 1) in the unsaturated zone and migrate by diffusion or mass transport mechanisms, as detailed in Section 1. At an unpaved site, infiltrating water may displace the contaminant film which may accumulate until it flows downward under greater-than-residual saturation conditions. Locus 2 is then displaced downward to locus 7 ahead of the advancing water front, or is dissolved in the advancing water front and becomes mobile in locus 8 or remains relatively immobile in locus 3. Contaminants moving from locus 7 into locus 3, by the mechanism described above, will be less mobile because of their association with the unsaturated pore space water held in place by capillary tension. The contaminants transferred into locus 3 can be, however, easily remobilized when the soil pore spaces are saturated by either infiltrating water, a rising groundwater table, or volatilization. Contaminants may be transferred from locus 7 to locus 5 by a rising water table, which isolates blobs of liquid contaminant from the main body of liquid contaminant. Once in locus 5, the contaminants are immobile and are held in the interstitial spaces by capillary tension.

These contaminants can be remobilized by dissolution (as described in Section 7.2.2.1) or mechanical dispersion of minute droplets. The overall effect is an increase in potential partitioning to locus 8 (mobile solute in groundwater). Should the water table again fall, locus 5 could be reabsorbed in locus 7. Contaminants can be transferred from locus 7 to locus 6 by the rising and lowering water table, so that liquid contaminant partially fills the void spaces of the unsaturated zone. This remaining liquid contaminant is termed residual saturation and may be different for water-dry and water-wet soil particles.

Adsorption. Contaminants in locus 7 may come into direct contact with water-dry soil particles or biological matter. As a result of contact, contaminants may either adsorb (mono-molecular layer film) or adhere (multi-molecular layer film) to particles in these loci and be less mobile than when in locus 7. For detailed discussions of these processes, see Sections 2, 4, and 9.

Dissolution (Detailed in Section 7.2.2). Dissolution of contaminants from locus 7 in the relatively immobile water film surrounding soil particles in the unsaturated zone (locus 3) could take place. As discussed above, however, there is great potential for contaminants to be remobilized from this temporary sink.

Diffusion (Details in Section 10.2.3). Contaminant species may diffuse from locus 7 into the water-dry surface of a mineral particle (locus 10). Although contaminants in locus 10 are believed to be immobile, the total impact of this process on reducing the mobility of contaminants in locus 7 and its impact on remediation of the locus is exceeding small.

7.2.3.2 Other Fixation Approaches

Hydrodynamic control (groundwater pumping or draining) and grout curtains are control options in some geologic settings that may effectively halt mass migration of locus 7 to a new physical location. There are, however, no other fixation approaches that could conceivably prevent migration of contaminants from their nearly-pure state in locus 7 to some other locus.

7.2.4 Transformation

7.2.4.1 Biodegradation (Details in Section 11)

Bulk liquid hydrocarbons are much less subject to biodegradation than organic contaminants dissolved in water. This is because of the limitation of oxygen in organic liquids which is requisite for aerobic biodegradation. Microorganisms which can survive in, and cause "spoilage" of, organic liquids are known. The rate of reaction is low, however, due to limitations of oxygen or other nutrients. It is likely that contaminants in locus 7 would disperse and diffuse to other loci, where they may degrade more quickly, before they degraded as a bulk liquid in locus 7.

Table 7–5. Total Porosities for Common Consolidated and Unconsolidated Sediments

Unconsolidated Sediments	Total Porosity (percent)	Consolidated Sediments	Total Porosity (percent)
Clay	45–55	Sandstone	5–30
Silt	35–50	Limestone	1–20
Sand	25–40	Shale	0–10
Gravel	25–40	Fractured crystalline rock	0–10
Sand & gravel	10–35	Vesicular basalt	10–50
Glacial till	10–25	Dense, solid rock	<1

Source: Driscoll, 1986.

7.2.4.2 Chemical Oxidations (Details in Section 3.2.4.2)

Contaminants floating on the water surface can undergo abiotic transformation as the constituent comes into contact with oxidized metal complexes.

7.3 STORAGE CAPACITY IN LOCUS

7.3.1 Introduction and Basic Equations

The storage capacity of locus 7 can be described by the total porosity (Driscoll, 1986), which is defined as:

$$\Theta_t = \frac{\text{Volume of total pore space}}{\text{Volume of bulk solid}} \tag{7.3}$$

where the difference between the pore space volume and that of the bulk solid is filled by geologic material. Table 7–5 presents ranges of percent total porosity for a variety of unconsolidated and consolidated sediments.

Those porosity values measured in nature may be lower than those values reported in Table 7–5 because additional phases may occupy void volume, reducing the volume available for contaminant storage. These other space-occupying factors are:

1) Entrapped water;
2) Biomass;
3) Entrapped soil gas.

The volumes of the factors listed above are site-specific and may vary with time. Entrapped water may be displaced, biomass may increase or decrease in response to nutrient levels, and entrapped soil gas may be displaced or absorbed into the contaminant or aqueous phases. Several of these factors are briefly discussed below as they relate to porosity-reduction.

The quantity of entrapped water in a porous medium depends on the size and uniformity of the grains. Presuming that the entrapped water is held by capillarity, the quantity of water held ranges from the field capacity to the wilting point of the porous medium. These terms are discussed in Section 12.2.2.2. Figure 12–4 gives ranges of values for the field capacity and wilting point of various soils with which the quantity of entrapped water may be estimated.

Entrapped soil gas may reduce the storage capacity of the porous medium for contaminant liquid. Air dissolution in a liquid contaminant, however, may be relatively rapid, making this consideration relatively unimportant in calculating the storage capacity of locus 7.

As a final note on factors which may offset the storage capacity, changes in the crystalline structure of clays due to incorporation of organic compounds have been reported. This may be a significant factor in calculating storage and transmission of liquid contaminants in some geologic settings. Certain clays also adsorb water, swelling and reducing void volume. The increased percent volume in clay due to structural changes may be as high as 400%.

Under these concepts, the storage capacity of locus 7 is:

$$m_c = \Theta_c \rho_c = (\Theta_t - \Theta_a - \Theta_w - \Theta_b)\ \rho_c \qquad (7.3)$$

where:
- m_c = storage capacity for contaminant (kg/L)
- Θ_t = total porosity (dim.)
- Θ_c = contaminant storage capacity (dim.)
- Θ_a = entrapped soil gas fraction (dim.)
- Θ_w = entrapped water fraction (dim.)
- Θ_b = biomass fraction (m^3/m^3 dim.)
- ρ_c = density of contaminants (g/cm^3 or kg/L)

7.3.2 Guidance on Impacts for, and Calculations of, Maximum Value

For the calculation of a maximum value for the contaminant storage with equation 7.3, the following assumptions were made:

1) no entrapped soil gas, $\Theta_a = 0$;
2) entrapped water is at wilting point;
3) no biomass, $\Theta_b = 0$.

As stated above, Figure 12–4 gives the fraction of water held in a soil at the wilting point. Under these assumptions, equation 7.3 reduces to:

$$m_c = (\Theta_t - \Theta_w)\,\rho_c \tag{7.4}$$

7.3.3 Guidance on Impacts for, and Calculations of, Average Values

To calculate an average value for contaminant storage capacity, some of the assumptions in the previous section will be altered. For lack of better information, it is assumed that the volume of biomass is negligible ($\Theta_b = 0$), and that the volume of entrapped air is also small ($\Theta_a = 0$). Instead of relying upon the wilting point of a soil for an estimation of entrapped water volume, the field capacity of the soil will be used.

7.4 EXAMPLE CALCULATIONS

7.4.1 Storage Capacity Calculation

Assume gasoline in a sand formation for these calculations. The maximum total porosity for sand from Table 7–5 is 0.40. The average porosity for sand from Table 7–5 is 0.33. The density of gasoline is 0.74 g/cm^3 (Table 1–1A).

7.4.1.1 Maximum Storage Capacity:

Assuming $\Theta_a = \Theta_b = 0$ and Θ_w = wilting point, which is 0.05 (Figure 12–4), and using equation 7.3:

$$m_c = (0.4 - 0.05)\ 0.74$$

$$= 0.26\ \text{g/cm}^3$$

7.4.1.2 Average Storage Capacity:

Assuming $\Theta_a = \Theta_b = 0$ and Θ_w = field capacity of 0.09 (Figure 12–4):

$$m_c = (0.33 - 0.09)\ 0.74$$
$$= 0.18\ \text{g/cm}^3$$

7.4.2 Transport Rate Calculations

Based on flow equations presented in Sections 2, 5, 6, 8 and 12, a simple calculation of liquid contaminant flow velocity is presented. Returning to the sandy soil of the previous section, the porosity available for contaminant storage ranges from 0.25 to 0.40. From Table 7–1, the kinematic viscosity of the unweathered liquid contaminant is 0.00636 cm^2/sec. For a clean sand, the saturated hydraulic conductivity may range from 10^{-4} to 10^0 cm/sec (Freeze and Cherry, 1979); a medium value of 10^{-2} cm/sec is chosen for the hydraulic conductivity. With equation 7.1, the saturated hydraulic conductivity of the sand for water may be scaled to a fluid conductivity of the sand for the synthetic gasoline:

$$K_o = (\nu_w / \nu_o) \cdot K_w$$

where ν_w = 0.01 cm^2/sec for water
ν_o = 0.00636 cm^2/sec for synthetic gasoline

The scaled fluid conductivity is 1.57×10^{-2} cm/sec.

Assuming a hydraulic gradient of 0.001 cm/cm, the flow velocity of the liquid contaminant (synthetic gasoline) is:

$$V_c = \left(\frac{K_o}{\Theta_c}\right)\left(\frac{dh}{dl}\right) \tag{7.5}$$

where dh/dl = hydraulic gradient (cm/cm)

From equation 7.5, the estimated flow velocity of synthetic gasoline in the sand ranges from 3.93×10^{-5} cm/sec to 6.33×10^{-5} cm/sec.

7.5 SUMMARY OF RELATIVE IMPORTANCE OF LOCUS

7.5.1 Remediation

The remediation of contaminants in locus 7 is essential because contaminants in this locus serve as a continual source of contamination for the vadose zone and the saturated zone. Remediation of other loci may be necessary as well to remove the health and/or environmental threat posed by the contaminants.

Remediation of contaminants in this locus has been by the direct

Table 7–6. Loci Interactions with Liquid Contaminant Floating on Groundwater

Interacting Locus	Phase Contacted	Transfer Process	Relative Importance
1	air	volatilization	high
2	dry solid	adhering	high
3	water	dissolution	moderate
5	water & solid	adhering	moderate
6	solid	adhering	high
8	water	dissolution	high
10	solid	adsorption	low
11	biota	sorption	moderate
12	water	dissolution	moderate
13	voids	bulk transfer	high

removal of liquid contaminants by active pumping. In the relatively recent past, other corrective actions have been utilized in conjunction with active removal. While there is no known evidence of any other corrective action being as effective or efficient as direct liquid removal, other alternatives, such as enhanced biodegradation, active soil venting (vacuum extraction, steam stripping, etc.), and in situ treatment should be evaluated before a corrective action is selected for a given site.

7.5.2 Loci Interaction

The fact that locus 7 consists of bulk liquid contaminant and is mobile in its bulk state in three dimensions means that the contaminants in this locus can, ultimately, affect all other loci. Table 7–6 summarizes the partitioning and relative importance of that partitioning between locus 7 and the other loci.

As is evident from the table, the loci most affected by locus 7 are nos. 1, 2, 6, 8, and 13. The ability of locus 7 to transfer pure contaminant to loci 1, 2, 6 and 13 make it a critical distribution locus. Of these, the largest mass of contaminants will probably be transferred to loci 1 and 13, with lesser potential partitioning to loci 2 and 6. The potential for partitioning to locus 8 is also relatively high because of the large potential surface area of contact at the bottom of locus 7 and the duration of contact.

Loci 3, 5, 11, and 12 are also potential receptors of contaminants partitioning from locus 7, but are of considerable less importance than those mentioned above because of the mass of contaminants that could be transferred there. Locus 10 has high potential to receive contaminants from locus 7, but appears to be a relatively small-volume sink.

7.5.3 Information Gaps

There is a considerable amount of research that has been completed and data generated that is relevant to the study of locus 7. Most of this information is, however, site-specific. The following is a brief list of information gaps that, if filled, would improve the understanding of the behavior of contaminants in locus 7:

(1) Solubility data of individual contaminants under various groundwater conditions;
(2) Data on the effects of gasoline additives on fundamental parameters (e.g., solubility, activity, volatility);
(3) Data on physical and chemical characteristics of "weathered" mixtures;
(4) Data on hydraulic properties of the site; and
(5) Areal distribution of the liquid contaminant in the subsurface.

7.6 LITERATURE CITED

Dietz, D.N. 1971. Pollution of Permeable Strata by Oil Components. *In*: Water Pollution by Oil. Institute of Petroleum, London, England.

Driscoll, F.G. 1986. Groundwater and Wells, Johnson Well Division, St. Paul, MN.

Freeze, R.A. and J.A. Cherry, 1979. Groundwater. Prentice-Hall, Inglewood Cliffs, NJ.

Lyman, W.J. 1987. Environmental Partitioning of Gasoline in Soil/Groundwater Compartments. *In*: Final Seminar Proceedings of the Underground Environment of an UST Motor Fuel Release, U.S. EPA, Edison, N.J.

Yaniga, P.M. 1984a. "Hydrocarbon Contamination of Groundwater: Assessment and Abatement," testimony presented to U.S. Senate Committee on Environmental and Public Works and Subcommittee on Toxic Substances and Environmental Oversite.

Yaniga, P.M. 1984b. Hydrocarbon Retrieval and Apparent Hydrocarbon Thickness: Interrelationships to Recharging/Discharging Aquifer Conditions. Proceedings of the NWWA/API Conference on Petroleum Hydrocarbons and Organic Chemicals in Groundwater, Houston, TX, pp. 299–325.

7.7 OTHER REFERENCES

Case Studies

Testa, S.M., D.M. Baker, and P.L. Avery. 1989. Field Studies and Occurrence, Recoverability, and Mitigation Strategy of Free Product Liquid Hydrocarbon. *In*: Environmental Concerns in the Petroleum Industry. American Association of Petroleum Geologists, pp. 57–81.

Lapham, S.M. 1986. Advanced Stages of a Hydrocarbon Recovery/ Closed Loop Hydrocarbon Flushing System Successful in a High Permeable Glacial Outwash in Apple Valley, MN. Proceedings of the NWWA/API Conference on Petroleum Hydrocarbons and Organic Chemicals in Groundwater, Houston, TEX, pp. 755–769.

General Ground Water Theory

Driscoll, F.G. 1986. Groundwater and Wells. Johnson Well Division, St. Paul, MN.

Fetter, C.W., Jr. 1980. Applied Hydrogeology. C.E. Merrill Publishing Co., Columbus, OH.

Immiscible-Phase Experimentation

Schiegg, H.O. 1988. Physical Simulation of Three-phase Immiscible Fluid Displacement on Porous Media. VAW No. 40(5), U.S. Department of Energy.

Schwille, F. 1988. Dense Chlorinated Solvents in Porous And Fractured Media, Lewis Publishers, Chelsea, MI.

Immiscible-Phase Simulation

Abriola, L.M. and G.F. Pinder. 1985. A Multiphase Approach to Modeling Porous Media Contamination by Organic Compounds, 1. Equation Development. Water Resources Research, 21(1):11–18.

Abriola, L.M. and G.F. Pinder. 1985. A Multiphase Approach to Modeling Porous Media Contamination by Organic Compounds, 2. Numerical Simulation. Water Resources Research, 21(1):19–26.

Aziz, K. and A. Settari. 1979. Petroleum Reservoir Simulation, Applied Science Publishers, London.

Falta, R.W. and I. Javandel. 1987. A Numerical Method for Multiphase Multicomponent Contaminant Transport in Groundwater Systems. Trans. American Geophysical Union, 68(44).

Faust, C.R. 1985. Transport of Immiscible Fluids Within and Below the Unsaturated Zone—A Numerical Model. Water Resources Research, 21:587–596.

Hochmuth, D.P. and D.K. Sunada. 1985. Ground-water Model of Two Phase Immiscible Flow in Coarse Material. Ground Water, 23(5):617–626.

Hunt, J.R. and N. Sitar. 1988. Nonaqueous Phase Liquid Transport and Cleanup, 1. Analysis of Mechanisms. Water Resources Research, 24(8):1247–1258.

Hunt, J.R., N. Sitar, and K.S. Udell. 1988. Nonaqueous Phase Liquid Transport and Cleanup, 2. Experimental Studies. Water Resources Research, 24(8):1247–1258.

Huyakorn, P.S. and G.F. Pinder. 1978. New Finite Element Technique for the Solution of Two-phase Flow through Porous Media. Advances in Water Resources, 1.

Kaluarachchi, J.J., J.C. Parker, and A.K. Katyal. 1988. A Numerical Model for Water and Light Hydrocarbon Migration in Unconfined Aquifers under Vertical Equilibrium. Advances in Water Resources, in review.

Kuppusamy, T., J. Sheng, J.C. Parker, and R.J. Lenhard. 1987. Finite Element Analysis of Multiphase Immiscible Flow through Soils. Water Resources Research, 23:625–631.

Osbourne, M. and J. Sykes. 1986. Numerical Modeling of Immiscible Organic Transport at the Hyde Park Landfill. Water Resources Research, 22(1):25–33.

Parker, J.C., J.J. Kaluarachchi, and A.K. Katyal. 1988. Areal Simulation of Free Product Recovery from a Gasoline Storage Tank Leak Site. Proceedings of the NWWA/API Conference on Petroleum Hydrocarbons and Organic Chemicals in Groundwater, Houston, TX, pp. 315–332.

Parker, J.C., R.J. Lenhard, and T. Kuppusamy. 1987. A Parametric Model for Constitutive Properties Governing Multiphase Flow in Porous Media. Water Resources Research, 23:618–624.

Immiscible-Phase Flow Theory

Cary, J.W., J.F. McBride, and C.S. Simmons. 1989. "Trichloroethylene Residuals in Capillary Fringe as Affected by Air-entry Pressures," J. Environ. Qual., 18(1):72–77.

Parker, J.C. R.J. Lenhard, and T. Kuppusamy. 1987. Physics of Immiscible Flow in Porous Media. EPA/600/2-87/101.

Stallman, R.W. 1964. "Multiphase Fluid Flow in Porous Media—A

Review of Theories Pertinent to Hydrological Studies," USGS Prof. Paper 411-E, Department of Energy.

Monitoring and Measurement Studies

Abdul, A.S., S.F. Kia, and T.L. Gibson. 1989. "Limitations of Monitoring Wells for the Detection and Quantification of Petroleum Products in Soils and Aquifers," Ground Water Monitoring Rev., 9(2), 90–99.

Wilson, J.L., S.H. Conrad, E. Hagan, W.R. Mason, and W. Peplinski. 1988. "The Pore Level Spatial Distribution and Saturation of Organic Liquids in Porous Media, "Proceedings of Petroleum Hydrocarbons and Organic Chemicals in Groundwater, 107–134, National Water Well Association.

Remediation

Camp Dresser & McKee Inc. 1988. Cleanup of Releases from Petroleum USTs: Selected Technologies. EPA/530/UST-88/001.

Hunt, J.R., N. Sitar, and K.S. Udell. 1986. "Organic Solvents and Petroleum Products in the Subsurface: Transport and Cleanup," U. Cal-Berkely, UCB-SEERHL Report No. 86–11, San. Eng. and Env. Health.

Somers, J.A. 1973. The Fate of Spilled Oil in the Soil. Hydrological Sciences, 19(4):501–521.

Testa, S.M. and M.T. Paczkowski. 1989. Volume Determination and Recoverability of Free Hydrocarbon. Ground Water Monitoring Review, 9(1):120–128.

Yaniga, P.M. 1984. Hydrocarbon Retrieval and Apparent Hydrocarbon Thickness: Interrelationships to Recharging/Discharging Aquifer Conditions. Proceedings of the NWWA/API Conference on Petroleum Hydrocarbons and Organic Chemicals in Groundwater, Houston, TX, pp. 299–325, National Water Well Association.

SECTION 8

Contaminants Dissolved in Groundwater

CONTENTS

List of Tables 208
List of Figures 208
8.1 Locus Description 209
8.1.1 Short Definition 209
8.1.2 Expanded Definition and Comments 209
8.2 Evaluation of Criteria for Remediation 209
8.2.1 Introduction 209
8.2.2 Mobilization/Remobilization 212
8.2.2.1 Partitioning into Mobile Phase 212
8.2.2.2 Transport with Mobile Phase 212
Transport Rate 216
Dispersion 217
Diffusion 217
Advection 218
Retardation 219
8.2.3 Fixation 223
8.2.3.1 Partitioning onto Immobile (Stationary) Phase 223
8.2.3.2 Other Fixation Approaches 224
8.2.4 Transformation 224
8.2.4.1 Biodegradation 224
8.2.4.2 Chemical Oxidation 226
8.3 Storage Capacity in Locus 226
8.3.1 Introduction and Basic Equations 226
8.3.2 Guidance on Inputs for, and Calculation of, Maximum Value 226
8.3.3 Guidance on Inputs for, and Calculation of, Average Values 228
8.4 Example Calculations 228
8.4.1 Storage Capacity Calculations 228
8.4.1.1 Maximum Value 228
8.4.1.2 Average Value 228

8.4.1.3 Example of Solubility Enhancement 229
8.4.2 Transport Rate Calculations 229
8.5 Summary of Relative Importance of Locus 232
8.5.1 Remediation 232
8.5.2 Loci Interactions 232
8.5.3 Information Gaps 233
8.6 Literature Cited 234

TABLES

8-1 Estimated Retardation Factors for Hydrocarbon Gasoline Constituents 224
8-2 Water Soluble Fractions for Three Gasolines 228
8-3 Estimated Solubility of Gasoline in Water 230
8-4 Loci Interactions with Hydrocarbon Dissolved in Groundwater 233

FIGURES

8-1 Schematic Cross-Sectional Diagram of Locus No. 8-Contaminants Dissolved in Groundwater (i.e., Water in the Saturated Zone) 210
8-2 Schematic Representation of Important Transformation and Transport Processes Affecting Other Loci 211
8-3 Processes of Dispersion on a Microscopic Scale 214
8-4 Concentration Distribution of Dissolved Components as a Result of Hydrodynamic Dispersion 214
8-5 Approximate Temperature of Groundwater in the Conterminous United States at Depths of 10 to 25 Meters 218
8-6 Relations Among Hydraulic Conductivity, Uniformity Coefficient, and Median Grain Diameter 220
8-7 Advance of Reactive and Nonreactive Contaminants through a Column 221
8-8 Effect of the Distribution Coefficient on Contaminant Retardation During Transport in a Shallow Groundwater Flow System 222
8-9 Relationship of Retardation Factors to Octanol-Water Partition Coefficient by Conventional and Revised Equations 225
8-10 Normalized Breakthrough Curves for Displacement of Benzene of 15-cm Lincoln Fine Sand Soil Column as a Function of Water Velocity 225

SECTION 8

Contaminants Dissolved in Groundwater

8.1 LOCUS DESCRIPTION

8.1.1 Short Definition

Contaminants dissolved in groundwater (i.e., water in the saturated zone).

8.1.2 Expanded Definition and Comments

Locus no. 8 consists of groundwater contaminated with dissolved pollutants. The dissolution process may have occurred in the unsaturated zone (from locus nos. 2 or 6), at the groundwater table (from locus no. 7), or from the saturated zone (locus no. 5). The locus forms a continuous phase in the saturated zone and is mobile. This saturated flow may be in the vertical or horizontal direction; the flow is more even than in the unsaturated zone because of the phase continuity and a more consistent water supply. The flowing groundwater may transport the dissolved contaminants (up to several kilometers) from the leak site. No air is in contact with locus no. 8 except at the surface of the groundwater table, which, in combination with the fact that groundwater is far from being a well-mixed system, essentially eliminates volatilization as a significant loss mechanism. Figures 8-1 and 8-2 present a schematic cross-sectional diagram of locus no. 8 and a schematic representation of the transformation and transport processes affecting other loci, respectively.

8.2 EVALUATION OF CRITERIA FOR REMEDIATION

8.2.1 Introduction

Contaminants dissolved in groundwater become part of a continuous, mobile phase. The contaminant's behavior is affected by the partitioning

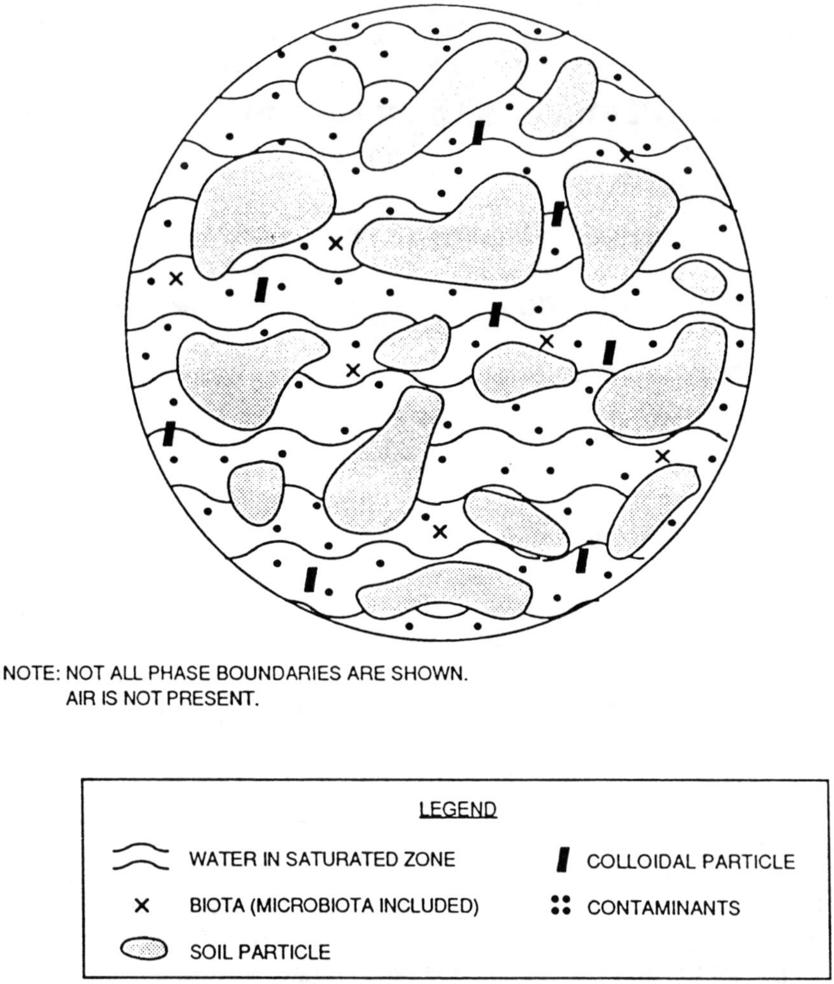

Figure 8–1. Schematic cross-sectional diagram of locus no. 8–contaminants dissolved in groundwater (i.e., water in the saturated zone).

characteristics, particularly the soil sorption partitioning, and by transformation processes, mainly biodegradation.

Remedial measures used for contaminants in this locus tend to fall into three categories: mobilization, immobilization, and transformation. Mobilization techniques, such as pump and treat schemes and artificial recharge, seek to move the contamination to an extraction point where it is captured and treated. During these processes, the contaminants in the groundwater become diluted in concentration. Immobi-

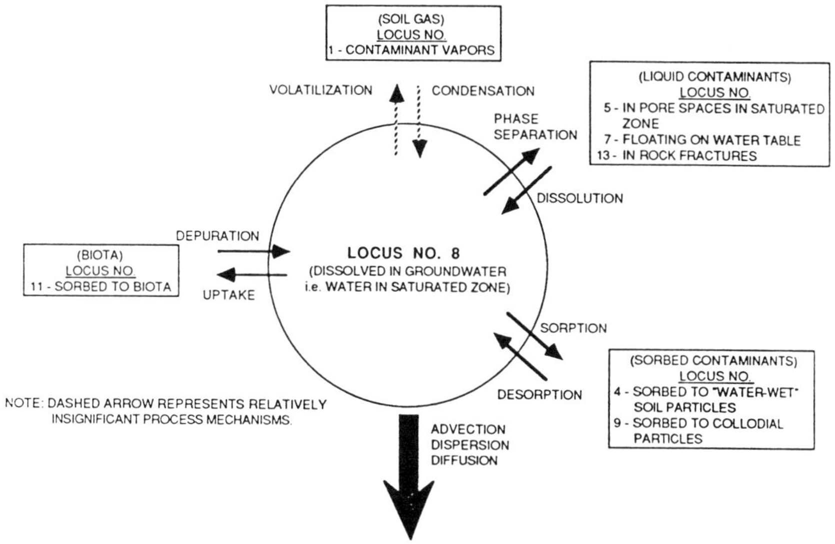

Figure 8–2. Schematic representation of important transformation and transport processes affecting other loci.

lization techniques (e.g., slurry walls) attempt to limit the migration of contaminants by reducing the mobility of a certain part of a groundwater flow regime. This technique alone does not treat the contaminants to dilute the concentrations, but may be combined with other techniques. Transformation processes generally refer to biodegradation, or the enhancement of the natural biological transformation processes, usually through the addition of nutrients, particularly oxygen.

The dissolved oxygen concentration is typically the rate-limiting factor for aerobic biodegradation. Groundwater typically contains very low concentrations of dissolved oxygen, although some systems do contain sufficient oxygen to allow limited aerobic degradation. The introduction of oxygen to groundwater, through air sparging, pure oxygen injection, or the addition of oxygenated compounds such as hydrogen peroxide, is often used to stimulate degradation. Other nutrients, including nitrogen and phosphorus, may also be added.

Anaerobic biodegradation seems to be far less important for hydrocarbon degradation. Far less is known about anaerobic degradation than aerobic degradation. Because the rates of anaerobic degradation are far slower, only aerobic degradation is considered for hydrocarbon remediation.

Remedial measures used to ameliorate groundwater that has been contaminated by dissolved hydrocarbon constituents must address the following questions:

(1) What is the extent of the contaminant plume, both areally and vertically?
(2) What contaminants are present in the plume, and what are the physical and chemical characteristics of those compounds?
(3) What is the rate of transport of the contaminant plume? How feasible would it be to change the rate of transport, either by immbolization or by mobilization?
(4) What is the partitioning behavior of the contaminants and how does that affect the potential removal (e.g., is much of the contaminant sorbed tightly in dead pore space, where it will be very difficult to remove)?

The following sections address the answers to these questions and especially discuss how these answers affect remediation decisions.

8.2.2 Mobilization/Remobilization

8.2.2.1 Partitioning into Mobile Phase

Locus no. 8 consists of contaminants dissolved in the groundwater. For this locus, the contaminant already exists in a mobile phase. The dissolution process may have occurred in the unsaturated zone (from locus no. 2 or no. 6), at the groundwater table (from locus no. 7), from the saturated zone (locus no. 5), or from desorption from water-wetted soil or rocks (locus no. 4) or colloids (locus no. 9). The dissolution of liquid contaminant into the water phase is described in detail in Section 7.

Partitioning of dissolved hydrocarbon into another mobile phase can result from volatilization into the soil gas in the unsaturated zone (locus no. 1) or phase separation from the dissolved hydrocarbon back into the liquid hydrocarbon (locus no. 5 and 7). Since air is only in contact with locus no. 8 at the water table and groundwater is not necessarily a well-mixed system, volatilization is not considered to be a significant loss mechanism. Phase separation, defined as the reverse of dissolution, would be insignificant in comparison to dissolution.

8.2.2.2 Transport with Mobile Phase

Transport of solutes in the groundwater is typically described by a conservation-of-mass statement, which states that the net change in an elemental volume over some time period equals the inflow minus the outflow, adjusted to account for any reactions that occur within the elemental volume during that time period. The flow into and out of the volume may be described by advection and hydrodynamic dispersion. (See Figures 8–3 and 8–4.) Reactions that occur within that volume (e.g., biodegradation) can be difficult to describe mathematically, and are often assumed negligible. Transport may then be described by the advection-dispersion equation, the basic groundwater flow equation. If we consider nonreactive (conservative) solutes in saturated, homogeneous, isotropic materials under steady state, uniform flow conditions, the one-dimensional form of the equation is (Freeze and Cherry, 1979):

$$\frac{\partial C_i}{\partial t} = \frac{1}{R}\left(D_l \frac{\partial^2 C_i}{\partial l^2} - V_l \frac{\partial C_i}{\partial l}\right) \tag{8.1}$$

where C_i = concentration of contaminant i dissolved in groundwater (mg/L)
D_l = hydrodynamic dispersion coefficient (cm^2/sec)
V_l = average linear groundwater velocity (cm/sec)
l = distance along flowline (cm)
R = retardation factor (dimensionless)
t = time (sec)

The first term on the right-hand side of equation 8.1 represents the diffusive flux, while the second term represents the advective flux. The coefficient of hydrodynamic dispersion is composed of two components: mechanical dispersion and molecular diffusion. The diffusion coefficient becomes significant in relatively quiescent systems. Dispersion can also be important when considering dilution or retardation of the contaminant. The dispersion coefficient can be calculated as follows:

$$D = \alpha_L V + D^* \tag{8.2}$$

where D = dispersion coefficient (cm^2/sec)
α_L = longitudinal dispersivity of the media (cm)
V = average linear groundwater velocity (cm/sec)
D^* = effective diffusion coefficient (cm^2/sec)

The average linear groundwater velocity, V_l, in equation 8.1, is equal to the specific discharge divided by the effective porosity (q/Θ). The influence of this transport mechanism dominates the effect of diffusion for most cases. Remedial measures that work by mobilizing the groundwater (e.g., extraction) depend on advective flow. Mathematically, this velocity may be represented:

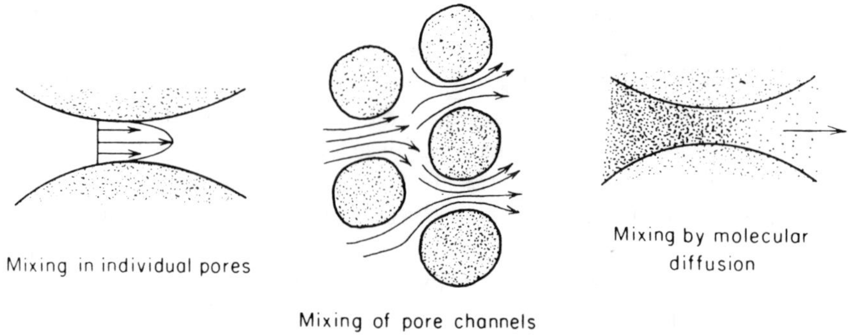

Figure 8–3. Processes of dispersion on a microscopic scale. *Source:* Freeze and Cherry, 1979. (Copyright Prentice-Hall Inc., 1979. Reprinted with permission.)

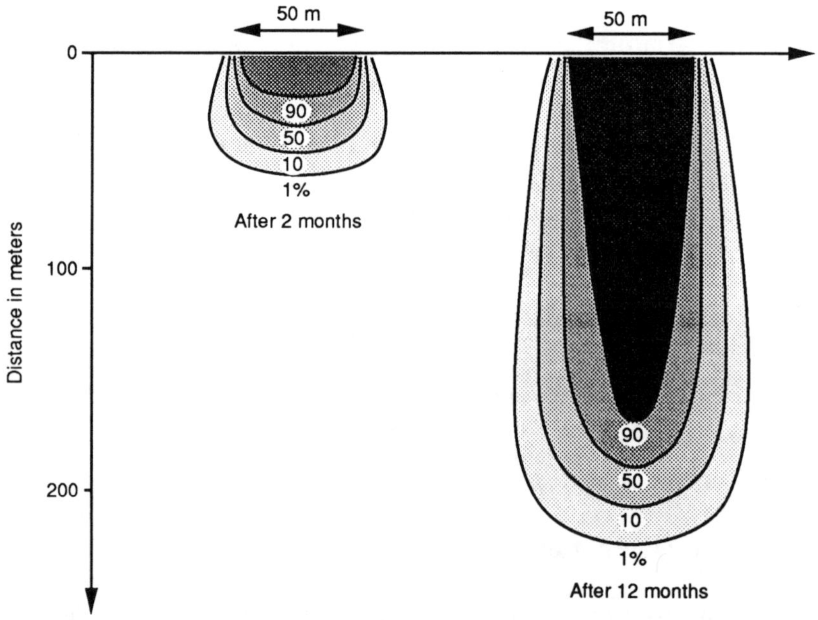

Figure 8–4. Concentration distribution of dissolved components as a result of hydrodynamic dispersion (Flow velocity: 0.5 m/day). *Source:* After Schwille, 1981.

$$V_l = q/\Theta = (1/\Theta)\,\frac{k\,\rho\,g}{\mu}\;\frac{dh}{dl} \tag{8.3}$$

where
v_l = average linear groundwater velocity (cm/sec)
q = volume flow per unit area (cm^3/sec cm^2)
k = intrinsic permeability of the media (cm^2)
ρ = density of the fluid (g/cm^3)
g = gravitational constant (cm/sec^2)
μ = dynamic viscosity (g/cm-sec)
Θ = effective porosity of the media (dimensionless)
$\frac{dh}{dl}$ = hydraulic gradient (cm/cm)

Retardation can have a significant influence on the spread of the dissolved contaminant plume with time. Retardation is simply the ratio of the velocity of a dissolved contaminant plume in relation to the bulk velocity of the groundwater. It can be described by the conventional retardation equation:

$$R = V_w/V_c = 1 + K_d\rho_b/\Theta \tag{8.4}$$

where
R = retardation factor (dimensionless)
V_w = average velocity of water (cm/sec)
V_c = average velocity of chemical pollutant (cm/sec)
K_d = sorption coefficient (cm^3/g) (see Section 4.2.2)
ρ_b = soil bulk density (g/cm^3) (see Section 3.3.2)
Θ = soil porosity (dimensionless) (see Section 3.3.2)

Equation 8.4 shows that the retardation factor moves inversely to the porosity. That is, at low porosity values the retardation increases, while retardation decreases as the porosity increases. The retardation factor is related to the sorption behavior of solutes in groundwater. Sorption will be discussed more fully in the following section.

Equation 8.1 considered only one-dimensional flow. In the field, however, dispersion occurs in three dimensions. For the three dimensions, the advective-diffusion equation reads:

$$\frac{\partial C_i}{\partial t} = \left(\frac{1}{R}\right)\left[Dx\,\frac{\partial^2 C_i}{\partial x^2} + Dy\,\frac{\partial^2 C_i}{\partial y^2} + Dz\,\frac{\partial^2 C_i}{\partial z^2}\right] - \left(\frac{1}{R}\right)\left[Vx\,\frac{\partial C_i}{\partial x} + Vy\,\frac{\partial C_i}{\partial y} + Vz\,\frac{\partial C_i}{\partial z}\right] \tag{8.5}$$

Notice that the dispersion and advective terms are basically the same as in equation 8.1, however, they are applicable to x, y, and z components in Cartesian coordinates.

A modified solution to the advective-dispersive equation based on a description of Baetsle (1969), consists of initial and boundary conditions where the contaminant is assumed to originate as an instantaneous point source at x = 0, y = 0, and z = 0 of mass, M. The mass of contaminant is transported away from the source in a steady-state, uniform flow field moving in the x-direction in a homogenous, isotropic media. The concentration distribution of contaminant mass is given by:

$$C_i\ (x, y, z, t) = \frac{m}{8\ (\pi t)3/2\ \sqrt{D'_x D'_y D'_z}} \exp\left(\frac{-x^2}{4D'_x t} - \frac{y^2}{4D'_y t} - \frac{z^2}{4D'_z t}\right) \quad (8.6)$$

where C_i = concentration of constituent i dissolved in water (g/cm^3)
m = mass of contaminant (g)
D′ = D/R = dispersion coefficient divided by the retardation factor (cm^2/sec)
x = X - (Vt/R)
X = distance traveled in the x direction (cm)
y = distance traveled in the y direction (cm)
z = distance traveled in the z direction (cm)
V = average linear groundwater velocity in the x direction (cm/sec)
t = elapsed time (sec)

If the density of the contaminant is less than that of water and the vertical component of transport is minor, the z direction can be ignored and the solution becomes:

$$C_i\ (x, y, t) = \frac{m}{4\ (\pi t)\sqrt{D'_x D'_y}} \exp\left(\frac{-x^2}{4D'_x t} - \frac{y^2}{4D'_y t}\right) \quad (8.7)$$

Other analytical solutions for equation 8.1 with different initial and boundary conditions are described by Rifai et al. (1956), Ebach and White (1958), Ogata and Banks (1961), Ogata (1970), and others.

Transport Rate. The rate of movement due to advection of the center of mass (i.e., the highest concentration) is simply stated as:

$$V_c = V/R \quad (8.8)$$

where: V_c = average velocity of the contaminant (cm/sec)
V = average linear groundwater velocity (cm/sec)
R = retardation factor (dimensionless)

Examples of transport calculations using equations 8.6 and 8.7 will be shown in Section 8.4.2.

Dispersion. In equation 8.1, the dispersion term

$$\frac{\partial}{\partial x}\left(D\frac{\partial C_i}{\partial x}\right)$$

is defined in accordance with Fick's laws. Near the source of contamination, however, dispersion is a non-Fickian phenomenon. In a porous media, dispersion from the source becomes Fickian at distances on the order of tens to hundreds of meters. This is due to the variation in dispersivity (α_L) in equation 8.2 from the source (usually increasing with distance). The change in dispersivity is considered to be a second-order effect relative to the heterogeneity of the media.

Dispersivities (α_L), which are dependent on grain size distribution but independent of grain shape, are easily measured in a laboratory. They are more difficult to measure in the field, however; here values range from 0.01 to 2 cm, higher than laboratory values due to heterogeneities in the macroscopic flow field.

Diffusion (Details in Section 7.2.2). Diffusion in solutions is the process whereby molecular constituents move under the influence of their kinetic activity in the direction of their concentration gradient (from high concentration to low). Diffusion occurs in the absence (or presence) of advection and ceases only when concentration gradients become nonexistent. In equation 8.2, the effective diffusion coefficient, D*, takes into account the tortuous path for nonadsorbed species in porous media.

$$D^* = \tau D \tag{8.9}$$

where: D^* = effective diffusion coefficient (cm^2/sec)
D = diffusion coefficient (cm^2/sec) (see Section 3.2.2)
τ = tortuosity factor ($0<\omega<1$) (dimensionless)

The tortuosity factor may be expressed as a function of porosity (Θ) of the media (Millington and Quirk, 1961):

$$\tau = \Theta^{1.33} \tag{8.10}$$

For neutral organic chemicals in water, typical values for D are in the range of 10^{-4} to 10^{-5} cm^2/sec (see Table 3–1). The diffusion coefficient is temperature dependent; for example, at 5°C D is only about half as

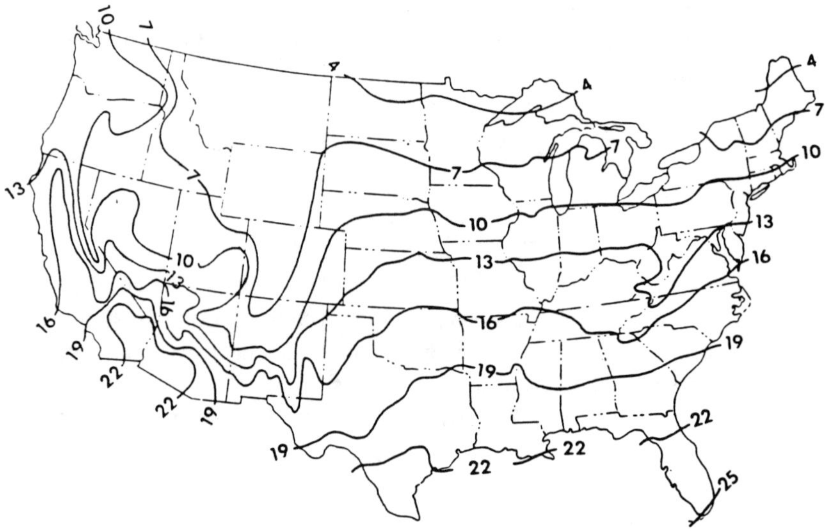

Figure 8–5. Approximate temperature of groundwater in the conterminous United States at depths of 10 to 25 meters (degrees Celsius). *Source:* Heath, 1983.

large as it is at 25°C. Groundwater temperature within any region is relatively stable, although it does vary across larger areas (Figure 8–5).

Advection. The average linear groundwater velocity (V), defined in equation 8.3, is the main component of the advective term in equation 8.1. The hydraulic conductivity (K) and hydraulic gradient (dh/dl) strongly influence the average linear groundwater velocity.

The saturated hydraulic conductivity is often measured rather than the intrinsic permeability of a media.

$$K = \frac{k \rho g}{\mu} \qquad \text{(see equation 8.3)}$$

where
- K = saturated hydraulic conductivity (cm/sec)
- k = intrinsic permeability (cm^2)
- ρ = density of fluid (g/cm^3)
- g = gravitational constant (cm/sec^2)
- μ = dynamic viscosity (g/cm-sec)

Hydraulic conductivity is related to the grain size of a porous media (see Figure 8–6). A simple empirical relationship to estimate the saturated hydraulic conductivity is based on the grain size distribution.

Hydraulic fracturing may also be useful for increasing the conductivity of an area surrounding a well (Murdoch et al., 1990). This method involves pumping fluid into a well until pressures exceed a critical value, beyond which fracturing begins. Sand is then pumped into the fractures to hold them open and provide a high permeability channel. This method, developed for enhanced oil recovery many years ago, is especially useful in regions of low permeability, and may help to improve remediation methods that depend on mobilizing contaminants (groundwater pumping, soil venting, etc.).

The main component of advection is the average linear groundwater velocity, which depends on the hydraulic conductivity and gradient of the media. The average linear groundwater velocity is greatest in homogeneous soils with high porosity, and with more soluble constituents. The conductivity can be increased for the purpose of remediation (e.g., flushing, extraction, etc.) by increasing the gradient.

The grain size distribution, easily measured in a laboratory from a soil sample, should be used to find the following parameters:

D_{60} = grain size diameter larger than 60% (by weight) of the soil's grains
D_{50} = median grain size
D_{10} = grain size diameter larger than 10% (by weight) of the soil's grains

The uniformity coefficient is equal to D_{60}/D_{10}. Once these values are known, they can be used with Figure 8–6 (Robson, 1978) to find the hydraulic conductivity. The range of values for hydraulic conductivity is very large (Table 8–1), from 100 to 10^{-11} cm/s. Hydraulic conductivity values can be measured in the field through a variety of aquifer test methods (e.g., slug test or pump test). For neutral organic compounds in water, typical groundwater velocities may range from tens to hundreds of meters/year. Naturally-occurring hydraulic gradients are typically on the order of 0.01 (dh/dl).

Retardation. Contaminants dissolved in groundwater can become retarded by sorption onto mineral grains in soils or rocks in the saturated zone. Sorption retards the movement of a contaminant compared to the groundwater movement (see locus no. 4 and no. 9). When partitioning of a contaminant between the liquid and solid phases is completely reversible, the retardation of the contaminant relative to the bulk mass of the groundwater can be described by equation 8.4.

Species in a mixture of reactive contaminants that are dissolved in groundwater will travel at different rates depending on their retardation factors (see Eq. 8.4). Thus, the contaminant plume will exhibit a chro-

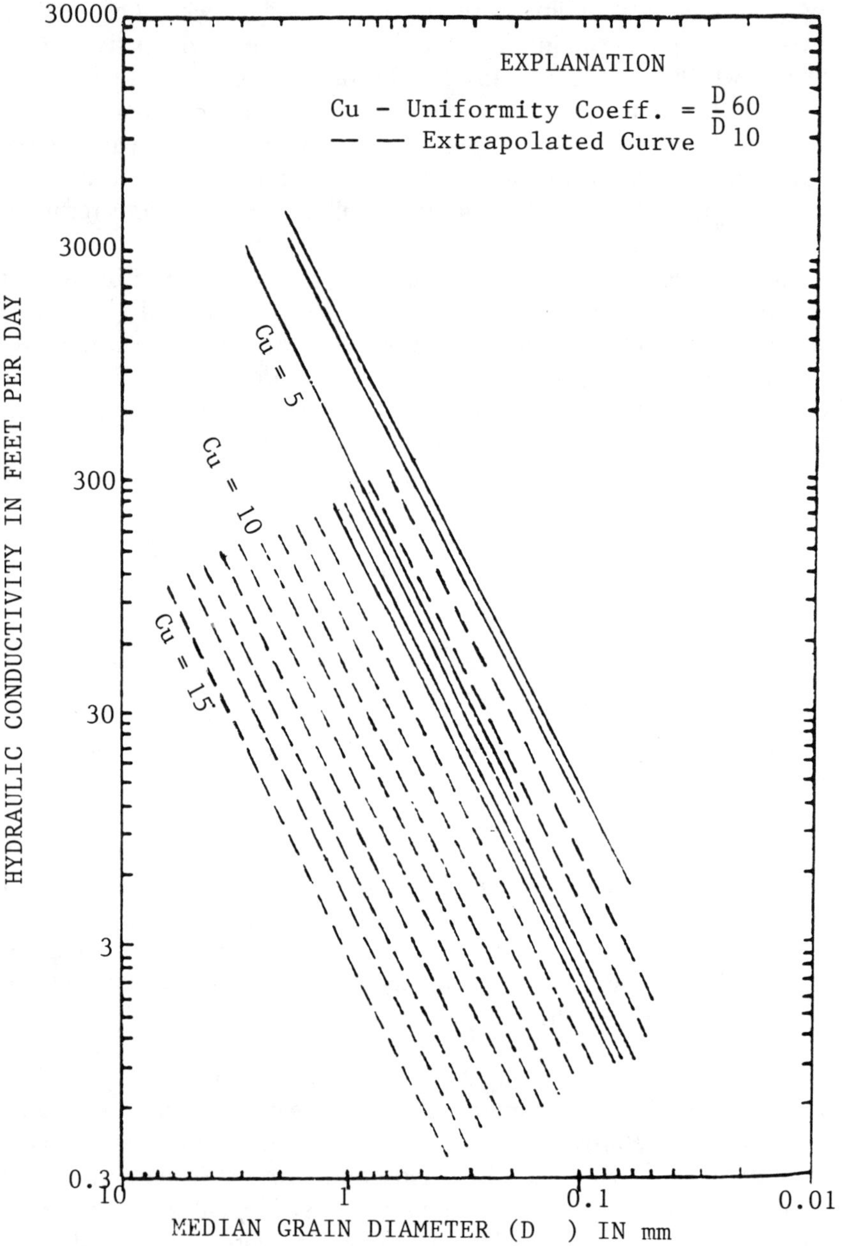

Figure 8–6. Relations among hydraulic conductivity, uniformity coefficient, and median grain diameter. *Source:* Robson, 1978.

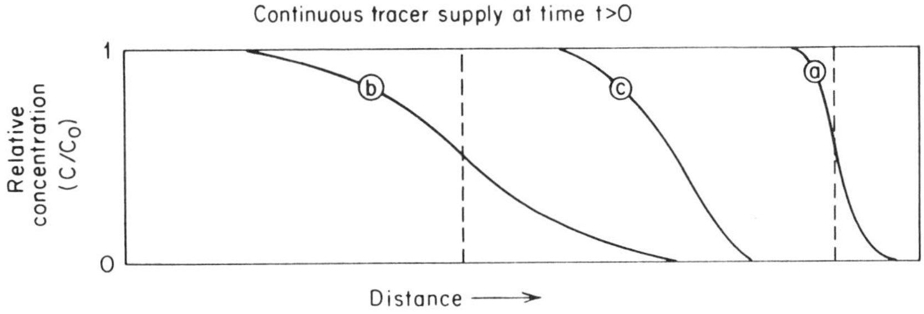

(a) Dispersed front of non-retarded solute
(b) Front of solute that undergoes equilibrium partitioning between liquid and solids
(c) Front of solute that undergoes slower rate of transfer to solids

Figure 8–7. Advance of reactive and nonreactive contaminants through a column. *Source:* Freeze and Cherry, 1979. (Copyright Prentice-Hall Inc., 1979. Reprinted with permission.)

matographic effect over time. As a contaminant plume advances along flow paths, the front is retarded as a portion of the contaminant mass sorbs to the solid phase (Figure 8–7). If the input of contaminant mass to the groundwater is discontinued, and the groundwater concentrations drop (for example, as a groundwater pump and treat scheme progresses), contaminants may desorb from the soil back into the groundwater. Thus, it is doubtful that all of the contaminant can be permanently immobilized, even though the retardation of the concentration front may be strong. However, a portion of the contaminants may be irreversibly fixed relative to the time scale of interest (see Figure 8–8).

Retardation is strongly affected by subsurface carbon content. The effect on retardation differs, however, based on the phase of the carbon. Dissolved organic carbon (DOC) (e.g., highly soluble compounds such as MTBE and tertiary butyl alcohol and colloidal particles of humic and fulvic acids) may significantly lower retardation. Organic matter in the solid phase increases sorption and hence retardation.

Garrett et al. (1986) report on how the presence of methyl tertiary butyl ether (MTBE) in gasoline spills affects the movement of the plume and the detection of a release. MTBE has multiple effects related to retardation and movement. First, MTBE, which may comprise up to 11% by volume of gasoline, is soluble in water to 43,000 mg/L (Garrett et al., 1986), about 25 times more soluble than benzene, otherwise the most soluble gasoline constituent at 1,780 mg/L. Generally, sorption of organic compounds is inversely proportional to their solubility. Thus, MTBE is likely to be minimally retarded, allowing it to travel faster than

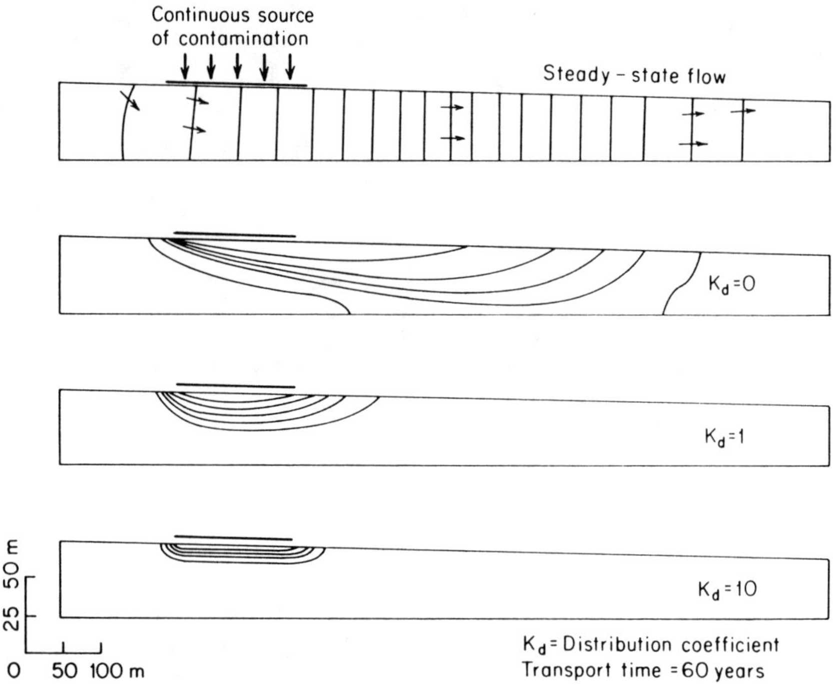

Porosity = 0.3
Hydraulic conductivity = 0.5 m/day
α_l = 10 m
α_t = 0.5m
Transport time = 60 years
Concentration contours at C/Co = 0.9, 0.7, 0.5, 0.3, and 0.1

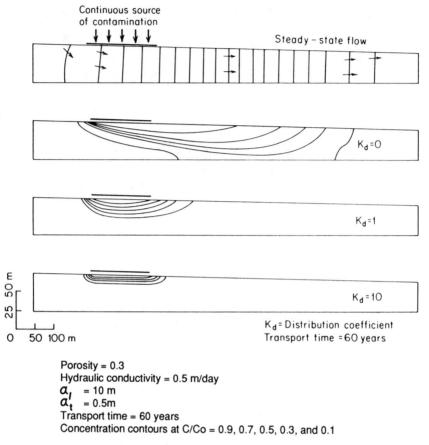

Figure 8–8. Effect of the distribution coefficient on contaminant retardation during transport in a shallow groundwater flow system. *Source:* Freeze and Cherry, 1979. (Copyright Prentice-Hall Inc., 1979. Reproduced with permission.)

other gasoline constituents. Indeed, Garrett et al. (1986) report that MTBE often forms a "halo" around a dissolved gasoline plume, where MTBE can be detected but other gasoline compounds are not detected.

MTBE also decreases the retardation of other gasoline constituents by increasing their dissolved concentrations in the groundwater. Benzene, toluene, ethylbenzene, and xylenes (BTEX) are more soluble in ether than in water. When MTBE is present in the gasoline plume, the BTEX concentrations in the groundwater are higher. Because the BTEX compounds are more soluble, they are likely to be more mobile (retarded less) in the groundwater.

For groundwaters that contain significant dissolved DOC, the conventional retardation equation (Eq. 8.4) can be modified to reflect the effect that DOC has:

$$R = 1 + \frac{K_d \rho_b / \Theta}{1 + K_{DOC}\,(DOC)\,10^{-6}} \tag{8.11}$$

where:
R = retardation coefficient (dimensionless)
K_d = sorption coefficient (L/kg)
ρ_b = soil bulk density (kg/L)
Θ = porosity (dimensionless)
K_{DOC} = sorption constant for compound in DOC (L/kg) (~Koc)
DOC = dissolved organic carbon concentration (mg/L)

This equation presumes that the solid organic matter is less than 0.1%, so Kd-type constants may replace Koc constants. Higher levels of organic carbon in the solid matter can increase sorption and elevate retardation effects.

Figure 8–9 shows that retardation factors are significantly lower at DOC concentrations of 100 ppm, particularly for high-molecular weight organic compounds. The variabilitiy in the retardation factors due to dissolved carbon adds to the inherent variance; Table 8–1 shows that retardation factors range over two orders of magnitude. The higher molecular weight aliphatic compounds are most strongly retarded.

8.2.3 Fixation (Details in Section 4.2.3)

8.2.3.1 Partitioning onto Immobile (Stationary) Phase

Sorption onto soil particles (locus no. 4) is a possible means of fixation. Soil sorption is an important factor in solute retardation, discussed in the preceding section. Sorption is not reliable for preventing transport with the ambient groundwater flow for all compounds. For other compounds, however, sorption to soil does virtually prevent significant transport (especially high molecular-weight compounds). Thus, sorption would probably not be an effective mechanism for permanent fixation of all of the contaminant, although a portion of the contaminant may be irreversibly fixed relative to a specific time scale.

Sorption could possibly be an effective mechanism for fixation of most of the contaminant in a relatively quiescent groundwater system. A slower groundwater velocity results in a greater amount of sorption and,

Table 8–1. Estimated Retardation Factors for Hydrocarbon Gasoline Constituents

	Estimated Soil Sorption Constants[a]		Estimated Retardation Factors[b]
	K_{oc} (L/kg)	K_d (f_{oc} = 0.1%)	R
n-Butane	490	0.49	4.5
Isobutane	420	0.42	4.0
n-Pentane	910	0.91	7.6
Isopentane	880	0.88[c]	7.3
1-Pentane	460	0.46[c]	4.3
n-Hexane	1,900	1.90[c]	14.7
1-Hexene	910	0.91	7.6
2-Methylpentane	1,500	1.50[c]	11.8
Cyclohexane	960	0.96	7.9
Benzene	190	0.19	2.4
n-Heptane	4,300	4.30[c]	32.0
2-Methylhexane	3,200	3.20	24.0
Methylcyclohexane	1,800	1.80	14.0
Toluene	380	0.38	2.7
n-Octane	8,200	8.20[c]	60.0
2,4-Dimethylhexane	5,200	5.20	38.4
Ethylbenzene	680	0.68	5.9
m-Xylene	720	0.72[c]	6.2
2,2,4-Trimethylhexane	8,700	8.70[c]	63.6
1,3,5-Trimethylbenzene	940	0.94	7.8
2,2,5,5-Tetramethylhexane	14,000	14.0 [c]	101.8
1,4-Diethylbenzene	2,900	2.9	21.9
Dodecane	88,000	88.0	634.0

a. $K_d = K_{oc}\, f_{oc}$
f_{oc} = Fraction of organic carbon
b. Using equation 8.4, a bulk density ρ_s = 1.8 L/kg and soil porosity Θ = 0.20
c. Nonlinearity is important for these constituents when their concentrations are close to their solubility limit.

thus, lower concentrations in the transmitted water. An example is given for benzene in Figure 8–10.

8.2.3.2 Other Fixation Approaches

None are immediately evident.

8.2.4 Transformation

8.2.4.1 Biodegradation (Details in Section 11)

Groundwater is largely devoid of oxygen due to its slow movement and lack of surficial contact with air. As a result, the aerobic microorga-

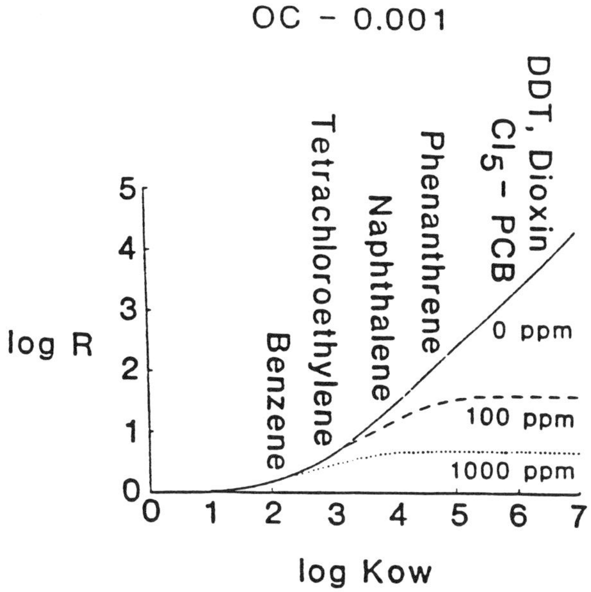

Figure 8–9. Relationship of retardation factors to octanol-water partition coefficient by conventional (———) and revised (---,---) equations. *Source:* Kan and Tomson, 1986. (Copyright Water Well Journal Publishing Co., 1986. Reprinted with permission.)

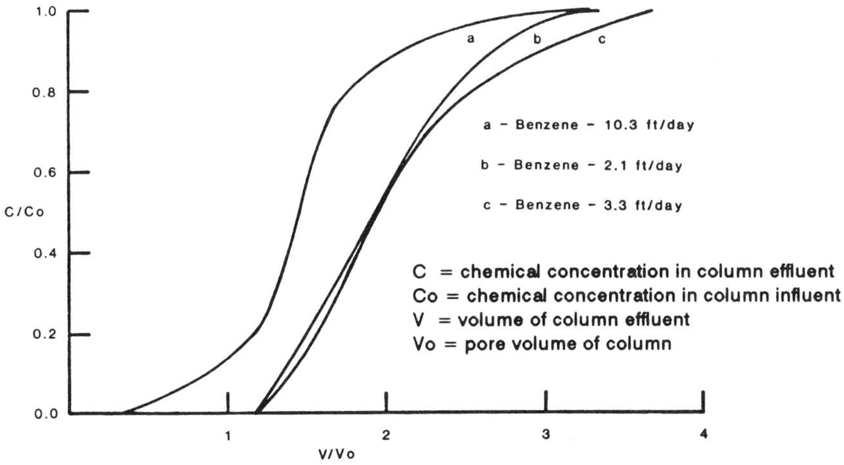

Figure 8–10. Normalized breakthrough curves for displacement of benzene of 15-cm Lincoln fine sand soil column as a function of water velocity. *Source:* Clark et al., 1986. (Copyright Water Well Journal Publishing Co., 1986. Reprinted with permission.)

nisms primarily responsible for biodegradation cannot survive in large numbers in this locus. Biodegradation of contaminants in the groundwater, therefore, is usually an insignificant means of removal. However, if contaminants are transported to the water table or vadose zone (due to fluctuations in the groundwater elevation) which are much more well oxygenated, biodegradation can occur much more rapidly.

8.2.4.2 Chemical Oxidation (Details in Section 3.2.4.2)

Contaminants in groundwater can undergo abiotic transformation (provided precursors such as phenol and catechol are present) as a result of aerobic microbial process.

8.3 STORAGE CAPACITY IN LOCUS

8.3.1 Introduction and Basic Equations

Some of the information below considers the case for gasoline, which is the product most commmonly held in an UST. The maximum concentration of an unadulterated gasoline that can be dissolved in groundwater is directly related to the solubilities of each constituent in a gasoline blend (for further discussion of dissolution, see Section 7).

For unoxidized hydrocarbon in contact with pure water the concentration of the solute in aqueous solution is proportional to the concentration of the constituent in gasoline and the solubility of the solute.

$$C_{iw} = X_{ig} S_{iw} \tag{8.12}$$

$i = 1, 2, 3 \ldots n$ constituents

where C_{iw} = concentration of constituent i in the water phase (mg/L)
X_{ig} = mole fraction of constituent i in gasoline (dimensionless)
S_{iw} = solubility of constituent i in pure water (mg/L)

- Calculations are more difficult if significant amounts of oxygenated additives are present.

8.3.2 Guidance on Inputs for, and Calculation of, Maximum Value

An approximate maximum for non-oxygenated gasoline blends or for hydrocarbon constituents of oxygenated fuels is on the order of 200 mg/L. The value for oxygenated fuels (i.e., those containing MTBE, TBA, ethanol, methanol, etc.) is poorly constrained at present, but is at least an order of magnitude greater than that for non-oxygenated fuels. The general effect of co-solvents on solubility can be stated as:

$$\ln S^{m} = \ln S^{w} + \sigma f \tag{8.13}$$

where S^{w} = mole fraction solubility in water (dimensionless)
S^{m} = mole fraction solubility in mixed solvents (dimensionless)
σ = parameter related to solutes' surface area and interfacial free energy
f = volume fraction of co-solvent ($0<f<1$) (dimensionless)

The value of 200 mg/L stated above for the maximum concentration of gasoline without additives dissolved in groundwater is an approximation based on a laboratory study (API, 1985) and solubility calculations from literature—reported values (Lyman, 1987). The actual concentrations found in the field are typically much lower than the maximum estimated value. The maximum values in the field typically diminish with time, as advection and dispersion serve to spread the plume and dilute the concentrations.

Hydrocarbon solubility in water is not significantly affected by modest changes in temperature, ionic strength, pH, or pressure. Temperature and pressure are nearly constant for most groundwaters. Solubility is reduced somewhat with increasing ionic strength in aqueous solution. Solubility may be affected, perhaps quite significantly, by co-solvent effects. Garrett et al. (1986) note that in gasolines that contain MTBE, the BTEX compounds are more soluble than in gasolines not containing MTBE, because BTEX are almost completely soluble in ether. Co-solvent values for σ (see equation 8.13) are empirically-derived; their availability is limited for gasoline products. In general, solubility increases with increasing dissolved organic carbon.

The composition of the dissolved gasoline is quite different from the composition of the bulk product (see Table 3-2). For non-oxygenated gasolines, the most soluble components are the light aromatics: benzene, toluene, xylene, and ethylbenzene (BTEX). The literature includes several studies of the water soluble fraction (WSF) of gasoline and other petroleum products often stored in underground storage tanks. The results show that BTX compounds comprise 23–55% of the bulk product but 42–74% of the WSF; benzene concentrations are enriched ten times in the WSF versus the bulk product (Guard et al., 1983). Coleman et al. (1984) performed solubilization experiments and found that the

Table 8–2. Water Soluble Fractions for Three Gasolines (PPM)

	Regular Leaded	Regular Unleaded	Super Unleaded
Benzene	30.5	28.1	67
Toluene	31.4	31.1	107.4
MTBE	43.7	35.1	966

unleaded gasoline product contained 49% aromatic compounds, but the WSF contained 95% after one-half hour of mixing.

Solubility data for gasoline that contains oxygenated additives is presented by Kramer and Hayes (1987). Partial results of their experimental work are shown in Table 8–2.

The concentration of benzene is roughly doubled and toluene, more than tripled in the super unleaded brand. The super unleaded MTBE concentration of 966 mg/L is 27 times regular unleaded. While Kramer and Hayes do not state the concentrations of benzene and toluene in the bulk product, the increased levels of benzene and toluene in the WSF of the super unleaded gasoline may be due to cosolvent effects from the MTBE.

8.3.3 Guidance on Inputs for, and Calculation of, Average Values

- Typical value for non-oxygenated gasoline blends is on the order of 10–20 mg/L.
- Oxygenated fuels are at least an order (if not several orders) of magnitude greater than for non-oxygenated fuels.

8.4 EXAMPLE CALCULATIONS

8.4.1 Storage Capacity Calculations

8.4.1.1 Maximum Value

- Use maximum value of 200 mg/L given in Section 8.3.2 for unadulterated gasoline.

8.4.1.2 Average Value

- Use average value of 20 mg/L given in Section 8.3.3.

These values are based, in part, on a weighted sum and average of calculated concentrations using equation 8.12:

$$C_i = X_{ig} S_{ig}$$

For illustrative purposes, an example is provided in Table 8-3.

8.4.1.3 Example of Solubility Enhancement

For oxygenated fuels, the effect of co-solvents on solubility can be shown by the following example. Consider the enhanced solubility of benzene in water due to the presence of ethanol using equation 8.13:

$$\begin{aligned} S^w &= 1780 \text{ mg/L } (25°C) \text{ in pure water (see Table 8–3)} \\ &= 4.11 \times 10^{-4} \text{ mole fraction} \end{aligned}$$

Assume ethanol present at 10% volume and assume o = 10, f = 0.1

$$\begin{aligned} \ln S^m &= \ln (4.11 \times 10^{-4}) + 10\,(0.1) = -7.797 + 1 \\ &= -6.797 \\ S^m &= 1.118 \times 10^{-3} \text{ mole fraction} \\ &= 4840 \text{ mg/L} \end{aligned}$$

As can be seen from the example, ethanol increases the solubility of benzene nearly three times.

8.4.2 Transport Rate Calculations

For unadulterated gasoline, the transport of each constituent needs to be considered in calculating the effect of transport on the concentration of the blend. The following example considers the transport of benzene dissolved in groundwater and uses equation 8.12.

- What is the concentration after one year of an instantaneous benzene-saturated point source of 1 m^3 in groundwater, with an average linear pore velocity in the x-direction of 100 m/yr and dispersivities of 1 m and 10 cm in the x and y directions, respectively? Steady-state groundwater flow in a homogeneous, isotropic media are implicitly assumed. Using equation 8.7:

$$C\,(x, y, t) = \frac{m}{4\,(\pi t)\sqrt{D_x D_y}} \exp\left(\frac{-x^2}{4D_x t} - \frac{y^2}{4D_y t}\right)$$

- Contaminant Mass, m, in system

$$m = (C_i)\,(V)$$

Where: C_i = benzene concentration at saturation
V = unit volume

Table 8–3. Estimated Solubility of Gasoline in Water

	X	S	C
Gasoline Component	Selected Conc. in Gasoline (Mol. Fr.)	Reported Solubility in Pure Water 25°C (mg/L)	Est. Conc. in Gasoline-Saturated Water (mg/L)
n-Butane	0.0163	61.4 (1 ATM)	1.00
Isobutane	0.0326	48.9 (1 ATM)	1.59
n-Pentane	0.0394	41.2	1.62
Isopentane	0.184	48.5	8.92
1-Pentene	0.0203	148	3.00
n-Hexane	0.0990	12.5	1.24
1-Hexene	0.0169	50	0.84
2-Methylpentane	0.0880	14.2	1.25
Cyclohexane	0.0338	59.7	2.02
Benzene	0.0364	1780	64.80
n-Heptane	0.0142	2.68	0.38
2-Methylhexane	0.0473	2.54	0.12
Methylcyclohexane	0.00966	15	0.14
Toluene	0.0515	537	27.6
n-Octane	0.00830	0.66	0.0055
2,4-Dimethylhexane	0.0664	1.5	0.10
Ethylbenzene	0.0179	167	2.99
m-Xylene	0.0625	162	10.1
2,2,4-Trimethylhexane	0.0148	0.8	0.012
1,3,5-Trimethylbenzene	0.0394	72.6	2.86
2,2,5,5-Tetramethylhexane	0.0100	0.13	0.0013
1,4-Diethylbenzene	0.0353	15	0.53
Dodecane	0.0558	0.005	3×10^{-4}
	Σ = 0.998	= 3240	= 131 219 (10°C)

Source: Lyman, 1987.

Note: No additives considered.

$$
\begin{aligned}
m &= (1.78 \times 10^6 \text{ mg/m}^3)(1 \text{ m}^3) \text{ (for benzene)} \\
&= 1.78 \times 10^6 \text{ mg}
\end{aligned}
$$

- Use equation 8.2 to determine dispersion coefficients, D_x' & D_y'

$$
\begin{aligned}
D^* &= 1 \times 10^{-5} \text{ cm}^2/\text{s (benzene diffusion in water)} \\
\alpha_x &= 100 \text{ cm} \\
\alpha_y &= 10 \text{ cm determined by measured extent of the plume} \\
V &= 100 \text{ m/yr}
\end{aligned}
$$

$$
\begin{aligned}
D_x' &= \alpha_x V + D^* \\
&= (100 \text{ cm})(10{,}000 \text{ cm}/31{,}536{,}000 \text{ sec}) + (10^{-5} \text{ cm}^2/\text{s}) \\
&= 0.03172 \text{ cm}^2/\text{s}
\end{aligned}
$$

$$D_y' = (10\ \text{cm})(10{,}000\ \text{cm}/31{,}536{,}000\ \text{sec}) + (10^{-5}\ \text{cm}^2/\text{s})$$
$$= 0.00318\ \text{cm}^2/\text{s}$$

Taking into account retardation (R = 1.5 for benzene)

$$D_x = D'_x/R = 0.03172/1.5 = 0.02115\ \text{cm}^2/\text{s}$$
$$D_y = 0.00318/1.5 = .001414\ \text{cm}^2/\text{s}$$

Since the mass, M, is defined as a point source that spreads with transport, the peak concentration will be at the center of the concentration profile (Gaussian distribution). Thus, in determining the maximum concentration at the centerline, we can let $y = 0$.

The concentration at any point at any time is at a maximum when the contaminant has traveled the least distance, that is, as $x \rightarrow 0$. This occurs when the distance traveled is equal to the average linear pore velocity at a particular time:

$$x \rightarrow 0 \text{ when } x = \frac{Vt}{R} \text{ (see definition of x for equation 8.6)}$$

Therefore, to determine the maximum concentration at time, t, we can set $x = 0$.

- Input values

$$t = 1\ \text{year} = 31{,}536{,}000\ \text{seconds}$$
$$m = (C_i)\ (V) = 1.78 \times 10^6\ \text{mg}$$
$$D^* = 1 \times 10^{-5}\ \text{cm}^2/\text{s}$$
$$\alpha_x = 100\ \text{cm}$$
$$\alpha_y = 10\ \text{cm}$$
$$V = 100\ \text{m/yr}$$
$$R = 1.5$$
$$D_x = 0.02115\ \text{cm}^2/\text{s}\ (1 \times 10^{-4}) = 2.1 \times 10^{-6}\ \text{m}^2/\text{s}$$
$$D_y = 0.001414\ \text{cm}^2/\text{s}\ (1 \times 10^{-4}) = 1.4 \times 10^{-7}\ \text{m}^2/\text{s}$$
$$x = 0$$
$$y = 0$$

$$C\ (x,y,t) = \frac{(1.78 \times 10^6)}{4\ (\pi)(31{,}536{,}000)\sqrt{(2.1 \times 10^{-6})(1.4 \times 10^{-7})}} \exp\ (0)$$

$$= 10{,}150\ \text{mg/m}^3$$
$$= 10.2\ \text{mg/L}$$

Thus, for 1,780 grams of benzene that traveled 66 meters in one year, the final concentration is 10.2 mg/L.

The rate of transport (using equation 8.8) is simply:

$$V_c = V/R$$
$$= 100/1.5 = 66.6 \text{ m/yr}$$

8.5 SUMMARY OF RELATIVE IMPORTANCE OF LOCUS

8.5.1 Remediation

Hydrocarbons that are dissolved in groundwater pose one of the more serious pollution problems. The solubility of the lighter hydrocarbon constituents greatly exceeds the concentration levels at which groundwater is considered to be seriously polluted. For example, the typical solubility of a gasoline might be approximately 20 mg/L, but it can be detected by taste and odor at concentrations of less than 0.005 mg/Liter.

This low taste and odor threshold and the toxicity of many hydrocarbons (even at low levels) argues against fixation as a remedial measure. Natural flushing, dispersion and biodegradation serve to reduce concentrations of hydrocarbons in the aquifer, but the time scale for this approach is sufficiently long to preclude this means (the "no-action" approach) from consideration for all but a few sites.

Many contaminated sites will require removal and treatment to restore the aquifer to a safer condition. Historically, this has meant pumping of the contaminated groundwater followed by above ground treatment, usually via air stripping or activated carbon. Biostimulation—the addition of oxygen and other nutrients to improve the natural biodegradation processes—is being used at an increasing number of sites but is still a far less popular option than either air stripping or activated carbon.

All three groundwater treatment methods have shown the ability to reduce aquifer concentrations to below regulated levels. Their application is site-specific, however, and some sites prove difficult to restore to pre-contaminated condition.

8.5.2 Loci Interactions

The loci that interact with hydrocarbon constituents dissolved in groundwater during partitioning and mass migration processes are summarized in Figure 8-2 and Table 8-4. In an open system, groundwater can act as a sink for mobile constituents from other loci. Dissolution of hydrocarbon constituents may have occurred in the unsaturated zone

Table 8–4. Loci Interactions with Hydrocarbon Dissolved in Groundwater

Process	Phases in Contact[a]	Interacting Loci	Relative Importance
Mobility			
Dissolution	wet soil	4	low
	liquid hydrocarbon	5,7	high
Bulk Transport (advection and dispersion/ diffusion)	liquid hydrocarbon	5,7	high
	wet soil	4	moderate
	rock	13	moderate-high
	air	1	low
Volatilization			
Immobility			
Sorption	wet soil	4	low
	rock	10,13	low

a. Biota (locus no. 11) are potentially in contact with all phases.

(loci nos. 2 or 6), at the groundwater table (locus no. 7), from the saturated zone (loci no. 4, 5, or 9). Bulk transport out of the locus is influenced by the media through which the groundwater flows (loci 4, 10, or 13) and the overburden. Aromatic constituents dissolved in groundwater can volatilize into the overlying soil gas (locus no. 1), although the relative amounts are probably not significant. The distribution of hydrocarbon in or out of the groundwater may depend on the nature of groundwater movement and proximity to discharge points. Of these loci, those containing liquid hydrocarbon are relatively more important.

8.5.3 Information Gaps

For a better understanding and definition of factors affecting hydrocarbon constituents dissolved in groundwater, the following information is needed:

(1) Fundamental data on solubility, especially for multi-component products like petroleum, in addition to co-solvent effects for additives.
(2) Composition and physicochemical properties of gasoline.
(3) Dissolution rates for product into groundwater, including recharge from unsaturated zone residual.
(4) Better definition of natural biodegradation rates in aerobic and anaerobic systems, and factors affecting these rates.
(5) Concentration of additives in different gasoline blends, their behavior

as co-solvents and in groundwater, and basic property data of additives.
(6) Sorption data (K_d, K_{oc}) representing natural soil systems that include nonequilibrium effects.
(7) The influence of media structure and heterogeneity on bulk transport and retardation.
(8) Further definition of dispersivities in relation to media heterogeneity and proximity to the source.

8.6 LITERATURE CITED

API. 1985. Subsurface Venting of Hydrocarbon Vapors from an Underground Aquifer. Publication No. 4410. Washington, D.C.

Baetsle, L.H. 1969. Migration of Radionuclides in Porous Media.

Progress in Nuclear Energy, Series XII, Health Physics. A.M.F. Duhamel (Ed.) Pergamon Press, Elmsford, NY. p. 707–730.

Clark, G.L., A.T. Kan, and M.B. Tomsom. 1986. Kinetic Interaction of Neutral Trace Level Organic Compound with Soil Organic Material. In: Proceedings of Petroleum Hydrocarbons in Ground Water: Prevention, Detection, and Restoration. NWWA/API, Houston, TX.

Coleman, W.E., J.W. Munch, R.P. Streicher, H.P. Ringhand, and F.C. Kopfler. 1984. The Identification and Measurement of Components in Gasoline, Kerosine, and No. 2 Fuel Oil that Partition into the Aqueous Phase After Mixing. Arch. Environ. Contam. Toxicol. 13: 171–178.

Ebach, E.A. and R.R. White. 1958. Mixing of Fluids Through Beds of Packed Solids. American Institute of Chemical Engineering Journal, Vol. 4, no. 2.

Freeze, R.A. and J.H. Cherry. 1979. Groundwater. Prentice-Hall Inc., New Jersey.

Garrett, P., M. Moreau, and J.D. Lowry. 1986. Methyl Tertiary Butyl Ether as a Ground Water Contaminant. In: Proceedings of Petroleum Hydrocarbons and Organic Chemicals in Ground Water Conference. NWWA-API.

Guard, H.E., J. Ng and R.B. Laughlin. 1983. Characterization of Gasolines, Diesel Fuels and their Water Soluble Fractions. Naval Biosciences Laboratory, Naval Supply Center, Oakland, CA.

Heath, R.C. 1983. Basic Ground-water Hydrology. USGS Water Supply Paper 2220. 84 pp.

Kan, A.T. and M.B. Tomson. 1986. Facilitated Transport of Naphthalene and Phenanthrene in a Sandy Soil Column with Dissolved Organic Matter—Macromolecules and Micelles. In: Proceedings of Petroleum Hydrocarbons and Organic Chemicals in Groundwater. NWWA/API, Houston, TX.

Kramer, W.H. and T.J. Hayes. 1987. Water Soluble Phase of Gasoline: Results of a Laboratory Mixing Experiment. New Jersey Geological Survey Technical Memorandum 87-5. Trenton, NJ.

Lyman, W.L. 1987. Environmental Partitioning of Gasoline in Soil/ Groundwater Compartments. Final Seminar Proceedings on Underground Environment of an UST Motor Fuel Release. USEPA RREL, sponsors, Edison, NJ.

Millington, R.J. and J.M. Quirk. 1961. Permeability of Porous Solids. Transactions Faraday Society, 57:1200-1207.

Murdoch, L.C., G. Losonsky, I. Klich, and P. Cluxton. 1990. Field Tests of Hydraulic Fracturing to Increase Fluid Flow in Soils. Poster Session, Sixteenth Annual EPA Hazardous Waste Research Symposium, Cincinnati, OH. April 3-5.

Ogata, A. 1970. Theory of Dispersion in a Granular Medium. USGS Professional Paper 411-I.

Ogata, A. and R.B. Banks. 1961. A Solution of the Differential Equation of Longitudinal Dispersion in Porous Media. USGS Professional Paper 411-A.

Rifai, M.N.E., W.J. Kaufman and D.K. Todd. 1956. Dispersion Phenomena in Laminar Flow through Porous Media. Report No. 3IER, Series 90, Sanitary Engineering Research Laboratory, University of California, Berkeley.

Robson, S. 1978. Application of Digital Profile Modeling Techniques to Groundwater Solute Transport at Barstow, CA. U.S.G.S. Water Supply Paper 2050.

Schwille, F. 1981. Groundwater Pollution in Porous Media by Fluids Immiscible with Water. *In*: Quality of Groundwater, Proceedings of an International Symposium, Noordwijkehout, The Netherlands. p. 451-463.

SECTION 9

Contaminants Sorbed onto Colloidal Particles in Water in Either the Unsaturated or Saturated Zone

CONTENTS

List of Tables 238
List of Figures 238
9.1 Locus Description 239
    9.1.1 Short Definition 239
    9.1.2 Expanded Definition and Comments 239
9.2 Evaluation of Criteria for Remediation 240
    9.2.1 Introduction 240
    9.2.2 Mobilization/Remobilization 243
        9.2.2.1 Partitioning onto Mobile Phase 243
        9.2.2.2 Transport with Mobile Phase 246
    9.2.3 Fixation 249
        9.2.3.1 Partitioning onto Immobile (Stationary) Phase 249
    9.2.4 Transformation 250
        9.2.4.1 Biodegradation 250
        9.2.4.2 Chemical Oxidation 250
9.3 Storage Capacity in Locus 251
    9.3.1 Introduction and Basic Equations 251
    9.3.2 Guidance on Inputs for, and Calculation of, Maximum Value 251
    9.3.3 Guidance on Inputs for, and Calculation of, Average Values 251
9.4 Example Calculations 252
    9.4.1 Storage Capacity Calculations 252
        9.4.1.1 Maximum Value Calculations 252
        9.4.1.2 Average Value Calculations 252
9.5 Summary of Relative Importance of Locus 253
    9.5.1 Remediation 253
    9.5.2 Loci Interactions 253
    9.5.3 Information Gaps 253

9.6 Literature Cited 254

TABLES

9–1 Relative Sizes of Particles 241
9–2 Factors Influencing Fixation of Colloidal Particles ... 250

FIGURES

9–1 Schematic Cross-Sectional Diagram of Locus No. 9—Contaminants Sorbed to Colloidal Particles in Water in Either the Unsaturated or Saturated Zone ... 240
9–2 Schematic Representation of Important Transformation and Transport Processes Affecting Other Loci 241
9–3 Continuum of Particulate and Dissolved Organic Carbon in Natural Waters 242
9–4 Typical Distribution of Dissolved Organic Carbon in Interstitial Water of Soils 246
9–5 Increase in Mobility of Contaminants Versus Relative Concentration of Colloidal Particles 247
9–6 Mobility of a Hydrophobic Compound Relative to the Mobility of the Same Compound Without the Presence of a Macromolecule as a Function of Octanol-Water Partition Coefficient and Amount of Organic Carbon in the Mobile Phase 248

SECTION 9

Contaminants Sorbed onto Colloidal Particles in Water in Either the Unsaturated or Saturated Zone

9.1 LOCUS DESCRIPTION

9.1.1 Short Definition

Contaminants sorbed onto colloidal particles in water in either the unsaturated or saturated zone.

9.1.2 Expanded Definition and Comments

Hydrophobic compounds in solution, which may preferentially sorb to the stationary soil matrix, may also sorb to "free" colloidal particles in water. This locus is recognized as being potentially important as it allows for the mobilization of strongly-sorbed contaminants that would otherwise remain immobile due to sorption on a stationary phase. The definition includes liquid contaminants "attached" to colloidal particles rather than just "sorbed". Water in both the saturated and unsaturated zones is included with regard to transport on colloids.

Research on this locus is somewhat limited. However, recent studies suggest that the process of deposition of dissolved contaminants onto the mobile colloidal phase is similar to that for locus 4 (sorption onto the stationary soil phase). Figures 9–1 and 9–2 present a schematic cross-sectional diagram of locus no. 9 and a schematic representation of the transformation and transport processes affecting other loci, respectively.

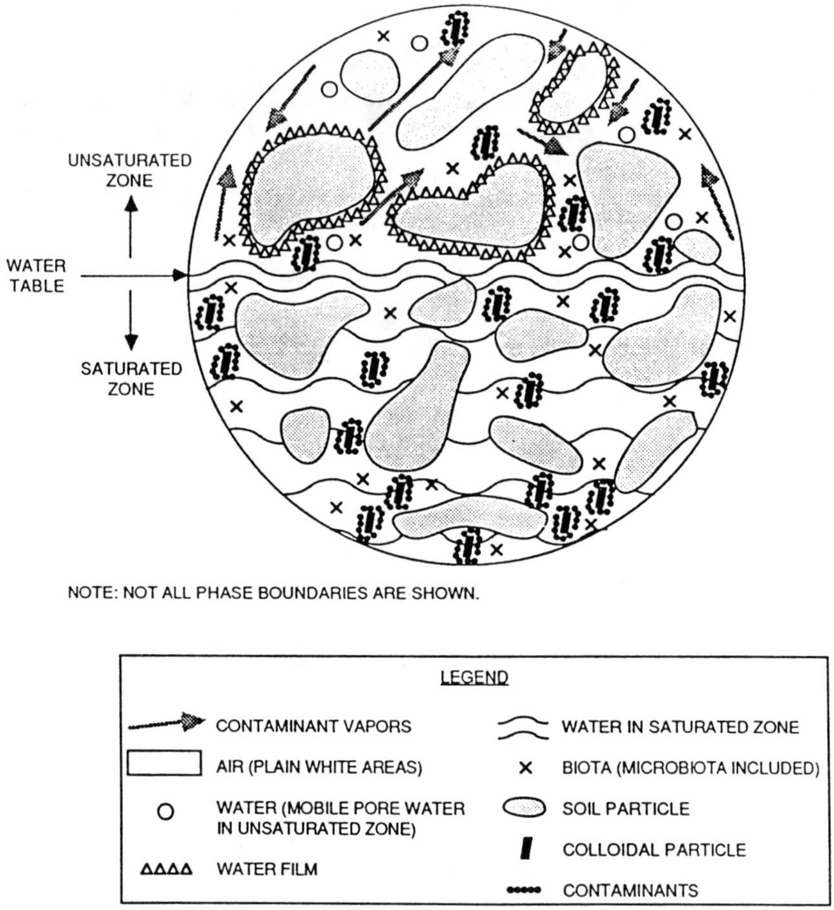

Figure 9–1. Schematic cross-sectional diagram of locus no. 9–contaminants sorbed to colloidal particles in water in either the unsaturated or saturated zone.

9.2 EVALUATION OF CRITERIA FOR REMEDIATION

9.2.1 Introduction

Colloidal particles form a homogeneous continuous phase dispersed in water. They are distinguished from truly dissolved materials of ionic or molecular size and larger suspended materials which are influenced by gravity. Colloids are electrically charged particles (usually negatively charged) that may be comprised of small solid particles, macromole-

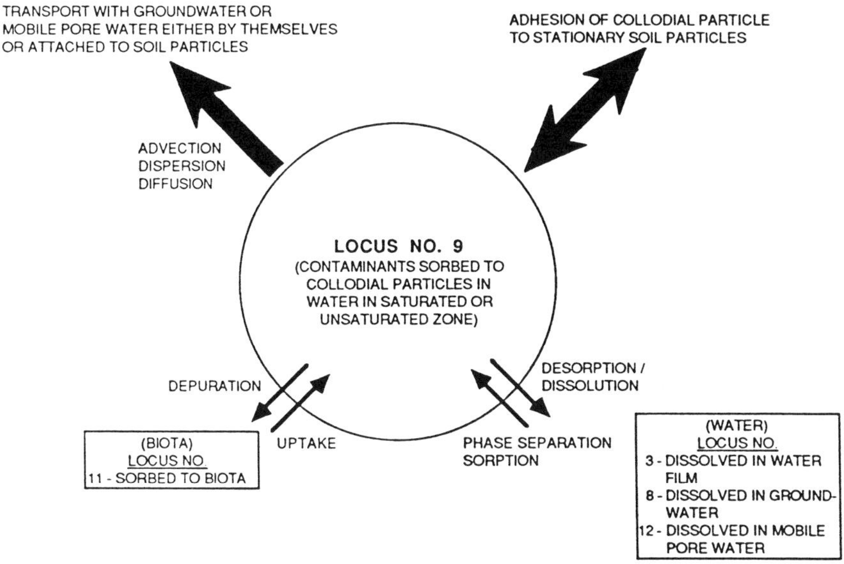

Figure 9–2. Schematic representation of important transformation and transport processes affecting other loci.

cules, small droplets of liquids, or small gas bubbles. Because colloidal particles are very small (0.45 to 0.001 μm), gravity is an insignificant force acting on them when compared with the electrical forces acting on the surface of the particles. Table 9–1 compares the size of colloidal particles to those of other particles. Figure 9–3 demonstrates the general colloidal particles size range in groundwater systems. Note that colloidal-sized particles can be "free" or "bound", and the colloidal phase discussed herein is associated only with free particles. Soils such as

Table 9–1. Relative Sizes of Particles

Particle	Approximate Diameter[a]
Atoms	0.1 to 0.6 mμ
Molecules	0.2 to 5.0 mμ
Molecular groups	0.5 to 10 mμ
Colloidal particles	0.001 to 10 mμ
Microscopically visible particles	>250 μ
Colloidal clay	<0.2 μ
Clay	<2 μ
Silt	0.05 to 0.002 mm
Very fine sand	0.10 to 0.05 mm
Fine sand	0.25 to 0.10 mm

a. 1Å (Angstrom) = 10^{-7} mm = 10^{-8} cm; 1 mμ (millimicron) = 10^{-6} mm = 10^{-7} cm; 1 μ (micron) = 10^{-3} mm = 10^{-4} cm
Source: Stevenson, 1982.

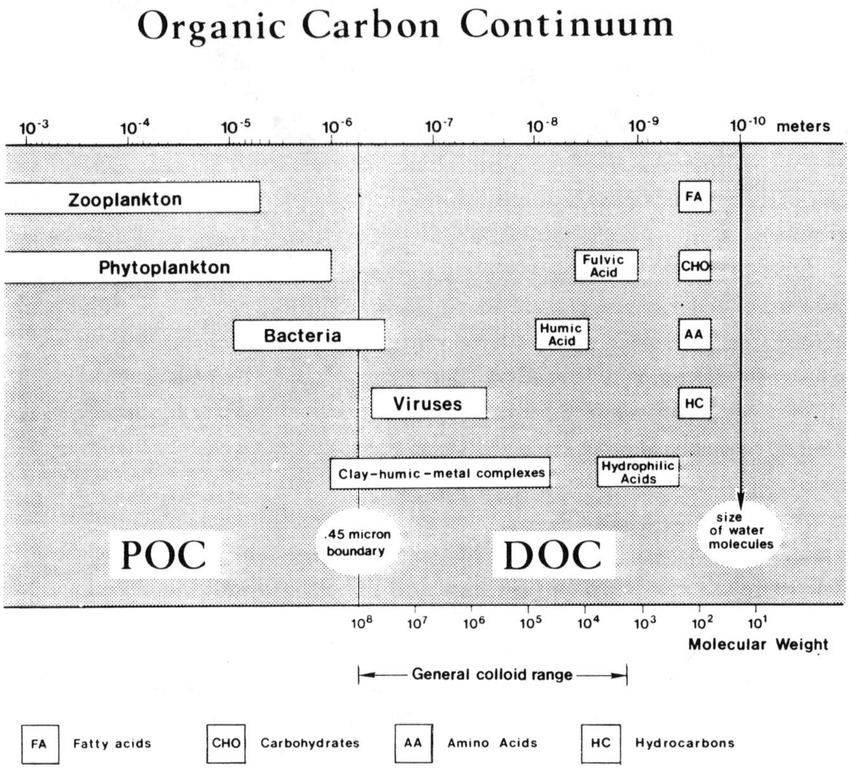

Figure 9–3. Continuum of particulate and dissolved organic carbon in natural waters. *Source:* Thurman, 1985. (Copyright Prentice Hall, Inc. Reprinted with permission.)

clays may be comprised to a large degree by immobile, or bound, colloidal-sized particles, but these are not associated with locus 9.

There are two general classes of colloidal particles: 1) hydrophobic colloidal particles, which repel water (i.e., clay particles) and 2) hydrophilic colloidal particles which have affinity for water (i.e., humic and fulvic substances). Organic colloidal particles like humic and fulvic substances are present in soils, sediments, and aquifer materials. Organic matter plays a large role in adsorption of dissolved contaminants to a solid phase. Humic substances constitute 40 to 60 percent of natural dissolved organic carbon (DOC) and are the largest fraction of organic matter in natural waters (Thurman, 1985). Inorganic colloidal particles include clays, metal oxides, and inorganic precipitates in the submicron size range.

Colloidal particles of interest here are mobile with water. They can adsorb organic and inorganic contaminants which have been dissolved in water and enhance their mobility. According to Nightingale and Bianchi (1977), colloidal clay particles, which were mobilized from the surface soils, were the cause of turbidity in wells several hundred meters distant from the recharge site.

Corrective action mechanisms that may immobilize colloidal particles by transformation from aqueous phase to solid phase are:

- Decrease in groundwater velocity which would enhance the attachment process of colloidal particles onto soil particles.
- Increase in ionic strength in groundwater. As ionic strength increases, the precipitation of organic acids increases.
- Decrease in pH of groundwater, which causes precipitation of colloidal humic substances.

Conversely, the preceding mechanisms may be reversed to facilitate continued mobilization of contaminants in locus 9. The following discussions present information about the factors affecting the mobilization, fixation and transformation of contaminants in locus 9.

9.2.2 Mobilization/Remobilization

9.2.2.1 Partitioning onto Mobile Phase

Contaminants are found in locus 9 when dissovled contaminants sorb to free colloidal particles. The contaminants remain in a mobile phase with the colloid. Immobilization occurs through desorption back to the dissolved phase (locus 8) and subsequent re-sorption to the immobile soil matrix (locus 4) or through immobilization of the colloidal particle itself.

As discussed in Section 9.2.1, there are two types of colloidal particles, organics and inorganics. Clays are inorganic colloidal particles which have layered structures of silicates and metal hydroxides, a planar geometry, and very large surface area with negative charges. Organic colloidal particles such as humic and fulvic substances are hydrophobic, organophilic, and have high cation exchange capacities because of negative charges at their surfaces.

In soils which contain more than 0.1 percent (by weight) of organic matter, sorption of neutral organics is controlled by this organic fraction; otherwise, sorption is slight and is controlled primarily by surface area of the soil particles. Because of some similarities between soils and colloidal particles, the processes involved in water:soil partitioning may be similar to those of water:colloid partitioning.

Colloidal particles may adsorb contaminants as a result of one of the following processes:

- Physical adsorption, which is related to van der Waals forces and is caused by a weak electrostatic interaction between atoms and molecules of the contaminants and the colloidal particles.
- Chemisorption, which involves direct formation of a chemical bond between the adsorbed contaminant molecule and the colloidal particle.
- Exchange adsorption, which occurs when contaminant ions concentrate at the surface of the colloidal particles as a result of electrostatic attraction between the contaminant and the negative surface of the colloidal particle.

Without considering water:colloid partitioning of contaminants, water:soil partitioning models may overestimate the mass of immobilized contaminants. Water:soil partitioning is often approximated by the following retardation equation:

$$R = 1 + K_d \rho_b/\Theta \tag{9.1}$$

where R = retardation of dissolved contaminant plume
K_d = partition coefficient for soil and water (L/kg)
ρ_b = bulk density of solid soil (g/cm^3)
Θ = soil porosity (dimensionless)

Water:soil partitioning of hydrophobic contaminants is covered in detail in Section 4.

Hutchins et al. (1985) presents a modified retardation equation based on water:soil:colloid partitioning of organic solutes:

$$R = 1 + \frac{K_d \rho_{b/\Theta}}{1 + K_p} \tag{9.2}$$

where K_p = colloid:water partition coefficient (L/kg) ; and other parameter as defined in equation (1).

The modified equation accounts for a decreased proportion of immobilized contaminant due to sorption onto mobile colloidal particles. Experimental data indicate that the modified retardation equation provides a reasonable estimation of this facilitated transport (Kan and Mason, 1986).

Typical values for soil density and porosity can be found in Section 4. The soil:water partition coefficient can be estimated from:

$$K_d = K_{oc} f_{oc} \tag{9.3}$$

where K_{oc} = organic carbon partition coefficient; and
f_{oc} = fractional organic content of soil by weight.

The colloid:water partition coefficient (K_p) can be estimated by:

$$K_p = K_{oc} (DOC)\ 10^{-6} \tag{9.4}$$

where K_{oc} = organic carbon partition coefficient; and
DOC = dissolved organic carbon content of water (mg/L).

Equations 9.3 and 9.4 point out the importance of organic carbon content in sorption of dissolved contaminants. Other key parameters affecting sorption were presented in Table 4–3. Although sorption on to colloidal particles is less well understood than sorption on to soil, many of the same parameters are likely to be involved.

The distribution of dissolved organic carbon (DOC) in the subsurface environment is especially important for locus 9. Figure 9–4 shows a typical distribution of DOC in interstitial waters of soils. DOC content is usually much higher in the pore water of the unsaturated zone, particularly near the surface. This suggests that facilitated transport of contaminants in locus 9 may be significantly greater vertically downward in mobile pore water of the unsaturated zone than in groundwater.

Among the factors that may be of importance in mobilizing/immobilizing locus 9 contaminants are temperature, salinity, and pH of the solution. As temperature increases, contaminant solubility generally increases, and the degree of partitioning on to colloidal matter is likely to decrease. However, control of mobilization by controlling temperature may be of limited practical value.

In general, waters low in specific conductivity will have lower concentrations of DOC, and therefore a smaller fraction of locus 9 contaminants than highly conductive waters. Increased specific conductivity, associated with increased salinity, has been shown to cause precipitation of DOC. Sholkovitz (1976) found a sudden removal of DOC at a salinity level of 0.5 to 1.5 percent. The maximum removal of DOC from solution appears to occur between 1.5 and 2.0 percent salinity. Humic substances, the major fraction of colloidal organic matter in natural waters, are particularly subject to precipitation when salinity increases are encountered. Artificially raising or lowering the salinity and/or conductivity of subsurface waters may help control mobility of locus 9 contaminants.

Sorption of organic acids, such as aquatic humic substances, is influenced by pH. For organic acids to adsorb completely, the pH of the solution should be 2 pH units below the pKa, the ionization constant for

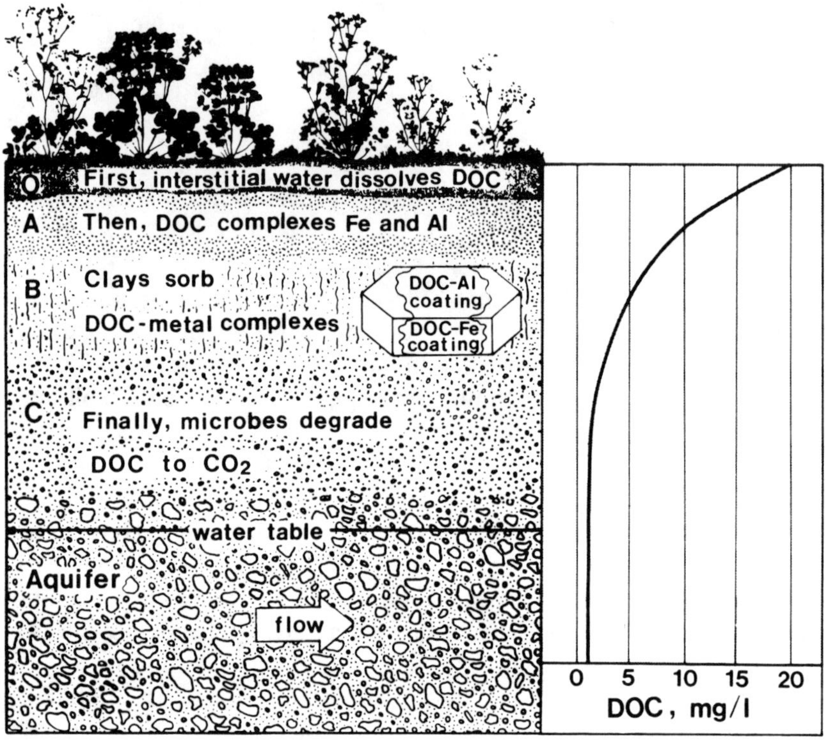

Figure 9–4. Typical distribution of dissolved organic carbon in interstitial water of soils. *Source:* Thurman, 1985. (Copyright Prentice Hall, Inc. Reprinted with permission.)

a weak acid (Thurman, 1985). Adsorption of organic acids is greatest at pH 2 or less, and desorption is greatest at pH 6 or more. Increasing the acidity of groundwater or pore water may result in immobilization of both colloidal humic substances and attached contaminants. Conversely, increasing alkalinity may result in increased mobilization of locus 9 contaminants.

9.2.2.2 Transport with Mobile Phase

Partitioning of hydrophobic contaminants from the dissolved phase to free colloidal particles as well as stationary soil has recently been demonstrated. The colloidal matter is transported with groundwater, carrying with it the sorbed contaminants. Enfield (1985) has noted that hydrophobic contaminants such as DDT have been moved farther in

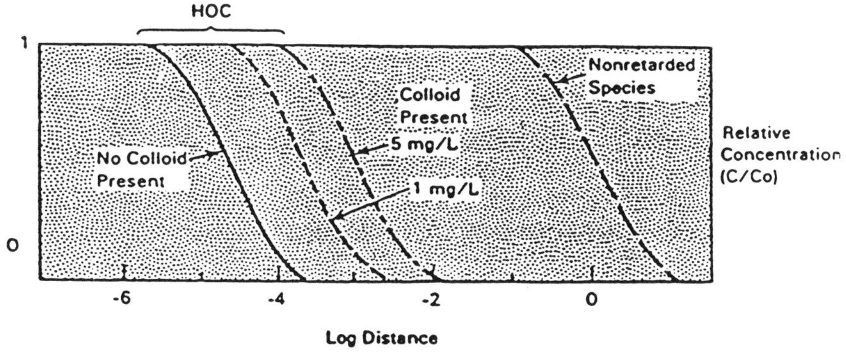

Figure 9–5. Increase in mobility of contaminants versus relative concentration of colloidal particles. *Source:* McCarthy and Wobber, 1986.

groundwater than laboratory models would suggest due to increased mobility on colloidal particles.

A fraction of the colloidal particles will "filter out" of solution during transport, as is discussed in Section 9.2.3, but the remaining free colloid moves at roughly the average velocity of groundwater. In some cases, the average free colloid velocity may be greater than that of groundwater. Enfield et al. (1989) estimated that the velocity of colloidal matter was 10 to 16 percent greater than groundwater during column studies using fine sand. They concluded that the relatively large-sized colloidal particles were excluded from the smaller pore spaces in the soil, reducing the effective void volume that can be occupied by the colloid.

The degree to which contaminant mobility is enhanced via locus 9 is influenced in large part by the sorptive properties of the contaminant, the concentration of free colloidal matter, and the organic carbon content of the colloidal matter. The average velocity of highly sorptive compounds may be increased significantly in the presence of waters with high DOC content in colloidal form. For less hydrophobic compounds, transport of contaminants via locus 9 is generally only a small fraction of the total mobile contaminant. Figure 9–5 demonstrates that the presence of 1 and 5 mg/L of organic colloidal particles in water increases mobility of contaminant by 10 and 50 times respectively; a significant increase but still several orders of magnitude less mobile than conservative compounds.

Figure 9–6 shows the relative change in mobility of hydrophobic contaminants due to transport with colloidal particles. As the figure shows, the importance of colloidal particles on mobility of contaminants diminishes as the octanol-water partition coefficient (K_{ow}) decreases. Higher

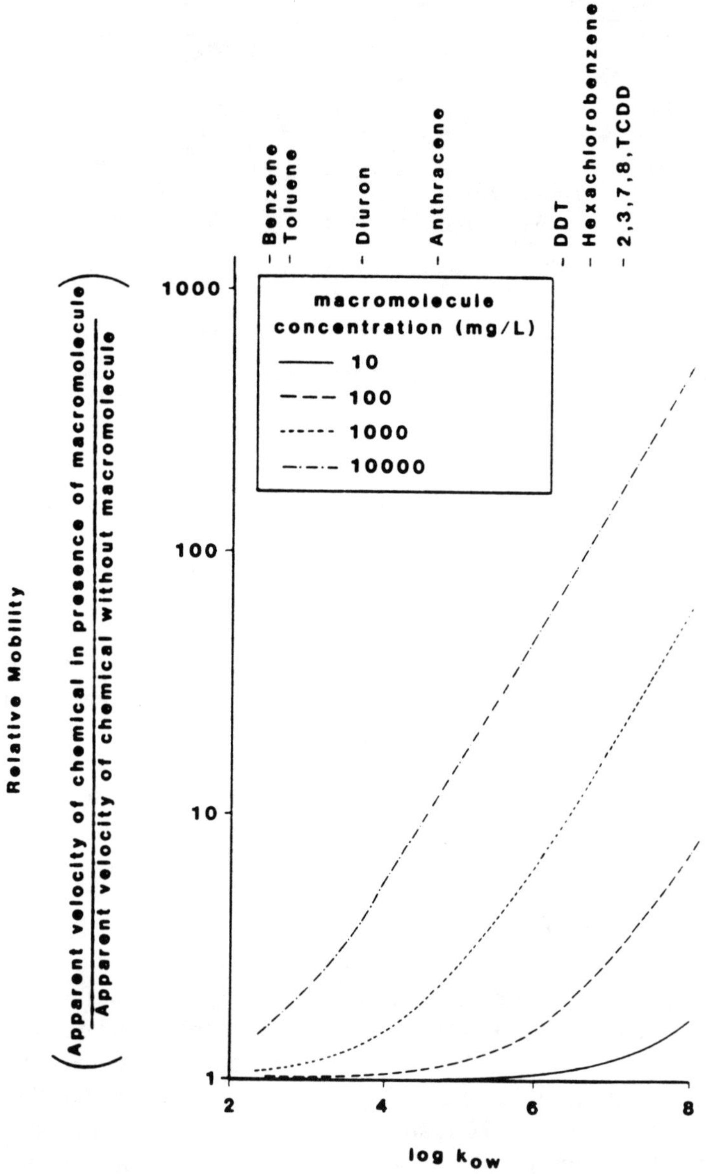

Figure 9–6. Mobility of a hydrophobic compound relative to the mobility of the same compound without the presence of a macromolecule as a function of octanol-water partition coefficient and amount of organic carbon in the mobile phase. *Source:* Enfield and Bengtsson, 1988. (Copyright Water Well Journal Publishing Co. Reproduced with permission.)

K_{ow} generally implies a greater preference for the sorbed rather than dissolved phase. Figure 9-6 shows that for colloidal concentrations less than 100 mg/L, typical for most groundwaters, only compounds with K_{ow} greater than 10,000 are significantly affected by this mechanism.

9.2.3 Fixation

9.2.3.1 Partitioning onto Immobile (Stationary) Phase

Fixation of locus 9 contaminants can occur through desorption to the mobile aqueous phase and subsequent re-sorption to the stationary soil matrix. There is currently little information available describing the process of desorption of contaminants from colloidal paraticles. Results of experiments by Enfield et al. (1989) suggest that water:colloid partitioning of organic solutes is a reversible reaction and therefore, some degree of fixation probably occurs via this process. However, the implications of this process on remediation are not well understood. Colloidal particles have similarities with soil particles; therefore, the processes of adsorption/desorption of contaminants and soil particles discussed in Sections 2 and 4 may be applicable to colloidal particles, and the materials in these sections may provide insight into this fixation mechanism.

Locus 9 contaminants may also be fixated through immobilization of the colloidal particles onto the stationary soil matrix. The particles can be "filtered out" of solution during transport as they come into contact with the soil. Brownian diffusion, gravitational sedimentation, and interception cause the particles to contact the soil matrix. The colloidal particles may then become attached to the soil particles by electrostatic, van der Waals, hydrodynamic, and specific chemical interaction forces between the respective particle surfaces. Electrostatic forces are the dominant forces that control the attachment of contaminated colloidal paraticles to soil particles. These forces result from isomorphic substitution within the crystal lattice, surface group ionization, and ion adsorption.

Table 9-2 shows some of the factors that influence fixation of colloidal particles onto immobile soil. A larger soil porosity and a larger soil particle size provide a lesser surface area exposure to the colloid, and would suggest a lower fraction of colloidal attachment to the soil matrix. A lower groundwater velocity would, in general, allow a greater frequency of contact between colloid and stationary soil, enhancing immobilization of the colloid.

Larger colloidal particle size and higher colloidal concentration generally indicate increased contact with stationary soil and, therefore,

Table 9–2. Factors Influencing Fixation of Colloidal Particles

Site Parameters	Solution Parameters
Soil Porosity	Colloidal Particle Size
Soil Particle Size	Colloidal Concentration
Groundwater Velocity	pH
	Ionic Strength

greater likelihood of attachment. Lowering of pH causes precipitation of the humic acid fraction of the colloid onto immobile soil. Ionic strength, which can be indirectly measured by specific conductivity or salinity, also influences colloidal fixation. A sudden increase in the ionic strength of water can cause precipitation of colloidal matter.

Of the parameters in Table 9–2, groundwater velocity, pH, and ionic strength are most easily controlled artificially, and therefore, of greatest interest in remediation. Decreasing groundwater velocity can result in lower mobility of contaminants in the free colloidal phase; however, the relative effect of this mechanism on overall contaminant mobility may be small. On the other hand, increased groundwater velocity could cause increased mobility of locus 9 contaminants under high hydraulic gradient conditions (i.e., groundwater pumping). The effect of pH and ionic strength on colloidal fixation are discussed in greater detail in Section 9.2.2.

9.2.4 Transformation

9.2.4.1 Biodegradation (Details in Section 11)

By comparison, the conditions for biodegradation in locus no. 9 closely resembled those of locus no. 4; the sorbed contaminants are in equilibrium with dissolved aqueous phase contaminants. Since nearly all microbial transformation likely occurs within the cell, the contaminants must first desorb to the liquid phase (loci nos. 8 and 12) prior to biodegradation. In the vadose zone, oxygen is much more available than below the water table, and biodegradation is therefore much more likely to occur there.

9.2.4.2 Chemical Oxidation (Details in Section 3.2.4.2)

Colloids serve as the major mechanism by which abiotic transformations occur. Chemical oxidation of contaminants is mediated by reduced clay-metal colloids and polymerization occurs as organic colloids are joined.

9.3 STORAGE CAPACITY IN LOCUS

9.3.1 Introduction and Basic Equations

The basic equation for calculating the storage capacity in locus no. 9 is as follows:

$$C_p = K_p C_w \tag{9.5}$$

Where C_p = concentration of contaminant on colloidal particles (μg/kg)
K_p = partition coefficient (L/kg)
C_w = concentration of contaminant in aqueous phase (μg/L)

9.3.2 Guidance on Inputs for, and Calculation of, Maximum Value

- The partition coefficient, K_p, can be estimated by equation 9.4:

$$K_p = K_{oc} (DOC) 10^{-6}$$

- DOC concentrations for groundwater range from 0.2–17 μg/L (Thurman, 1985). Use maximum value of 17 μg/L.
- K_{oc} values for gasoline constituents can be found in Table 4–4. For bulk gasoline, use K_{oc} = 191 L/kg, as discussed in Section 4.4.1.
- The pure water solubility of a compound will generally be higher than found in situ. Use solubilities of gasoline constituents in gasoline-saturated water found in Table 8–4. Use bulk gasoline maximum solubility of 200 mg/L as discussed in Section 8.3.

9.3.3 Guidance on Inputs for, and Calculation of, Average Value

- Determine K_p as discussed in Section 9.3.2.
- Use average DOC concentration of 0.7 μg/L (Thurman, 1985).
- Use K_{oc} values as discussed in Section 9.3.2.
- Use average C_w value of 20 mg/L for gasoline, as discussed in Section 8.4.

9.4 EXAMPLE CALCULATIONS

9.4.1 Storage Capacity Calculations

9.4.1.1 Maximum Value Calculations

- Storage capacity (C_p) for gasoline:

Using equation 9.4 to estimate K_p:

$$K_p = 191\ (17)\ 10^{-6} = 0.0032\ \text{L/kg}$$
$$C_w = 200\ \text{mg/L}$$

Then using equation 9.5:

$$C_p = (0.0032)(200) = 0.64\ \text{mg/kg}$$

- Maximum storage capacity for benzene:

From Table 4.4, K_{oc} for benzene = 62
Then for equation 9.4:

$$K_p = 62\ (17)\ 10^{-6} = 0.001\ \text{L/kg}$$

From Table 8–4, gasoline-saturated water solubility of benzene = 64.8 mg/L.
Then using equation 9.5:

$$C_p = (0.001)\ (64.8) = 0.065\ \text{mg/kg}$$

9.4.1.2 Average Value Calculations

- Average storage capacity for gasoline:

$$K_p = 0.0032\ \text{L/kg}$$
$$C_w = 20\ \text{mg/L}$$

Then using equation 9.5:

$$C_p = (0.0032)\ (20) = 0.064\ \text{mg/kg}$$

9.5 SUMMARY OF RELATIVE IMPORTANCE OF LOCUS

9.5.1 Remediation

Under most natural conditions, the significance of contaminated colloidal particles in groundwater contaminant is very low. Remediation efforts targeting locus 9 contaminants may be warrranted for highly hydrophobic compounds ($K_{ow} > 10{,}000$) and/or for waters very high in DOC ($> 10\ \mu g/L$). Under these conditions, contaminant mobility may be greater than normally expected.

For the saturated zone, immobilization techniques would likely focus on enhanced precipitation of free colloidal matter through adjustments in pH or ionic strength. Removal of locus 9 contaminants from groundwater by traditional pumping methods should be effective.

In the unsaturated zone, enhanced fixation of colloidal particles through pH or ionic strength adjustments will introduce considerable amounts of water to the subsurface. This is likely to facilitate mobility in this zone rather than retard it, and will generally not be feasible. Traditional soil venting methods may serve to immobilize colloidal particles in pore water as water vapors are extracted and the soil "dried out".

9.5.2 Loci Interactions

Loci 3, 5, 6, 7, 8 and 12 interact with attachment of contaminants onto colloidal particles in groundwater. For the contaminated colloidal particles, the partitioning process into groundwater (aqueous phase) and soil particles (solid phase) is not clearly defined and not enough information is available. Bulk transport into the locus occurs from locus nos. 3, 5, 6, 7, 8 and 12.

Currently little is known about the desorption process of contaminants from the colloidal particles in groundwater; therefore, bulk transport out of locus no. 9 is not well defined. It may be related to locus nos. 2, 3, 4, 5, 8, 11 and 12.

9.5.3 Information Gaps

There is limited information available on the subject of colloidal particle behavior and the influence of these particles on the mobility of contaminants in the subsurface environment. Critical areas in which additional research is needed to understand the role of colloidal particles in the rate and transport of contaminants are as follows:

1. Factors affecting the process of adsorption/desorption of contaminants on and off colloidal particles.
2. Effects of colloidal particles on the solubility of contaminants in groundwater (i.e., do these particles increase solubility). Additional qualitative and quantitative data are needed.
3. Factors affecting the mass transport of contaminated colloidal particles with groundwater.
4. Effect of precipitation (filtration) of free colloidal particles on the permeability of the soil and the effect of sorbed contaminant on precipitation/filtration of free colloidal particles.
5. Concentrations, properties and constituents of inorganic colloidal particles in the groundwater.
6. Importance of organic and inorganic colloidal particles in contaminant transport.
7. Type of contaminants most influenced by locus 9 transport.

9.6 LITERATURE CITED

Enfield, C.G. 1985. Chemical Transport Facilitated by Multiphase Flow Systems. Water Sci. Technol., 17:1–12.

Enfield, C.G. and G. Bengtsson. 1988. Macromolecular Transport of Hydrophobic Contaminants in Aqueous Environments. Ground Water, 26(1):64–70.

Enfield, C.G., G. Bengtsson, and R. Lindquist. 1989. Influence of Macromolecules on Chemical Transport. Environ. Sci. Technol. Vol. 23, no. 10.

Hutchins, S.R., M.B. Tomson, P.B. Bedient, and C.H. Ward. 1985. Fate of Trace Organics During Land Application of Municipal Wastewater. CRC Critical Reviews of Environmental Control. Vol. 15. pp. 355–416.

Kan, A.T. and B.T. Mason. 1986. Facilitated Transport of Naphthalene and Phenanthrene in a Sandy Soil Column with Dissolved Organic Matter. Proceedings of the NWWA/API Conference on Petroluem Hydrocarbons and Organic Chemicals in Groundwater. Houston, TX. pp. 93–106.

McCarthy, J.F. and F.J. Wobber. 1986. Transport of Contaminants in the Subsurface: The Role of Organic and Inorganic Colloidal Particles. U.S. Department of Energy.

Nightingale, H.I. and W.C. Bianchi. 1977. Groundwater Turbidity Resulting from Artificial Recharge. Groundwater, 15(2): 146–152.

Sholkovitz, E.R. 1976. Flocculation of Dissolved Organic and Inorganic

Matter During the Mixing of River Water and Seawater. Geochemica et Cosmochemica Acta. 40:831–845.
Stevenson, F.J. 1982. Humus Chemistry. John Wiley & Sons, New York.
Thurman, E.M. 1985. Organic Geochemistry of Natural Waters. M. Nijhoff and W. Junk Publishers, Boston. pp. 1–17.

SECTION 10

Contaminants That Have Diffused into Mineral Grains or Rocks in Either the Unsaturated or Saturated Zone

CONTENTS

List of Tables 258
List of Figures 258
10.1 Locus Description 259
10.1.1 Short Definition 259
10.1.2 Expanded Definition and Comments 259
10.2 Evaluation of Criteria for Remediation 259
10.2.1 Introduction 259
10.2.1.1 Description of Pore Spaces 260
10.2.1.2 Description of Contaminant Phases . 260
10.2.1.3 Summary of Corrective Action Potential 261
10.2.2 Mobilization/Remobilization 262
10.2.2.1 Partitioning onto Mobile Phase 262
Diffusion 263
Diffusion Between Primary and Secondary Porosity 265
Diffusion Between Strata of Different Hydraulic Conductivities 268
Diffusion in the Unsaturated Zone .. 270
Sorption 271
Advection 272
10.2.2.2 Transport of/with Mobile Phase 273
10.2.3 Fixation 273
10.2.3.1 Partitioning onto Immobile (Stationary) Phase 273
10.2.4 Transformation 273
10.2.4.1 Biodegradation 273
10.2.4.2 Chemical Oxidation 274
10.3 Storage Capacity in Locus 274
10.3.1 Introduction and Basic Equations 274

10.3.2 Guidance on Inputs for, and Calculation of, Maximum Values 274
10.3.2.1 Concentration of Contaminants in Secondary Pore Water 274
10.3.2.2 Total Primary Porosity 274
10.3.3 Guidance on Inputs for, and Calculation of, Average Values 275
10.3.3.1 Concentration of Contaminants in Secondary Pore Water 275
10.3.3.2 Total Primary Porosity 275
10.4 Example Calculations 275
10.4.1 Storage Capacity Calculations 275
10.4.1.1 Maximum Values 275
10.4.1.2 Average Values 276
10.4.2 Transport Rate Calculations 276
10.5 Summary of Relative Importance of Locus 277
10.5.1 Remediation 277
10.5.2 Loci Interactions 277
10.5.3 Information Gaps 278
10.6 Literature Cited 278

TABLES

10–1 Typical Primary Porosities 267
10–2 Typical Effective Diffusivities for Nonsorbing Species 268

FIGURES

10–1 Schematic Cross-Sectional Diagram of Locus No. 10–Contaminants That Have Diffused into Mineral Grains or Rocks in Either the Unsaturated or Saturated Zone 261
10–2 Schematic Representation of Important Transformation and Transport Processes Affecting Other Loci 262
10–3 Schematic Diagram of Rock Matrix 263
10–4 Schematic Diagram of Diffusion Across a Fissure-Matrix Boundary 264
10–5 Physical Representation of Equation 10.1 for Mass Diffusing Through a Rock Slab 267
10–6 Comparison Between Measurements and Simulations for Low-Velocity Case 269
10–7 Comparison Between Measurements and Simulations for High-Velocity Case 270
10–8 Diagram of Diffusion Between Strata 271

SECTION 10

Contaminants That Have Diffused into Mineral Grains or Rocks in Either the Unsaturated or Saturated Zone

10.1 LOCUS DESCRIPTION

10.1.1 Short Definition

Contaminants that have diffused into mineral grains or rocks in either the unsaturated or saturated zone.

10.1.2 Expanded Definition and Comments

This locus focuses on contaminants that have gone beyond surface sorption (e.g., locus no. 2 and 4) and migrated or diffused into the inorganic mineral grains or rocks. While somme of the contaminants might penetrate intact mineral crystals or amorphous solids by pure diffusion, a larger fraction is likely to have entered through microfractures, micropores or the thin spaces between mineral (e.g., clay platelet) layers. In the latter case, capillary tension would be the major driving force. In both cases, the contaminants are very immobile because they are trapped within a stationary phase and would have to be extracted from the minerals (a slow process) in many remediation schemes. Figures 10–1 and 10–2 present a schematic cross-sectional diagram of locus no. 10 and a schematic representation of the transformation and transport processes affecting other loci, respectively.

10.2 EVALUATION OF CRITERIA FOR REMEDIATION

10.2.1 Introduction

As indicated in the definition of this locus, there are essentially two types of locations within a rock into which an organic contaminant might move: 1) intergranular pore spaces, and 2) between intramineral layers. Contaminants could enter these spaces in the gaseous, aqueous or pure liquid phases, or as a combination of these. The migration of these phases could occur in the unsaturated or the saturated zone according to processes of diffusion, sorption and advection.

10.2.1.1 Description of Pore Spaces

The pore spaces addressed in this discussion encompass the primary porosity of a rock, or those intergranular and intramineral spaces which were created during the rock formation process. Diameters of these pore spaces may range from many angstroms (10^{-8}m) to a few microns (10^{-6}m). Secondary porosity, or spaces formed in fracturing or geochemical weathering, is addressed in locus no. 13, yet there is some relationship between primary and secondary porosity within the context of diffusion into the rock matrix: higher levels of secondary porosity increase the accessibility of primary pore spaces. Thus, secondary porosity is discussed here inasmuch as it relates to primary porosity, while transport and remediation within the secondary porosity is discussed in locus no. 13.

The minerals of primary concern regarding intramineral contaminant migration are the phyllosilicates, especially micas such as biotite and the clay minerals, including kaolinite, illite, vermiculites, and montmorillonite. Phyllosilicates have the special feature that they are composed of tetrahedral and octahedral silicate layers, which allow ions, water molecules, and, potentially, small organic molecules to move between those layers. Although the concentration of these minerals is usually low in comparison to the one or two major constituents of a rock, their presence is occasionally important in some rocks as they are the only minerals which might sorb or otherwise take up enough of the diffused contaminant to have ramifications for corrective action. Sorption to the surfaces of mineral grains within a rock occurs much like the sorption processes discussed in locus no. 4.

10.2.1.2 Description of Contaminant Phases

The contaminant phase that will be addressed in this section is the aqueous phase. Although multi-phase contaminant/groundwater systems are likely to occur in the vicinity of this locus, such combinations are extremely difficult to model and will not be considered here. Furthermore, information concerning gas phase and liquid phase diffusion into the rock matrix is very sparse, with most studies, both theoretical and experimental, focusing on solute diffusion.

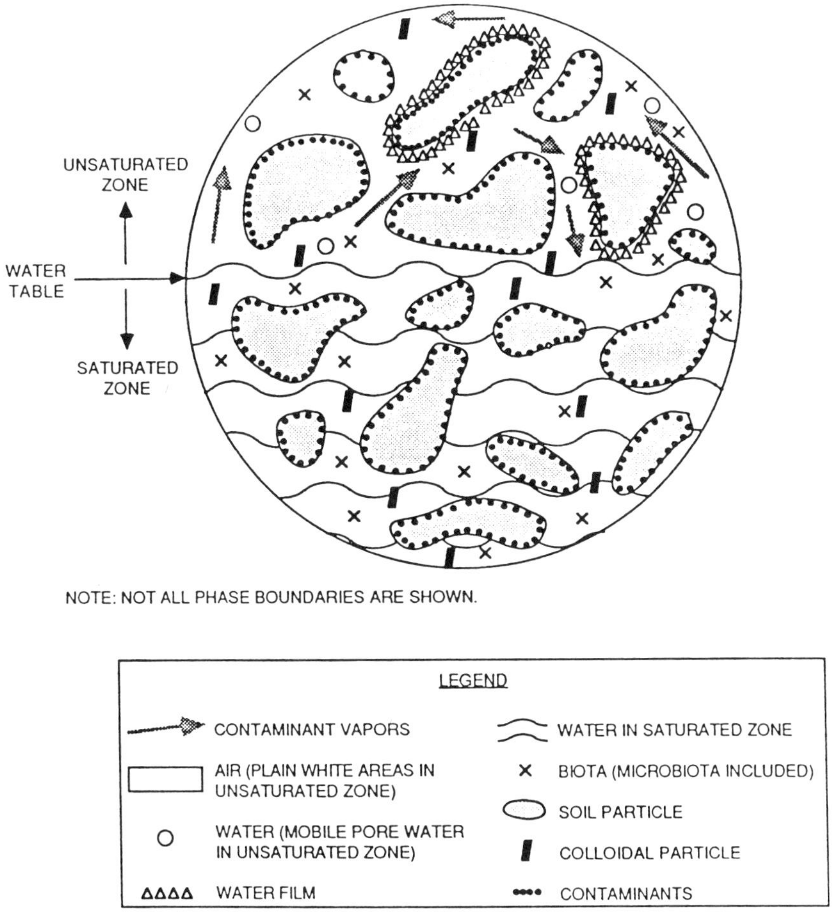

Figure 10–1. Schematic cross-sectional diagram of locus no. 10–contaminants that have diffused into mineral grains or rocks in either the unsaturated or saturated zone.

10.2.1.3 Summary of Corrective Action Potential

The three processes which will be discussed are diffusion, sorption and advection. As might be expected, each of these processes holds different implications for remediation. In a very general sense, the most important remediation mechanisms involve controlling the concentration gradient and affecting the sorption potential of the petroleum derivatives. Contaminants can be mobilized from the rock matrix by reversing the concentration gradient, encouraging the chemicals to diffuse out

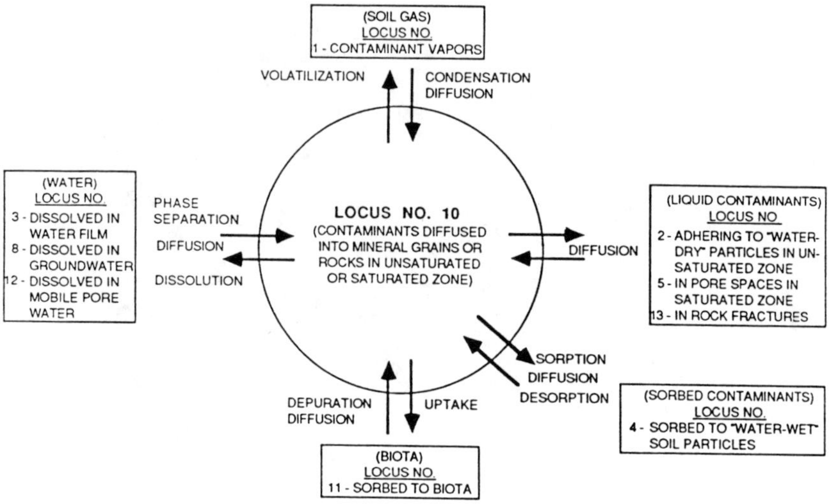

Figure 10–2. Schematic representation of important transformation and transport processes affecting other loci.

of the rock. Such diffusion may be reduced if the minerals composing the rock matrix have a high capacity to sorb organic chemicals. On the other hand, in the absence of advection, if the direction of diffusion can be maintained, then the contaminant may be contained in a limited area. Regardless of diffusive processes, minerals with a high sorption capacity will tend to immoblize the contaminants unless those organic molecules are displaced by substances with higher affinities for the minerals, in which case the contaminant may be remoblized. The discussion that follows will provide the important details of the processes (mainly diffusion and sorption) affecting the migration of organic chemicals in the rock matrix and investigate, where possible, remediation alternatives for this locus.

10.2.2 Mobilization/Remobilization

10.2.2.1 Partitioning onto Mobile Phase

The most important partitioning mechanisms that could move contaminants from the rock matrix to a mobile phase are diffusion and desorption. (Contaminants being transported by advection are already in the mobile phase). Figure 10–3 shows an enlarged portion of Figure 10–1 in order to clarify what the consolidated matrix looks like, and Figure 10–4 shows a schematic representation of the various phases

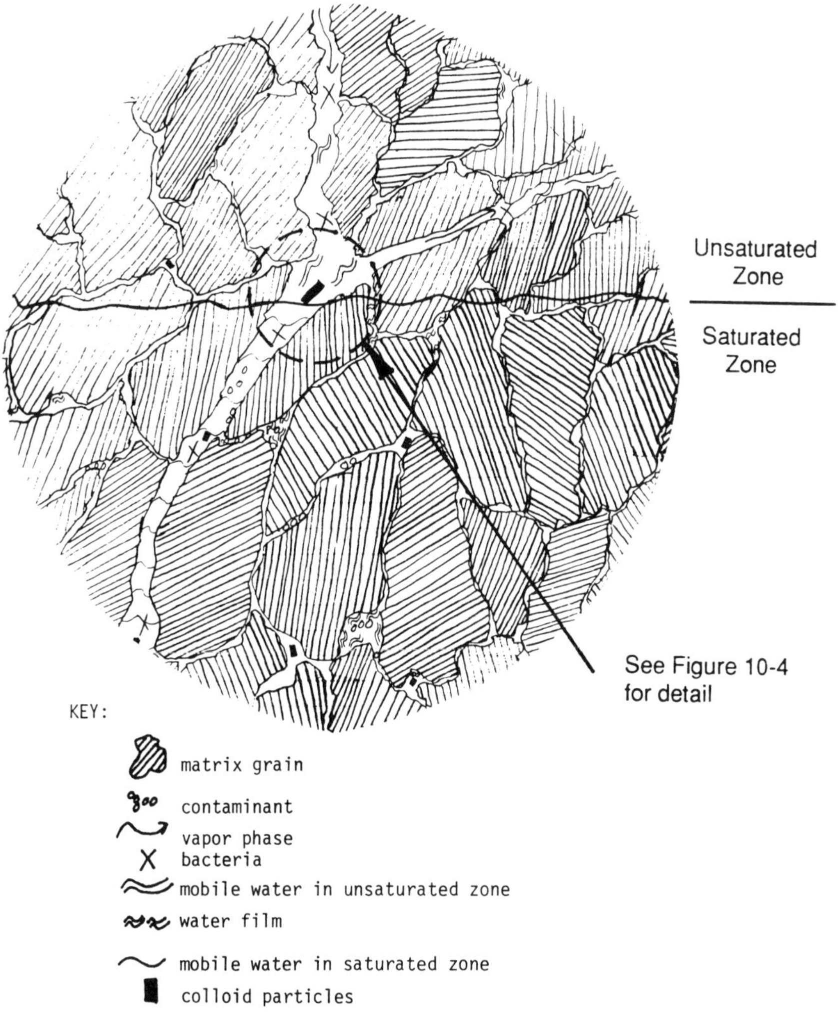

Figure 10–3. Schematic diagram of rock matrix.

through which a diffusing chemical might pass as it moves into or out of the matrix. Within the rock matrix, the chemical could be in either the gaseous, aqueous or the liquid contaminant phase.

Diffusion. Molecular diffusion is one of a number of dispersive processes. It occurs as a result of Brownian motion, with net migration of contaminant molecules from areas of higher to lower concentration. In the case of a completely inert contaminant, this process is theoretically

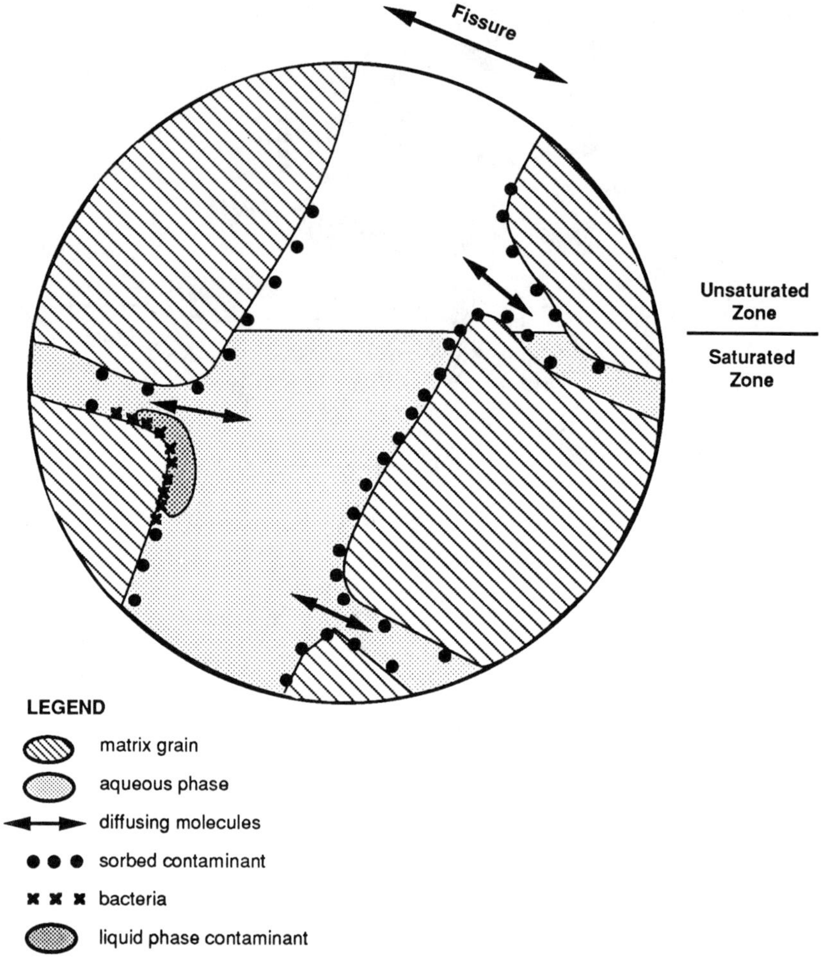

Figure 10–4. Schematic diagram of diffusion across a fissure-matrix boundary. (Detail from Figure 10–3.)

reversible so that a reversed concentration gradient will draw the contaminant back out of the rock matrix.

Diffusion differs from most other dispersive processes in that it does not require advection to occur but rather is driven by concentration gradients. In fact, higher advective velocities increase the viability of other dispersion mechanisms so that the relative importance of diffusion is lessened. One study has suggested that for average advective velocities of less than 1 m/day, diffusion is the dominant transverse dispersion mechanism (Grane and Gardner, 1961, cited by Sudicky et al., 1985).

Little work has been published regarding liquid or gas phase diffusion, so this section will address only the aqueous phase. Liquid phase and aqueous phase diffusion are likely to be quite similar. As shown in Figure 10–4, an organic chemical in the aqueous phase that has diffused into the rock may diffuse out by moving through several phases. In the saturated zone, the diffusing hydrocarbon may move between water-wet or contaminant-wet matrix grains into the aqueous or the liquid phase, while in the unsaturated zone, the diffusing hydrocarbon may move additionally between dry grains and into the vapor phase. In this figure, diffusion is depicted as movement into a secondary macropore, such as a fracture, because the bulk of diffusion into a rock will occur in zones of high secondary porosity, which provides a relatively high level of available surface area for diffusion. At any rate, this discussion will be limited to aqueous phase diffusion with no attention to the moisture characteristics of the particles, as there is little theoretical information and no experimental data available at this level of detail.

Diffusion Between Primary and Secondary Porosity. The type of diffusion depicted in Figure 10–4 (matrix to secondary pore channel) can be described by the following equation, derived from Fick's first and second laws, which describes the mass of chemical expected to diffuse into the rock matrix from a macropore per unit area (Skagius and Neretnieks, 1986a and 1986b):

$$Q = \frac{C_1 D_e}{1} t - \frac{C_1 1 \alpha}{6} \qquad (10.1)$$

where Q = mass of diffused substance, at time t, per unit diffusive surface area (mg/m^2)
1 = distance of diffusion (normal to macropore surface)(m)
C_1 = concentration in macropore water (mg/m^3)
t = time (s)
D_e = effective diffusion coefficient, or effective diffusivity (m^2/s)
α = =Jrock capacity factor (unitless)

The rock capacity factor measures a rock's capacity to store, instead of transmit, as a function of both porosity and sorption:

$$\alpha = \Theta_t + K_d \rho \qquad (10.2)$$

where Θ_t = total (primary) porosity (dimensionless)
K_d = sorption coefficient(m^3/kg)

ρ = rock density (kg/m^3)

The effective diffusion coefficient measures a rock's capacity to transmit by diffusion, and is a combination of diffusivity in the primary porosity and the portion of that porosity that is devoted to transport:

$$D_e = n^+ D_p \tag{10.3}$$

where D_e = effective diffusion coefficient
Θ_{tr} = transport porosity (dimensionless)
D_p = pore diffusion coefficient (m^2/s)

It is worth noting that the total porosity may be differentiated into the effective porosity, including those pores that are interconnected and serve as conduits, and storage porosity, including those pores that have a dead end. With these definitions, the rock capacity factor might better be defined in terms of the effective porosity instead of the total porosity. It is also interesting to note that, in the derivation of equation 10.1, the generic diffusion coefficient of Fick's second law, D, was defined as $D = D_e/\alpha$, with D_e and α being independent of one another. The importance of this observation is discussed below in the portion of this section that addresses sorption.

Figure 10–5 shows a schematic picture of the physical setting where this equation applies. This equation applies for diffusion through a rock slab that has an initial contaminant concentration of zero and an outlet concentration that is effectively zero.

Table 10–1 gives some examples of primary porosity values for various rock types, and Table 10–2 gives examples of effective diffusion coefficients. For sorption coefficients, discussed in locus no. 4, few data are available. Rock densities will vary considerably with mineral composition, but they are relatively simple to determine. Thus, for a given macropore concentration and diffusional length of interest, such as the distance between macropores, the flux rate of an organic chemical into the rock matrix may be calculated. The time frame of interest should be chosen carefully because the effective diffusion coefficient has been assumed to be constant in time, which may be inaccurate over short periods of time (Sudicky et al., 1985).

Assuming that the petroleum derivatives considered here are nonreactive, then the reverse process, diffusion out of the rock matrix, ought to be described by the same equation in the saturated zone, where no significant discontinuity would exist between the matrix pore water phase and the fracture aqueous phase. The implication for remediation is that contaminants that have diffused into the rock matrix will re-enter

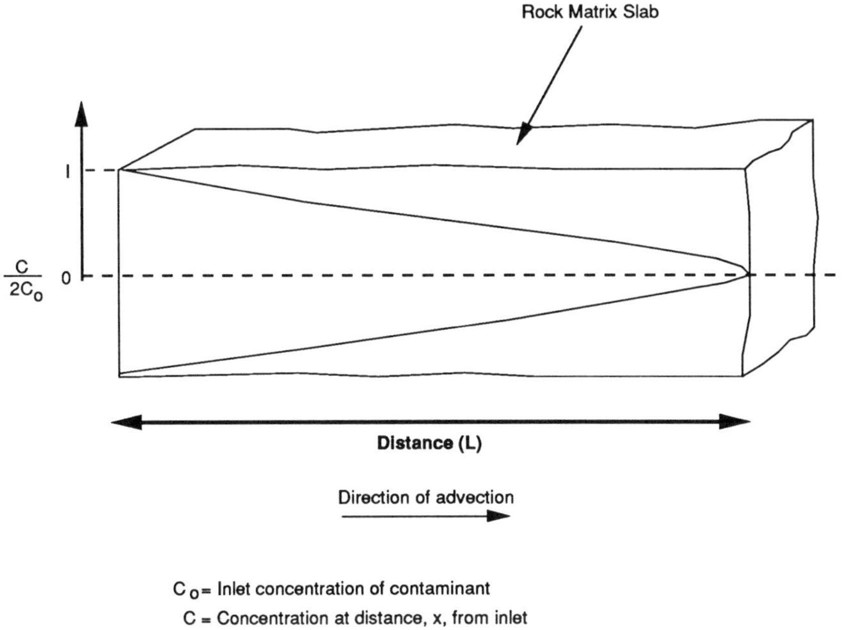

Figure 10–5. Physical representation of equation 10.1 for mass diffusing through a rock slab.

Table 10–1. Typical Primary Porosities

Rock Type	Typical Value or Range of Porosities (%)	Source
Granite	0.02–1.14	1
Gneiss	0.02–0.74	1
Limestone	10	2
Sandstone (semiconsolidated)	10	2
Basalt	10	2
Limestone	3.79–34	3
Granite	0.05–0.09	4
Clay-loam till	30–35	5
Equal sized spheres		
loosest packing	48	2
tightest packing	26	2

Sources: 1. Skagius and Neretnieks, 1986a.
2. Heath, 1983.
3. Garrels et al., 1949.
4. Neretnieks, 1980.
5. Grisak et al., 1980.

Table 10–2. Typical Effective Diffusivities for Nonsorbing Species

Rock Type	Typical Value or Range Diffusivities (cm^2/s)	Diffusing Substance*	Source
Granite	$4.1–66 \times 10^{-10}$	I	1
Granite	$0.22–6.9 \times 10^{-10}$	O	1
Gneiss	$1.8–13 \times 10^{-10}$	I	1
Gneiss	$0.01–7.8 \times 10^{-10}$	O	1
Unconsolidated sand, silt	1.21×10^{-5}	I	2
Granite	$5–36.3 \times 10^{-10}$	I	3
Gneiss	$3.3–28.7 \times 10^{-10}$	I	3
Limestone	2×10^{-5}	I	4
Granite	$0.25–10 \times 10^{-8}$	I	5
Clay-loam till	$1.9–5.0 \times 10^{-7}$	I	6

*I = ion
O = organic molecule

Sources: 1. Skagius and Neretnieks, 1986a.
2. Sudicky et al., 1985.
3. Skagius and Neretnieks, 1986b.
4. Garrels, et al., 1949.
5. Neretnieks, 1980.
6. Grisak et al., 1980.

the mobile phase as natural flushing or dilution changes the concentration gradient between primary and secondary porosity. Examining the rock capacity term reveals that the addition of a desorbing agent will increase the effectiveness of a reversed concentration gradient, provided that the desorbing agent is also able to move into the rock matrix.

The information provided combined with some simple field measurements of rock bulk densities allow the mass flux rate per unit area to be calculated readily. The usefulness of this flux rate is limited, however, by the ability to determine accurately the fracture extent or advective velocities within the fractures, because these matrix characteristics contribute to fracture diffusion and thus, to the fracture mass flux of contaminants (see Section 13 for a discussion of fracture characterization). If an order of magnitude estimate of this contribution is acceptable, then this mass flux rate may be useful. However, further difficulties arise in determining whether, for example, diffusivities determined from this equation for non-sorbing species may be applied to sorbing species or whether this analysis may be applied in the unsaturated zone.

Diffusion Between Strata of Different Hydraulic Conductivities. Diffusion occurs not only between primary and secondary porosity but also in porous media, especially in stratified regions. Sudicky et al. (1985) describe dispersion of a solute from an advective sand layer into the surrounding, lower conductivity silt layers. For a seven-day, continuous

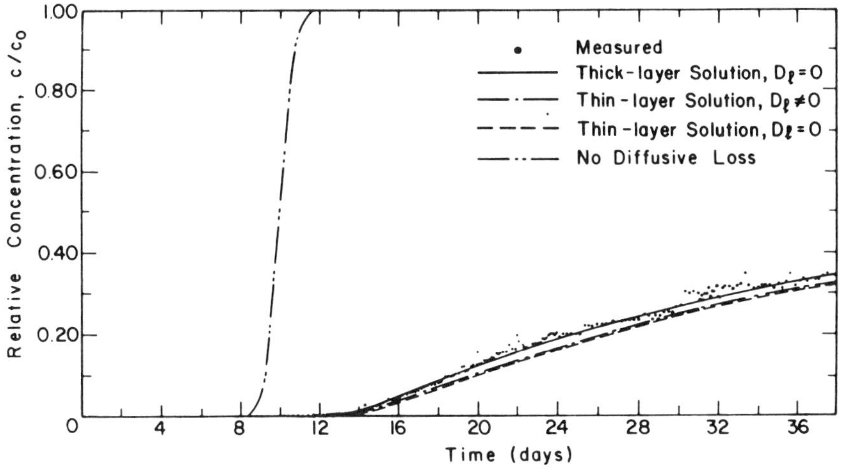

D_ℓ = Longitudinal hydrodynamic dispersion coefficient.

Figure 10–6. Comparison between measurements and simulations for low-velocity case. *Source:* Sudicky et al., 1985.

input of a nonsorbing tracer (NaCl), diffusion was found to be the main component of transverse dispersion. The experiments were not allowed to run long enough for the effluent concentration to approach zero; however, some differences were discovered between the high and low velocity cases. For a low advective velocity (0.1 m/day) in the sand layer, approximately 73 percent of the input mass had been recovered in 36 days, or 23 days after first appearance. On the other hand, in the high velocity case (0.5 m/day), approximately 63 percent of the input mass had been recovered in 12 days, or 10 days after first appearance. Concentration profiles for the low and high velocity case are shown in Figures 10–6 and 10–7. Although the leveling off of the concentration profile in Figure 10–7 suggests that some time might pass before the level of recovery reached that depicted in Figure 10–6, the high level of recovery up to this point supports the notion that higher velocities allow less diffusion into the silt layer. In addition, the fact that the curve in Figure 10–7 does level off instead of continuing to go quickly downward shows that some of the material that had diffused into the silt layer subsequently diffused back out. Sudicky et al. (1985) assume that this diffusion could be described by Fick's first law as illustrated in Figure 10–8, (at y = b):

$$J = \Theta D^* \frac{dc'}{dy} \tag{10.4}$$

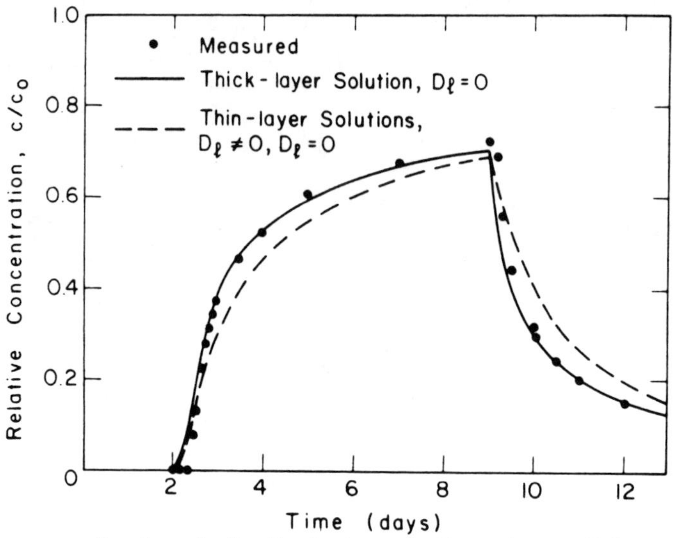

D_ℓ = Longitudinal hydrodynamic dispersion coefficient.

Figure 10–7. Comparison between measurements and simulations for high-velocity case. *Source:* Sudicky et al., 1985.

where J = flux across sand-silt interface ($kg/m^2/s$)
Θ = silt porosity (dimensionless)
D* = diffusivity (m^2/s)
c′ = concentration of solute in silt (kg/m^3)
y = distance from center of sand layer (m)
b = distance from center of sand layer to sand-silt interface (m)

Had the experiments been allowed to run until the relative concentration reached (nearly) zero, some simple mass balance calculations could have been performed to check the assumed values for porosity (Θ) and D*, which are reported here in Tables 10–1 and 10–2. Values for these parameters have been determined using unconsolidated media and should not be used in reference to consolidated media (rocks), however, the principles governing their derivation are the same, and the values are provided here for comparison.

Diffusion in the Unsaturated Zone. Little is known about diffusion in the unsaturated zone. Wang and Narasimhan (1985) have proposed a model in which flow in an unsaturated, fractured porous medium takes place mostly within the rock matrix, with fractures being bridged at points where asperities allow the sides of the fracture to touch and hold

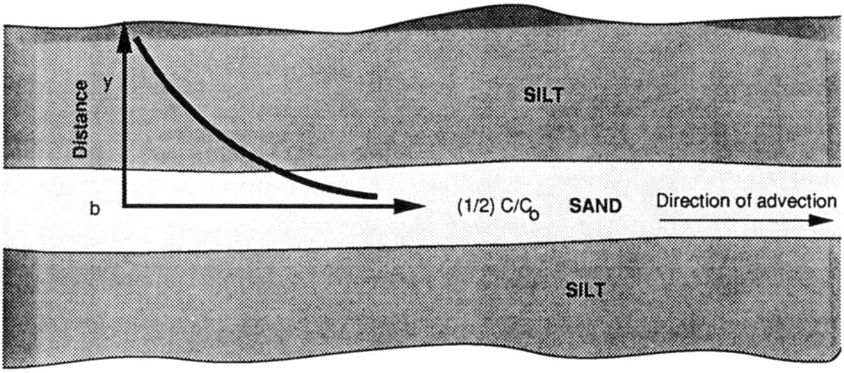

C_0 = Inlet concentration of contaminant
C = Concentration at distance, y, from center of sand stratum

Figure 10–8. Diagram of diffusion between strata.

a small amount of water. In fact, it is expected that any water residing in non-touching areas of fractures will have some (small) velocity into the matrix. The Darcian advective velocities predicted for various model configurations are on the order of only 10^{-3} to 10^{-4} m/day, so that diffusion is the major component of solute movement in the unsaturated zone, according to this model. As the authors themselves point out, however, more work is needed beyond the study of fluid flow to the study of solute movement.

Sorption (Details in Section 4.2). Sorption is the other main process governing the movement of a contaminant in the rock matrix. The process of surface sorption is discussed in locus no. 4. The sorption of organic chemicals to mineral surfaces, and especially to clay minerals, is controlled by van der Waal's forces and depends upon both the surface area and the solubility of the sorbing chemical in water. The sorption process described in Section 4 deals with sorption in soils, yet some of the principles discussed may be applicable for sorption to rock minerals. The principles of soil sorption may not apply to rock mineral sorption, however, because of the potentially considerable differences between the two loci.

While diffusion determines whether a solute may move into the matrix, sorption is largely responsible for the degree to which that contaminant may continue to move from one matrix block to the next and whether the contaminant will be able to diffuse back out. In fact, as indicated in equations 10.1 and 10.2, the mass diffused depends on not only the diffusivity but also the sorption coefficient which, together with primary porosity, determines the rock capacity factor. The process of sorption in soils is discussed in detail in Section 4; suffice it to say here that the principles are likely to be the same for minerals residing in a rock matrix, and the amount of contaminant which is able to sorb to the minerals will likely be limited by the amount of contaminant that is able to diffuse into the mineral grains. Of the two loci dealing with contaminants sorbed to or in close proximity with soil particles (4 and 9), locus no. 4 is probably the more similar physically to locus no. 10, although there are differences such as the degree of organic humus coating on, and surface area to volume ratio of, the particles of locus no. 4.

Clearly, there is a relationship between the degree of sorption and the diffusivity. As mentioned above in the discussion of diffusion into the rock matrix, this relationship is stated in the derivation of equation 10.1, where the generic diffusion coefficient of Fick's second law, D, is defined as $D = D_e/\alpha$ (Skagius and Neretnieks, 1986a). Thus, the generic diffusion coefficient depends upon both the capacity to transmit, reflected by D_e, and the capacity to store, reflected by α. These two parameters (D_e and α) have been considered independent of one another. Although no correlation was demonstrated by Skagius and Neretnieks (1986a) for the nonsorbing species tested, it is conceivable that a relationship does exist if sorption takes place in such a manner as to remove the sorbed molecules from the concentration gradient which is "felt" by subsequently diffusing molecules. In the context of diffusion between strata, experiments to examine a diffusivity retardation factor, which is a function of a solute-solid partitioning coefficient, have been performed (Starr et al., 1985). However, the models thus obtained did not fit observed data, with discrepancies increasing with decreased velocity in the higher conductivity stratum.

Advection. Advection affects this locus both directly and indirectly, but with less impact than either diffusion or sorption. As indicated in the discussion of diffusion, the degree of diffusive flux of a contaminant into the rock matrix is largely dependent upon the degree of secondary porosity. Thus, the advective transport of a contaminant into the secondary porosity (fractures and dissolution channels) is indirectly important to this locus. Advective transport in secondary porosity is discussed in Section 13. On the other hand, contaminant transport in an unfractured, consolidated, porous media is governed by an interplay of two processes: longitudinal (in the direction of flow) advective transport in higher conductivity strata, and tranverse diffusion into the surrounding lower conductivity strata.

In cases of consolidated and unfractured media, the modeling of advective transport is essentially the same as modeling in unconsolidated media, except that the actual values of the horizontal and vertical conductivity parameters will be different. Thus, in this unfractured case, the discussion of advective transport given in Section 8 is applicable here as well. The difficulty lies in the fact that there is a continuum between "fractured" and "unfractured" consolidated media. The level of fracturing for which unconsolidated media models are still accurate is not well defined. For the purpose of this locus, advection is only discussed qualitatively (in Section 10.2.2.1) as it directly affects diffusion processes.

10.2.2.2 Transport of/with Mobile Phase

Once the contaminant moves into the mobile phase, which occurs mainly in high conductivity strata or fractures in the saturated zone and is restricted by sorption, it may be transported with that phase. Transport is discussed in a number of other potentially applicable loci including loci 13 and 8.

10.2.3 Fixation

10.2.3.1 Partitioning onto Immobile (Stationary) Phase

The most important mechanism that could immobilize contaminants in the rock matrix is sorption. Although diffusion may be very slow, it will continue to move the contaminant for as long as a concentration gradient exists. Furthermore, as seen in the discussion of diffusion in stratified porous media, diffusion may significantly retard but not halt the advective movement of a non-sorbing contaminant. Thus, diffusion followed by sorption within the rock matrix will be the most significant retardation mechanism. Once again, the details of surface sorption are discussed in Section 4.2.3.

10.2.4 Transformation

10.2.4.1 Biodegradation (Details in Section 11)

Bacteria are the primary agents of biodegradation in the subsurface environment. Since they range in size from approximately 1 to 10 microns, they are excluded from any pore or crack with a smaller diameter of width. It is therefore unlikely that significant biodegradation occurs in the interior of minerals or rocks. Transport of necessary substrates such as oxygen and mineral nitrogen to the microbes may also be severely limiting.

10.2.4.2 Chemical Oxidation (Details in Section 3.2.4.2)

Chemical oxidation is not expected to occur within crystalline mineral lattices. No data are, however, available.

10.3 STORAGE CAPACITY IN LOCUS

10.3.1 Introduction and Basic Equations

The mass of contaminant per unit volume of unfractured rock is simply the following:

$$m_c = C\Theta_t \tag{10.5}$$

where m_c = mass of contaminant per unit volume of rock (mg/L)
C = concentration of aqueous phase contaminant in secondary pore (fractures and solution channels) waters (mg/L)
Θ_t = total primary porosity of rock (dimensionless)

Some values of Θ_t for both crystalline and porous media were given in Table 10–2. The total primary porosity includes both effective and storage porosity.

10.3.2 Guidance on Inputs for, and Calculation of, Maximum Values

10.3.2.1 Concentration of Contaminants in Secondary Pore Water, C

As indicated in Section 3.3.2, the maximum concentration may be taken as the contaminant solubility in water. The hydrocarbons of a typical gasoline without additives have an approximate solubility of 200 mg/L at 20°C. A gasoline containing oxygenated additives, such as ethanol or MTBE, has an approximate solubility of 2,000 mg/L.

10.3.2.2 Total Primary Porosity, Θ_t

For the calculation of a maximum storge capacity, the higher typical values of porosity (Table 10–1) should be used. Since porosities differ for different rock types, a porosity appropriate to the rock type of interest should be used.

10.3.3 Guidance on Inputs for, and Calculation of, Average Values

10.3.3.1 Concentration of Contaminants in Secondary Pore Water, C

An arbitrary but reasonable estimate of the average concentration is 10 percent of the maximum solubility. A factor of 10 allows for the dilution of contaminated water with clean pore water as a result of transport, as well as other factors which may reduce the contaminant level in time, such as volatilization or biodegradation. For a gasoline without additives, the average concentration is 20 mg/L, and for a gasoline with oxygenated additives, the average concentration is 200 mg/L.

10.3.3.2 Total Primary Porosity, Θ_t

A reasonable estimate of average primary porosity may be made by again consulting Table 10–2 and choosing values in the middle of the range of values shown. As mentioned before, average values should be chosen appropriately for the specific rock type of interest.

10.4 EXAMPLE CALCULATIONS

10.4.1 Storage Capacity Calculations

10.4.1.1 Maximum Values

To estimate the maximum storage capacity in a limestone for a gasoline containing oxygenated additives, use equation 10.5. As recommended in Section 10.3.2, the concentration in secondary pore water (C) should be 2,000 mg/L. From Table 10–1, the maximum limestone porosity given is 34 percent. Thus, the maximum storage capacity is:

$$\begin{aligned} M_c = C\Theta_t &= (2000 \text{ mg/L})(0.34) \\ &= 680 \text{ mg/L, or} \\ &= 680 \text{ g/m}^3 \end{aligned}$$

10.4.1.2 Average Values

To estimate the average storage capacity in a granite for a gasoline containing no additives, use equation 10.5 again. As recommended in Section 10.3.3, the concentration in secondary pore waters should be 20 mg/L. From Table 10–1, a reasonable value of average granite porosity is 0.3 percent. Thus, the average storage capacity is:

$$\begin{aligned} M_c = C\Theta_t &= (20\ \text{mg/L})(0.003) \\ &= 0.06\ \text{mg/L, or} \\ &= 60\ \text{mg/m}^3 \end{aligned}$$

10.4.2 Transport Rate Calculations

As indicated in Section 10.2.2.1, the mass of a pollutant diffusing into the matrix from the secondary porosity can be calculated using the following equations:

$$Q = \frac{C_i\, D_e}{1}\, t - \frac{C_i\, l\, \alpha}{6} \tag{10.1}$$

$$\alpha = \Theta_t + K_d\rho \tag{10.2}$$

with parameters as defined above. Maximum values for diffusive capacity can be determined by making some generous estimates of the parameters involved. A generous estimate of the slab length (l) for example, would assume a highly fractured system so that the matrix blocks would be very small. Such a system might yield a matrix block size of 0.25 m (see Section 13). A generous estimate of the rock capacity factor (α) would assume a non-sorbing contaminant ($K_d = 0$) and a low porosity (Θ_t). From Table 10–1, a low estimate of porosity is 0.02 percent, which corresponds to a crystalline rock, or 4 percent corresponding to a porous media. A generous estimate of the effective diffusivity (D_e) would lie on the high end, or, from Table 10–2, 6.6×10^{-9} cm²/s for crystalline and 2×10^{-5} cm²/s for porous rock. A high estimate of the concentration (C_1) may be obtained as recommended in Section 10.3.2. Finally, an estimate of time (t) could be chosen as 1 year, to make the calculation for a unit time frame. Then, for crystalline and porous rocks, the maximum flux parameter Q would be 7.5×10^{-6} mg/m² and 2.5×10^{-2} mg/m², respectively.

From these numbers, it is clear that the combination of D_e and t would have to increase by four (4) orders of magnitude for crystalline rocks to gain the same storage capacity as porous rocks. Since an increase in the diffusivity of more than two (2) orders of magnitude is highly unlikely, this means that a time scale of at least 100 years would have to be of interest in order for diffusion into crystalline rocks to be as significant as the diffusion which would take place in porous media in just one year.

10.5 SUMMARY OF RELATIVE IMPORTANCE OF LOCUS

10.5.1 Remediation

This discussion has indicated that the major mechanisms governing contaminant migration into or out of the rock matrix are diffusion and sorption. In fact, it has been indicated that the diffusive flux into even a porous medium is quite low relative to the presence of contaminants in other loci. Although sorption is an important process, it was not dealt with qualitatively in this section for two reasons: first, sorption is discussed in detail elsewhere; and second, and more importantly, sorption cannot provide a truly significant degree of fixation in this locus for two reasons. Although the sorbing minerals are often listed as common constituents of various rock types, they rarely comprise a significant fraction of any rock. In addition, even in cases where they do comprise significant portions of a rock, the low diffusion rate of contaminants into the rock limits the degree of sorption.

The real significance of findings for this locus lies in the fact that once contaminants are introduced to the locus they are not easily removed. This demonstrates the difficulty of cleaning a release site to zero-level concentrations. Although this locus poses neither a significant help nor a significant impedence to most methods of remediation, the fact that minor amounts of contaminant will diffuse into the rock indicates that zero-level remediation may not be possible. In cases where such a zero-level remediation effort is deemed desirable, the drastic measure of vitrification as a means of driving off the contaminants into the air by literally melting the rock is probably the only option which will truly work.

10.5.2 Loci Interactions

The loci that interact with locus no. 10 are summarized in Figure 10.2. The first, essential step in interaction between locus no. 10 and other loci is aqueous phase diffusion out of the rock matrix. Then, the diffused contaminants in the unsaturated zone may: partition out of water and into the vapor phase (locus no. 1), partition out of water and onto water-dry soil particles or rock surfaces (locus no. 2), remain dissolved in the relatively immobile water film surrounding well drained soil particles or rock surfaces (locus no. 3), remain dissolved in relatively mobile pore waters just above the capillary fringe or associated with heavy rainfall (locus no. 12), or undergo biodegradation (locus no. 11). In the saturated zone, on the other hand, the diffused contaminants may: become sorbed to water-wet soil particles or rock surfaces (locus no. 4); remain dissolved in mobile pore waters (locus no. 8) or in rock fracture and dissolution channels (locus no. 13), in which cases the contaminant may alternate continuously between diffusive and advective transport; or undergo biodegradation (locus no. 11). As has been indicated throughout this section, diffusion into the rock matrix from any other locus is so very slow that the other loci are more likely to control contaminant transport, with diffusion into and out of the rock matrix providing some small degree of retardation.

10.5.3 Information Gaps

In order to better understand the movement of contaminants into and through the rock matrix, the following data are needed:

1. Additional measurements of diffusivity and primary porosity;
2. A better understanding of fracture densities and what level of fracturing may still be modeled as an unfractured porous media;
3. A better understanding of sorption processes within the rock matrix, and especially of the capacity, if any, for sorption to effectively remove molecules from the total concentration gradient which governs diffusion; and
4. A better characterization of advective transport in stratified porous media, with special attention given to the extent of diffusion as the mechanism controlling transverse dispersion and to the relation between advective longitudinal velocity and transverse diffusion.

10.6 LITERATURE CITED

Garrels, R.M., Dreyer, R.M. and Howland, A.L. 1949. Diffusion of Ions Through Intergranular Spaces in Water-Saturated Rocks. Bull. Geol. Soc. Am., No. 60. pp. 1809–1828.

Grisak, G.E., J.F. Pickens, and J.A. Cherry. 1980. Solute Transport Through Fractured Media 2. Column Study of Fractured Till. Water Resources Research, 16(4):731–739.

Heath, R.C. 1983. Basic Groundwater Hydrology. USGS Water-supply Paper No. 2220. Alexandria, VA: USGS.

Neretnieks, I. 1980. Diffusion in the Rock Matrix: An Important Factor in Radionuclide Retardation? Journal of Geophysical Research, 85(B8):4379–4397.

Skagius, K. and I. Neretnieks. 1986a. Porosities and Diffusivities of Some Nonsorbing Specifies in Crystalline Rocks. Water Resource Research, 22(3):389–398.

Skagius, K. and I. Neretnieks. 1986b. Diffusivity Measurements and Electrical Resistivity Measurements in Rock Samples Under Mechanical Stress. Water Resource Research, 22(4):570–580.

Starr, R.C., R.W. Gillham, and E.A. Sudicky. 1985. Experimental Investigation of Solute Transport in Stratified Porous Media 2. The Reactive Case. Water Resources Research, 21(7):1035–1041.

Sudicky, E.A., R.W. Gillhan, and E.D. Frind. 1985. Experimental Investigation of Solute Transport in Stratified Porous Media 1. The Nonreactive Case. Water Resources Res., 21(7):1035–1041.

Wang, J.S.Y. and T.N. Narasimhan. 1985. Hydrologic Mechanisms Governing Fluid Flow in a Partially Saturated, Fractured, Porous Medium. Water Resources Research, 21(12):1861–1874.

SECTION 11

Contaminants Sorbed onto or into Soil Microbiota in Either the Saturated or Unsaturated Zone

CONTENTS

List of Tables 282
List of Figures 282
11.1 Locus Description 283
    11.1.1 Short Definition 283
    11.1.2 Expanded Definition and Comments 283
11.2 Evaluation of Criteria for Remediation 284
    11.2.1 Introduction 284
    11.2.2 Microbial Uptake of Hydrocarbons 286
    11.2.3 Biodegradation 291
        Mechanisms 291
        Kinetics 291
        Additional Factors Affecting Biodegradation ... 297
    11.2.4 Mobility 300
    11.2.5 Viability 301
11.3 Storage Capacity in Locus 302
    11.3.1 Introduction and Basic Equations 302
    11.3.2 Guidance on Inputs for, and Calculation of, Maximum Value 304
    11.3.3 Guidance on Inputs for, and Calculation of, Average Value 304
11.4 Example Calculations 304
    11.4.1 Storage Capacity Calculations 304
        11.4.1.1 Maximum Value 304
        11.4.1.2 Average Value 305
    11.4.2 Transformation Rate Calculations 305
11.5 Summary of Relative Importance of Locus 306
11.6 Literature Cited 306

TABLES

11–1 Factors Affecting Bioremoval 287
11–2 Bioremedial Actions 288
11–3 Bacteria-Water and Octanol-Water Partition Coefficients for Several Polyaromatic Hydrocarbons 290
11–4 Six Models for Mineralization Kinetics with Only the Variables of Substrate Concentration and Cell Density 295
11–5 Parameters and Asymptotic Standard Deviations of Four Models of Substrate Disappearance Kinetics Fit to Data on Metabolism of Nine Concentrations of [U-ring-^{14}C] Benzoate by *Pseudomonas* Sp. 299
11–6 Relative Biodegradability of Hydrocarbons 300

FIGURES

11–1 Schematic Cross-Sectional Diagram of Locus No. 11–Contaminants Sorbed onto or into Soil Microbiota in Either the Unsaturated or Saturated Zone 285
11–2 Schematic Representation of Important Transformation and Transport Processes Affecting Other Loci 286
11–3 Representation of Hydrocarbon Inclusion Body, "H," in *Acinetobacter sp.* Grown on Hexadecane 287
11–4 Pathways of Biodegradation for Several Hydrocarbons 292
11–5 Kinetic Models as a Function of Initial Substrate Concentration and Bacterial Cell Density 296
11–6 Kinetic Disappearance Curves for Six Kinetic Models of Biodegradation 298
11–7 Breakthrough Curves for *Escherichia Coli* and Coliphage F2 303

SECTION 11

Contaminants Sorbed onto or into Soil Microbiota in Either the Saturated or Unsaturated Zone

11.1 LOCUS DESCRIPTION

11.1.1 Short Definition

Contaminants sorbed onto or into soil microbiota in either the saturated or unsaturated zone.

11.1.2 Expanded Definition and Comments

Organic chemicals can be taken up by soil microbiota by several mechanisms: 1) sorption to the exterior cell membranes or to exuded extracellular material; 2) molecular transport across the cell membranes into the cells' cytoplasm; or 3) ingestion of particles or liquid droplets. In the latter case, inclusions of nearly pure liquid contaminant may exist inside the biota. While uptake is possible from air, liquid contaminant, or aqueous solution, uptake from aqueous solution or liquid contaminant are by far the more important. Soil biota include bacteria, fungi, and yeasts, but bacteria are by far the most numerous and, therefore, the most important. The mobility of microbes is proportional to the degree of moisture saturation in the soil, and the sorbed contaminants possess a similar degree of mobility. The potential importance of this locus as a temporary reservoir for contaminants has never been assessed.

Because only aerobic degradation is considered important for hydrocarbons, and the concentration of biota in soil drop off sharply with increasing depth, the unsaturated zone is probably much more important than the saturated zone for biodegradation of hydrocarbons. However, some groundwater (in the saturated zone) does contain dissolved oxygen sufficient to allow the aerobic biodegradation of a few milligrams/liter of dissolved hydrocarbons. Figures 11-1 and 11-2 present a schematic cross-sectional diagram of locus no. 11 and a schematic representation of the transformation and transport processes affecting other loci, respectively.

11.2 EVALUATION OF CRITERIA FOR REMEDIATION

11.2.1 Introduction

Locus no. 11 is by nature unique among the thirteen loci described in this report. Since microbes may exist inside or adjacent to each of the other loci, there is close interaction between locus no. 11 and each of the others. Also, being animate organisms, microbes are chemically complex and dynamic. They act as agents of physical and chemical change to leaked hydrocarbons in the subsurface. As a population, microbes can increase or decrease in number, adapt, and be transported from locus to locus, depending on environmental conditions. For these reasons, this section is structured differently than the others. Emphasis is placed on the thermodynamic and kinetic aspects of "bioremoval," which involves multiple mechanisms. The importance of locus no. 11 clearly lies in its potential to remove leaked hydrocarbons from the subsurface.

Biological removal of contaminant compounds from soil consists of two distinct, yet somewhat interdependent, mechanisms: microbial uptake and biodegradation. Microbial uptake is the physical-chemical removal of compounds by microbes from the surroundings. It may take the form of either adsorption to the microbe's surface or transport through the microbe's membrane covering into its interior. Transport through the cell membrane can itself take the form of either diffusion-like flow or bulk transport. Diffusion occurs at the molecular level and is strongly influenced by hydrophobic attraction between the contaminant and the cell. Bulk transport, also called inclusion, resembles "swallowing" by the cell of a volume of solution or pure product. Figure 11–3 shows several inclusion bodies of hexadecane inside a bacterium.

Biodegradation is the microbially mediated chemical transformation of organic compounds to form new product compounds. Biodegradation is largely attributed to reactions occurring within the cell, although many microbes excrete enzymes that catalyze transformations outside the cell. In the case of reactions within the cell, uptake is a necessary prerequisite to biodegradation.

Since microbes, consisting largely of bacteria, exist and thrive abundantly in surface soils, biodegradation is a spontaneous and "natural" process. The bacteria adapt to, and interact with, organic contaminants as they would other organic compounds. The rates of uptake and trans-

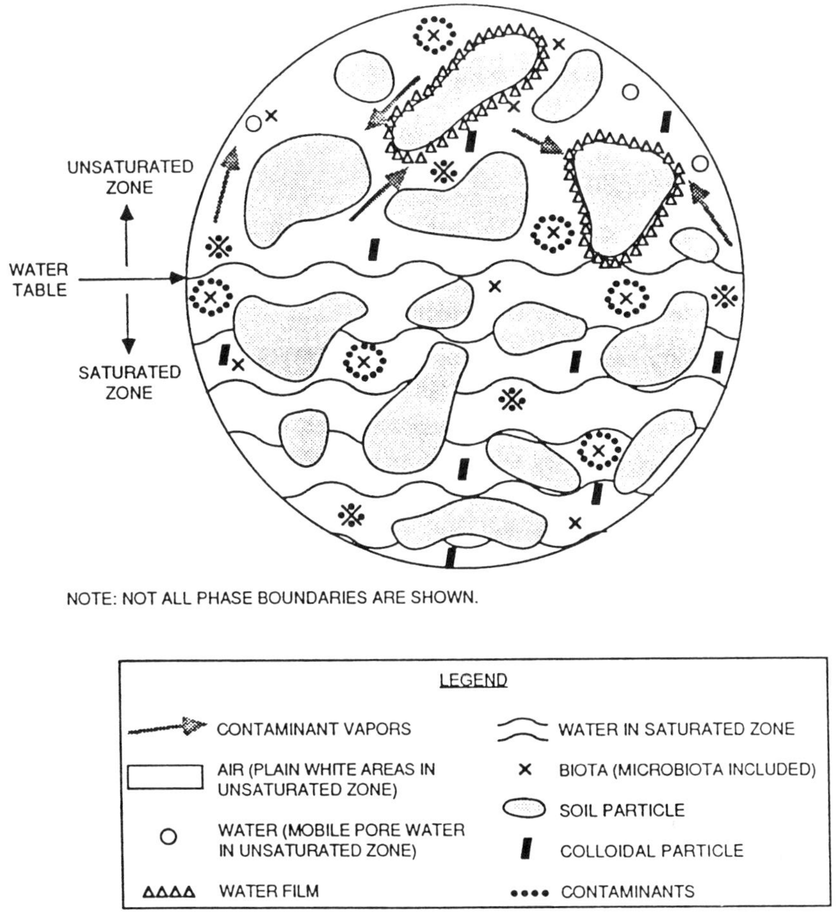

Figure 11–1. Schematic cross-sectional diagram of locus no. 11–contaminants sorbed onto or into soil microbiota in either the unsaturated or saturated zone.

formation can be artificially altered by manipulation of various factors. Bioremediation, the term applied to manipulating conditions to facilitate biodegradation, is a viable option that requires a clear understanding of the mechanisms of transport and adsorption and the kinetics of microbial metabolism and growth. An examination of the factors that limit the rates of these processes is thus in order. The rate limiting factors, which may be grouped according to their effects on uptake or metabolism, are listed in Table 11–1 and will be addressed in the following sections.

Various modes of bioremediation are listed in Table 11–2. All focus

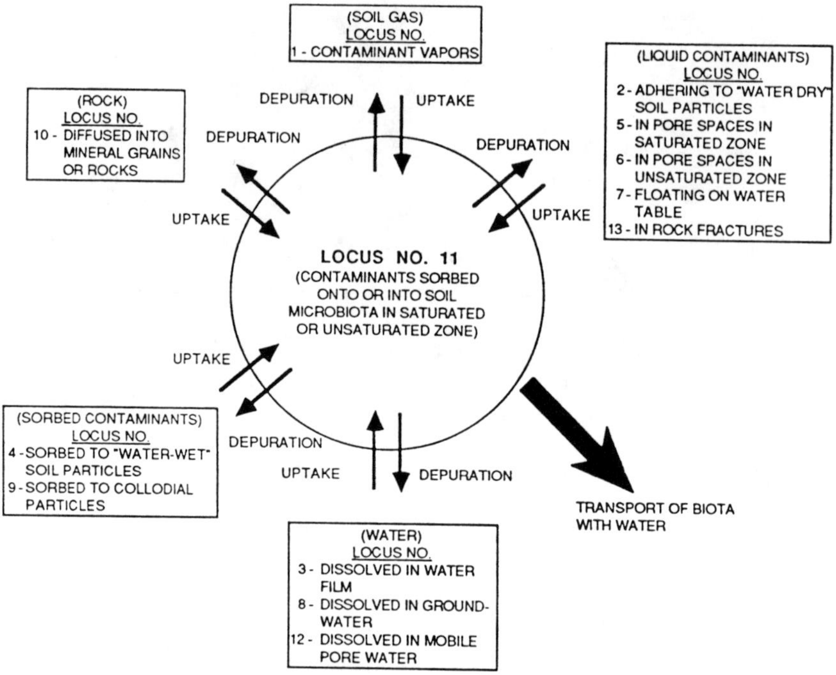

Figure 11–2. Schematic representation of important transformation and transport processes affecting other loci.

on the transformation of contaminants and not removal via sorption. This is because sorption is reversible and, therefore, not permanent whereas transformation is essentially irreversible. Composting and groundwater recirculation, with and without the addition of specially selected bacteria, are the more common and are proven remedial techniques. The vapor phase bioreactor and slurry reactors are both in development, but hold the promise of unique advantages for specific remedial requirements.

11.2.2 Microbial Uptake of Hydrocarbons

As stated previously, the two major bacterial uptake mechanisms are adsorption to the exterior membrane surface of the bacteria and absorption through the membrane to the cell's interior (cytoplasm). A review by Baughman and Paris (1981) revealed that researchers have tended to ignore the distinction between the two. However, some have assumed adsorption (or "uptake" in their terms) while only one suggested adsorp-

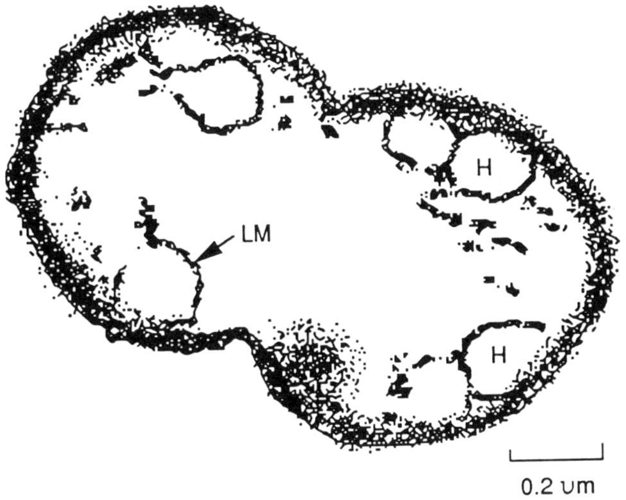

Figure 11–3. Representation of hydrocarbon inclusion body, "H", in *Acinetobacter sp.* grown on hexadecane. *Source:* Scott and Finnerty, 1976. (Copyright American Society for Microbiology. 1979. Reprinted with permission.)

tion as the sole mechanism. One group proposed sequential adsorption and absorption.

All experimental work reported in the literature has been performed in mixed aqueous solution. Transport limitations make it nearly impossible to measure the bulk partitioning of organic compounds between bacterial cells and soil. Thus, the available equilibrium partitioning data probably are only applicable in the immediate vicinity of soil microbes, especially in the vadose zone.

Table 11–1. Factors Affecting Bioremoval

A. Uptake
- bacterial concentration (population density)
- bacterial species
- contaminant concentration
- degree of microbial and contaminant dispersion in soil

B. Biodegradation
- molecular structure of contaminant compounds
- contaminant concentration
- bacterial concentration
- bacterial species
- nutrient (carbon, oxygen, nitrogen, phosphorus) concentrations
- temperature
- pH

Table 11–2. Bioremedial Actions

Mode	Soil Handling	Addition	Comments
1. Composting	Excavated Mixing	Direct Soil	– Simple direct method; "low tech" – requires control of volatilized compounds – similar to land treatment of refinery wastes
2. Groundwater recirculation	In situ	Delivered via Recirculation H_2O	– dependent on site hydrogeology – dependent on soil transmissivity
3. Selected microbes	Excavated or in situ	Either of the above	– may be used with any other mode – most effective in new spill cleanup – limited or no additional effect for old spills where indigenous bacteria are well acclimated.
4. Vapor Phase bioreactor	In situ	Mixed directly into bioreactor	– used downstream of vapor extraction systems – bioreactor may be solid phase (soil) or submerged (aqueous) culture – still experimental
5. Slurry reactor	Excavated	Mixed directly into bioreactor	– high energy input required – high reaction rate – capital intensive – still experimental

The relevant mathematical expressions applicable to uptake are the same as those for other equilibrium partitioning phenomena. They are the partition coefficient and the Freundlich equation. The partition coefficient, K_B, is defined as the ratio of the contaminant concentration in the cell, $[C]_{cell}$, to the concentration in the bulk medium, $[C]_{med}$, at equilibrium.

$$K_B = \frac{[C]_{cell}}{[C]_{med}} \tag{11.1}$$

where: $[C]_{cell}$ = grams hydrocarbon on or in microbes per gram dry soil
$[C]_{med}$ = grams hydrocarbon per gram dry soil

The Freundlich equation, proposed in 1909, adds an exponential parameter:

$$[C]_{cell} = K_f \, [C]_{med}^{1/n} \tag{11.2}$$

where: K_f = Freundlich coefficient (dimensionless)
$1/n$ = linearity constant (dimensionless)

Although both equations have been used, it is important to note that 1/n approaches unity for many systems at low concentration. Therefore, K_f will equal K_B in many instances.

There is a severe lack of partitioning data for hydrocarbons. Nearly all available data are for pesticides and other chlorinated compounds (Baughman and Paris, 1981; Bell and Tsezos, 1987). Some values of K_B for polyaromatic hydrocarbons (PAHs) are given in Table 11–3. Baughman and Paris (1981) found no data for non-cyclic hydrocarbons and no Freundlich parameters for any hydrocarbons.

In the absence of experimentally-derived K_B values, an alternative method of assessing potential partitioning behavior involves octanol-water coefficients. As previously mentioned, uptake by microbes is strongly dependent upon the hydrophobicity of the contaminant. Hydrophobicity is strongly a function of chemical structure. Specifically, for hydrocarbons, the higher the molecular weight, the more hydrophobic is the compound. For example, n-octane, C_8H_{18}, is more hydrophobic than n-hexane, C_6H_{14}. It could be expected, then, that the partition coefficient, K_B, is greater for octane than for hexane. This "hydrophobic effect" is confirmed when the partitioning of a compound between octanol and water (expressed as the octanol-water coefficient, K_{ow}) is compared with K_B.

Table 11–3 includes the octanol-water partition coefficients, for several PAHs. The relationship between K_B and K_{ow} is quite linear considering that several orders of magnitude are spanned. This is consistent with previous correlations made between K_{ow} and K_p values for higher organisms (Baughman and Paris, 1981). Thus, K_{ow} may be a useful surrogate for predicting K_B.

No detailed kinetic studies are available in the literature although

Table 11–3. Bacteria-Water and Octanol-Water Partition Coefficients for Several Polyaromatic Hydrocarbons

Compound	Organisms	K_B	Reference	K_{ow}	Reference
Benz [a] anthracene	Viable bacteria Killed bacteria	$(3.6 + 0.3) \times 10^4$ $(9.9 + 0.7) \times 10^4$	(1)	4.1×10^5	(3)
Benz [a] pyrene	Killed bacteria	$(3.5) \times 10^5$	(1)	2.2×10^6	(3)
Benz [f] quinoline	Viable bacteria Heat killed bacteria	149 540	(1)	1.6×10^3	(3)
Pyrene	Unspecified bacteria	2.7×10^4	(2)	1.5×10^5	(2)
Phenanthrene	Unspecified bacteria	6.3×10^3	(2)	3.3×10^3	(2)

References: (1) Smith et al., 1978.
(2) Karickhoff et al., 1979.
(3) Leo, A.J., 1975.

some workers have made measurements over time to verify equilibrium (Baughman and Paris, 1981). Generally, it appears that equilibrium is reached in a matter of minutes to several hours. There have been no exceptions involving bacteria under experimental conditions.

The four principle factors affecting uptake were listed in Table 11–1. Two important factors are the bacterial population density and the contaminant concentrations. For uptake to occur, a hydrocarbon molecule must come in close contact with a bacterium. The higher the concentrations of both bacteria and contaminants, the more frequent will be the contacts between the two. Assuming uniform dispersion of bacteria in the volume of soil, a high degree of dispersion of the contaminant within the soil is desirable to facilitate microbial/contaminant contact. It should be noted, however, that some minimum contaminant concentration is required for any biodegradation to occur (Alexander, 1985).

Baughman and Paris (1981) reviewed microbial uptake studies in which various bacterial species were contacted with the same compound. This review showed that the partition coefficient, K_B, was somewhat dependent upon the species of bacteria in the test. There are no data available, however, addressing the dependence of K_B on bacterial species for hydrocarbons. The number of species of bacteria present in a soil is dependent on the site history, especially with regard to previous contamination.

11.2.3 Biodegradation

Mechanisms. Examining the mechanisms of molecular transformation allows one to judge the effectiveness of detoxification by biodegradation. A series of biodegradation reactions, called the biodegradation pathway, typically occurs as more complex substances are transformed into simpler compounds. For most hydrocarbons, these simpler compounds are less toxic than the more complex hydrocarbons. For most cases, degradation results in carbon dioxide and water as final end products; this is termed mineralization. Other times, however, the hydrocarbons do not reach a mineralized endpoint but rather result in relatively stable aliphatic and aromatic compounds. These compounds may become integral parts of the soil humus (Alexander, 1977) or may be flushed out of the immediate vicinity in the form of highly soluble acids, ketones and alcohols (e.g., acetic acid, acetone). In the case of chlorinated hydrocarbons, mineralization is less often the biodegradation pathway. Chlorinated compounds may be transformed into recalcitrant and sometimes more toxic compounds.

The precise pathways of hydrocarbon oxidation are debated in the literature. However, transformations likely occur stepwise from end carbons, producing alcohols, aldehydes and fatty acids in sequence. Several examples are presented in Figure 11–4 (Singer and Finnerty, 1984).

Kinetics. Although a large body of literature exists on the kinetics of microbial growth, relatively little is available on biodegradation rates. Simkins and Alexander (1984) performed examination of degradation kinetics. Several assumptions are made for this analysis:

1. There is no transport limitation of any substrate.
2. The only substrate which may be limiting is carbon; O_2, N, and P are always in excess.
3. The microbial population consists of a single species and may be rate-limiting.
4. All organic contaminants present are equally metabolizable.

The implication of these assumptions when applying the kinetic models in subsurface conditions will be discussed later.

Although several models of cell growth have been proposed that depend upon only a single substrate, the classic model of kinetics was proposed by Monod:

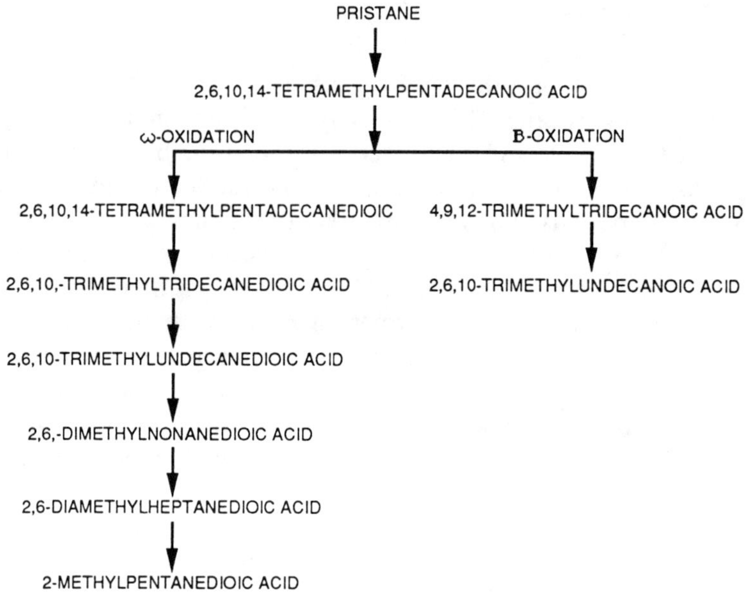

Pathway of pristane oxidation by *Brevibacterium erythrogenes*.

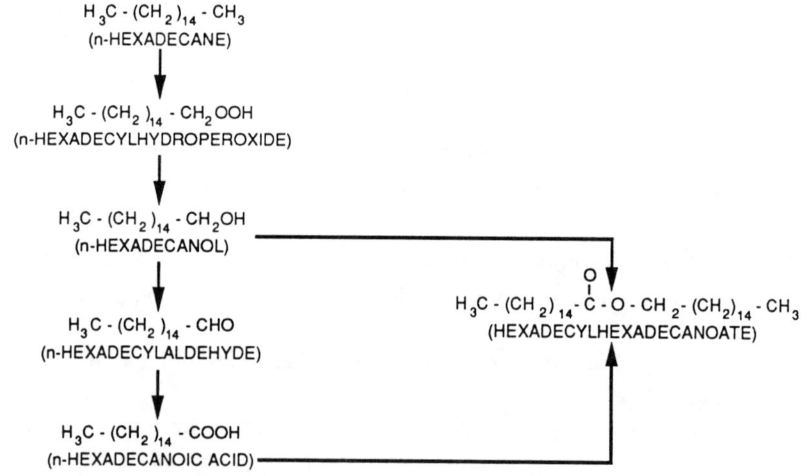

Proposed pathway of hexadecane metabolism in *Acinetobacter* sp. H01-N.

Figure 11–4. Pathways of biodegradation for several hydrocarbons. *Source:* Adapted from Singer and Finnerty, 1984.

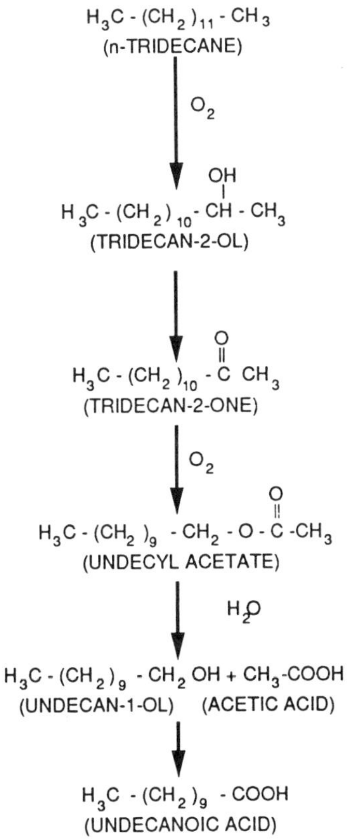

Pathway of Subterminal Alkane Oxidation in *Pseudomonas.*

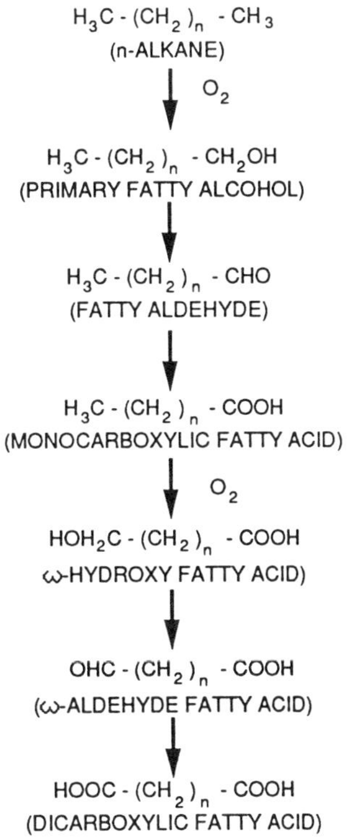

Pathway of Diterminal Alkane Oxidation

Figure 11–4. (continued).

$$\mu = \frac{\mu_{max} C}{(K_S + C)} \tag{11.3}$$

where

$\mu = dB/dt\ (1/B) =$ specific growth rate (day^{-1})

$B =$ microbial population density $\left(\frac{\text{bacteria cell count}}{\text{g soil}}\right)$

$\mu_{max} =$ maximum specific growth rate (day^{-1})

$C =$ substrate concentration $\left(\frac{\text{g hydrocarbon}}{\text{g soil}}\right)$

K_s = half velocity constant for growth* $\left(\frac{\text{g hydrocarbon}}{\text{g soil}}\right)$

t = time (day)

A mass balance equation must account for changes in both substrate concentration and microbial population density

$$C_o + qB_o = C + qB \qquad (11.4)$$

where C_o = initial substrate concentration
B_o = initial population density
q = cell quota = the amount of substrate consumed per cell produced

C and B are the substrate concentration and population density at any time after time zero. The cell quota, q, is reasonably constant when the carbon source is the limiting substrate. qB may then be replaced by X, the concentration of substrate required to produce the population density of cells, B. X_o is the analogous initial concentration. With this substitution the cell growth equation and the mass balance equation may be rewritten as follows:

$$\frac{dX}{dt} \cdot \frac{1}{X} = \frac{\mu_{max} C}{K_s + C} \qquad (11.5)$$

$$C_o + X_o = C + X \qquad (11.6)$$

If the mass balance equation and its time derivative are solved for X and dX/dt, respectively, these expressions can be substituted into equation 11.5. The result is:

$$\frac{-dC}{dt} = \frac{\mu_{max} C (C_o + X_o - C)}{K_s + C} \qquad (11.7)$$

Equation 11.7 is a general expression of substrate disappearance for systems in which only cell densities and substrate concentration determine the rate of substrate removal.

Simkins and Alexander (1984) identified six special cases which may be expressed as simplifications of the general expression. These are presented in Table 11-4. Figure 11-5 illustrates the approximate relative values of X_o, C_o and K_s required for each of these cases. Figure 11-6

*K_x = the substrate concentration at which $\mu = 1/2\ \mu_{max}$.

Table 11–4. Six Models for Mineralization Kinetics with Only the Variables of Substrate Concentration and Cell Density

Model and Characteristics	Equation and Inequalities
I. Zero order	
Differential form	$-dS/dt = k_1$
Integral form	$S = S_0 - k_1 t$
Derived parameter	$k_1 = \mu_{max} X_0$
Necessary conditions	$X_0 >> S_0$ and $S_0 >> K_s$
II. Monod, no growth	
Differential form	$-dS/dt = k_1 S/(K_s + S)$
Integral form	$K_s \ln(S/S_0) + S - S_0 = -k_1 t$
Derived parameter	$k_1 = \mu_{max} X_0$
Necessary condition	$X_0 >> S_0$
III. First order	
Differential form	$-dS/dt = k_3 S$
Integral form	$S = S_0 \exp(-k_3 t)$
Derived parameter	$k_3 = \mu_{max} X_0 / K_s$
Necessary conditions	$X_0 >> S_0$ and $X_0 << K_s$
IV. Logistic	
Differential form	$-dS/dt = k_4 S(S_0 + X_0 - S)$
Integral form	$S = \dfrac{S_0 + X_0}{1 + (X_0/S_0)\exp[k_4(S_0 + X_0)t]}$
Derived parameter	$k_4 = \mu_{max}/K_s$
Necessary condition	$S_0 << K_s$
V. Monod with growth	
Differential form	$-dS/dt = [\mu_{max} S(S_0 + X_0 - S)]/(K_s + S)$
Integral form	$K_s \ln(S/S_0) = (S_0 + X_0 + K_s)\ln(X/X_0) - (S_0 + X_0)\mu_{max} t$
Derived parameter	None
Necessary condition	None
VI. Logarithmic	
Differential form	$-dS/dt = \mu_{max}(S_0 + X_0 - S)$
Integral form	$S = S_0 + X_0[1 - \exp(\mu_{max} t)]$
Derived parameter	None
Necessary condition	$S_0 >> K.$

Source: Simkins and Aldexander (1984).

shows the shapes of all six kinetic curves. Table 11–5 gives parameter values for disappearance of benzoate with *Pseudomonas* bacteria using four of these models.

The assumption that transport of O_2, N, or P is not rate-limiting is not always robust. The assumption becomes more reasonable, however, as the size of the unit volume decreases and is most reasonable at the microscale level of locus no. 11 (i.e., a single cell which, if alive, is surrounded by water film). Also, transport of both microbes and substrates, including contaminants, increases with moisture content, espe-

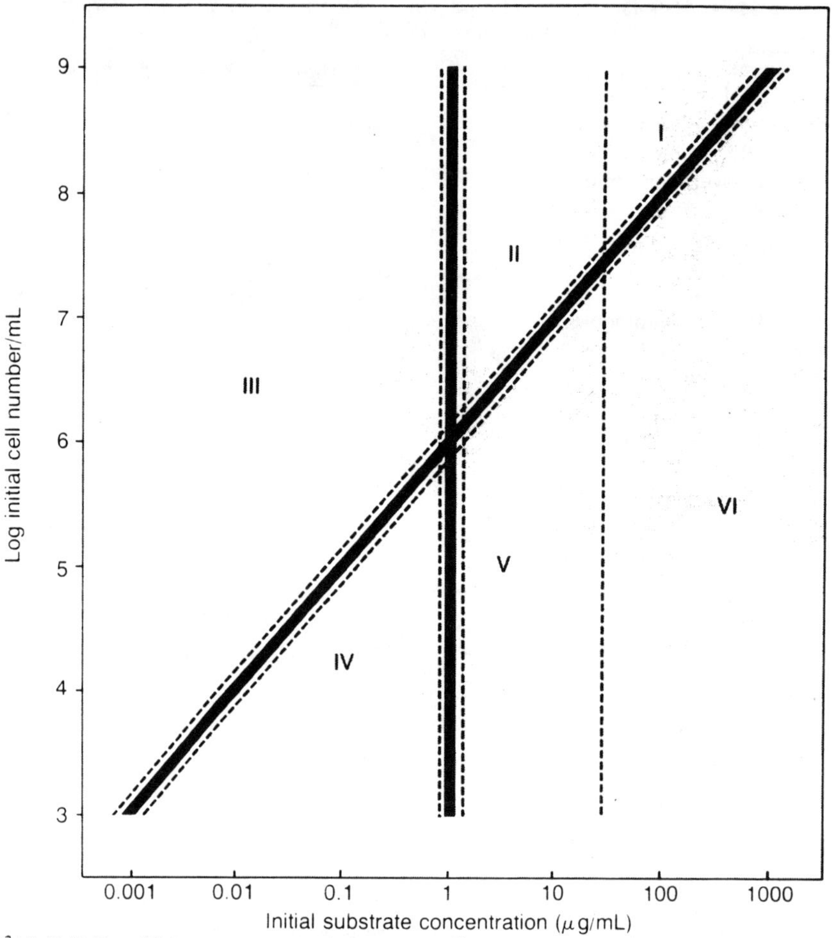

[a] I, II, III, IV, V, and VI designate regions in which zero-order, Monod (without growth), first-order, logistic, Monod (with growth), and logarithmic kinetics are expected (31)

Figure 11–5. Kinetic models as a function of initial substrate concentration and bacterial cell density. *Source:* Alexander, 1985. (Copyright American Chemical Society, 1985. Reprinted with permission.)

cially under saturation conditions. Saturation exists below the water table at all times, as well as above the water table, to some extent, after heavy rainfall.

The carbon limitation assumption is also open to question. Oxygen is probably most often the limiting substrate, except perhaps after a heavy rainfall. However, soluble carbon sources such as liquid hydrocarbons can be rate limiting when their concentration is sufficiently low. In fact, for some compounds, a minimum "threshold" concentration is necessary in order to detect any mineralization at all (Alexander, 1985). No threshold values for hydrocarbons were found but they are likely in the parts per trillion (ppt) range. The threshold for chlorinated species are generally higher, in the parts per billion (ppb) range, but vary from soil to aqueous media.

It was assumed that both a single bacterial species and a single carbon source are operative in the development of the kinetic equations. In fact, multiple species and multiple contaminant compounds are likely. This presents a problem since u_{max} is different for each species-compound combination at a given temperature. The precise kinetics of biodegradation would be extremely complex if all of the many combinations were considered. The approximation that all degradation is due to a single species is a good one if the contamination is old, however, because the population of hydrocarbons degraders will predominate by natural selection. Physiologically, the population will be sufficiently uniform that u_{max} may be used to approximate the growth for all active microbes.

Table 11–1 noted that molecular structure is an important variable in determining rates of biodegradation. This contradicts the fourth assumption, that all organic contaminants are equally degradable. Table 11–6 lists qualitatively the relative order of hydrocarbon biodegradability (Bartha and Atlas, 1977). Unfortunately, a quantitative compilation of μ_{max} values is not available. Microbe-specific μ_{max} values for individual compounds may be found in the literature. A "characteristic" $_{max}$ for theá mixture with the microbial population may be used for calculations.

Nevertheless, the multiplicity of contaminant compounds decreases with time as the more soluble and more volatile compounds are physically removed. This process is a part of "weathering" that all contaminants in the subsurface undergo. Also, for narrow "cuts" of petroleum products (e.g., gasoline) the biodegradability is similar for the majority of the volume of the mixture.

These considerations must be accounted for to practically apply theoretical kinetics to locus no. 11. This is difficult given the variability in subsurface conditions, variability of contamination from point to point, and ever-changing weather conditions. Order-of-magnitude rate estimates may be made by measuring cell density and hydrocarbon concentration and determining u_{max} of the system.

Additional Factors Affecting Biodegradation. The effects of molecular structure, contaminant concentration, bacterial number and species,

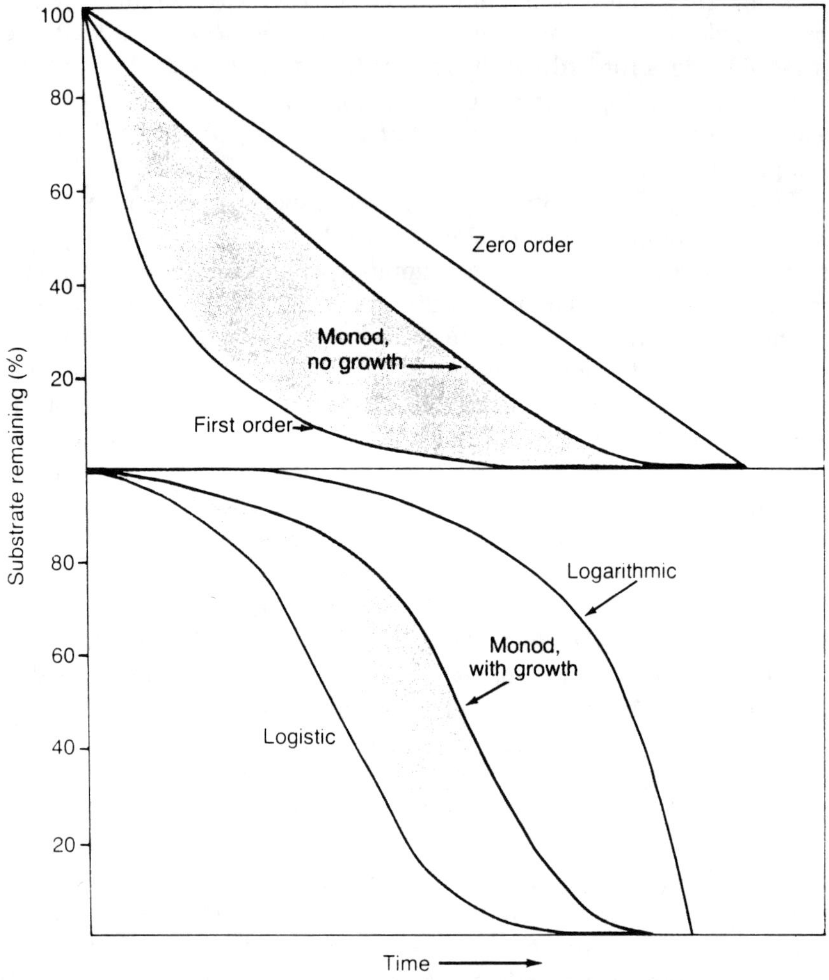

Figure 11–6. Kinetic disappearance curves for six kinetic models of biodegradation. *Source:* Alexander, 1985. (Copyright American Chemical Society. Reprinted with permission.)

and nutrient concentrations have been discussed. It is worth noting that temperature and pH also affect rates of biodegradation.

Optimum temperatures are definitely between 20°C and 40°C (Atlas, 1981). The soil temperatures of most areas of the United States fall within this range. Therefore, soil temperatures in the continental U.S. are not limiting except perhaps during winter months in northern lati-

Table 11–5. Parameters and Asymptotic Standard Deviations of Four Models of Substrate Disappearance Kinetics Fit to Data on Metabolism of Nine Concentrations of [*U*-ring-^{14}C] Benzoate by *Pseudomonas* Sp.

Actual C_o (μg/mL)	Model	Rate Constant	K_s (μg/mL)	X_o (μg/mL)	Estimated C_o (dpm/mL)	
0.010	First order	k_3 = 0.0290 ± 0.0009[a]	NA[b]	NA	1.310 ± 10	0.218 ± 0.006
0.032	First order	k_3 = 0.0293 ± 0.0015[a]	NA	NA	1.260 ± 20	0.244 ± 0.010
0.10	First order	k_3 = 0.0288 ± 0.0016[a]	NA	NA	1.270 ± 30	0.265 ± 0.008
0.32	Monod, no growth	k_1 = 0.008 ± 0.002[c]	0.13 ± 0.07	NA	1.320 ± 30	0.229 ± 0.009
1.0	Monod, no growth	k_1 = 0.017 ± 0.005[c]	0.38 ± 0.25	NA	1.320 ± 30	0.255 ± 0.013
3.2	Monod, growth	μ_{max} = 7.2 ± 0.6[d]	0.45 ± 0.58	2.1 ± 2.1	1.280 ± 30	0.232 ± 0.0001
10	Logarithmic	μ_{max} = 5.6 ± 0.6[d]	NA	3.0 ± 0.3	1.270 ± 40	NA
32	Logarithmic	μ_{max} = 7.2 ± 0.5[d]	NA	1.8 ± 0.3	1.300 ± 20	NA
100	Logarithmic	μ_{max} = 7.3 ± 2.3[d]	NA	1.8 ± 0.6	1.300 ± 10	NA

a. First-order rate constant, k_3, with units of minutes^{-1}.
b. Not applicable.
c. The constant $k_1 = \mu_{max} X_o$, having units of micrograms per milliliter per minute.
d. Rate constant for these models is μ_{max} with units of 10^{-3} min^{-1}.

Source: Simkins and Alexander (1984).

Table 11–6. Relative Biodegradability of Hydrocarbons (decreasing order)

1.	n-alkanes, C_{10}–C_{25}
2.	iso-alkanes; biodegradability decreases as degree of branching increases
3.	olefins
4.	low-molecular-weight aromatic hydrocarbons; may be toxic to microbes at high concentrations
5.	polycyclic aromatic hydrocarbons
6.	cycloalkanes; degraded by very few organisms

Source: Bartha and Atlas, 1977.

tudes. Optimum pH for biodegradation is near the neutral range or slightly basic.

11.2.4 Mobility

The mobility of organisms in the subsurface is highly dependent on the hydrology of the site of interest. Microbes move through entrainment in aqueous streams either independently or while attached to colloidal soil particles (see also locus no. 9). Thus, in the vadose zone, bacterial movement occurs almost exclusively by surface water percolation. In the saturated zone, transport occurs via groundwater movement. In both zones, microbial movement is attenuated due to the filtering effect of the soil.

Experiments have been performed by Krone et al. (1958) to model bacterial movement in the vadose zone. By applying *E. coli* to sand columns, the filtration or screening effect was measured. At first, there was a rise in bacterial content in the column effluent owing to detention of bacteria by sand. A maximum was reached followed by an asymptotic decrease to a constant effluent bacterial count. The decrease is due to the enhanced filtering effect by trapped microbes. As microbes are trapped in and clog soil pores, the effective porosity of the soil decreases. Fewer and fewer bacterial particles are able to pass until only larger sized pores remain which cannot retain microbes. If the rates of bacterial and water loading remain constant, the rate of bacterial percolation will thus also become constant.

The foregoing observations were made after applying a large homogeneous population of bacteria to the soil surface. The situation of a heterogeneous population of indigenous bacteria being mobilized by percolating groundwater is somewhat different. The bacterial concentration is likely smaller. Therefore, the enhanced filtering effect will be less evident. The size and shape of the bacteria also vary, which complicates the overall percolation rate determination. Also, there is not a constant application of new microbes. A somewhat different model is needed reflecting the intermittent nature of applied surface water and a relatively small population of microbes. Nevertheless, the model does lend some general insight into the resistance to movement of soil bacteria.

The soil structure of the site of interest also has a strong impact on microbial mobility. Smith et al. (1983) compared the bacterial filtering of intact and repacked soil columns. Up to 96 percent of bacteria applied to intact columns was recovered while only 40 to 60 percent was recovered from repacked columns. The authors concluded that flow through soil "macropores," whereby the bacteria bypass most of the adsorptive and retentive actions of the soil grains, is a common phenomenon.

Harvey et al. (1989) suggest that microbial growth may play an important role in transport through aquifers. Growth of bacteria, as they are transported through the aquifer, would partially compensate for the removal of bacteria due to sorption, biological adhesion, filtration ("straining"), predation, and lysis. An example presented by those researchers showed that without growth, the attenuation measured in one experiment would preclude bacteria from traveling more than 100 meters through a Cape Cod aquifer.

Movement below the water table may best be illustrated by the work of Bitton et al. (1974). Figure 11-7 is a "breakthrough" curve depicting the arrival of a bacterium and a much smaller virus at an observation well 500 feet from the injection pint. Filtration and dilution reduced the concentration by two orders of magnitude. The largest bacterial cells arrived at the observation point earlier than the smaller virus because they are less subject to filtration.

11.2.5 Viability

Soil microbes capable of degrading petroleum products include *Pseudomonas*, *Flavobacterium*, *Achromobacte*r, *Arthrobacter*, *Micrococcus* and *Acinetobacter*, among others. In fact, more than 200 soil microbial species have been identified that can assimilate hydrocarbon substrates (Savage et al., 1985). Total microbial counts of fertile soils range from 10^7 to 10^9 per gram of dry soil, and hydrocarbon degrader counts range from 10^5 to 10^6 per gram in soils with no history of pollution (Bossert and Bartha, 1984). Soils which have been exposed to petroleum products have bacteria counts on the order of 10^6 to 10^8 per gram. As previously mentioned, this effect is due to the selective pressure placed on a microbial population after a hydrocarbon release. Bacteria that readily utilize hydrocarbons will grow rapidly while those that do not will remain stable or decrease in numbers. Growth may be described by the Monod equation found in Section 11.2.3.

Survival times of bacterial cells have been studied for populations of pathogenic (disease causing) bacteria applied to the soil surface such as *S. typhimurium*, *E. coli* and *S. faecalis* (Bitton et al., 1983). Population decay is logarithmic with time after surface application. For indigenous microbial geni such as *Pseudomonas* and *Arthrobacter*, populations rise and fall with available substrate concentrations, temperature and moisture content. The individual lifespan of a bacterium probably ranges between two weeks and three months (Corapcioglu, 1984).

11.3 STORAGE CAPACITY IN LOCUS

11.3.1 Introduction and Basic Equations

The storage capacity of hydrocarbons by microbes in the subsurface is relatively small as compared with other loci. Equation 11.8 is derived intuitively from the foregoing discussions.

$$C_{HC}, \text{microbes} = C_{HC}, \text{soil} \cdot K_B \cdot B_{soil} \cdot M_{cell} \quad (11.8)$$

where C_{HC}, microbes = microbial hydrocarbon storage capacity; the grams of hydrocarbons in or on microbes per gram of dry soil (g/g)

C_{HC}, soil = the weight fraction of hydrocarbons in the dry soil (g/g)

K_B = the equilibrium partition coefficient of hydrocarbons between microbes and soil. It is assumed, for calculation purposes, that K_B values in Table 11–3 for bacteria and water apply equally well between bacteria and soil. Dimensions:

$$\frac{\text{(grams bacterial hydrocarbon)/(grams soil hydrocarbon)}}{\text{(gram bacterial)/(gram soil)}}$$

B_{soil} = the soil population count of hydrocarbon-degrading bacteria. It is assumed that only hydrocarbon degraders can take up (absorb or adsorb) hydrocarbons. Typical range is 10^5-10^8 cells per gram of dry soil.

M_{cell} = cellular mass; the mass of one bacterium is assumed to be a constant, estimated by Alexander (1977) to be 1.5×10^{-12} grams per cell.

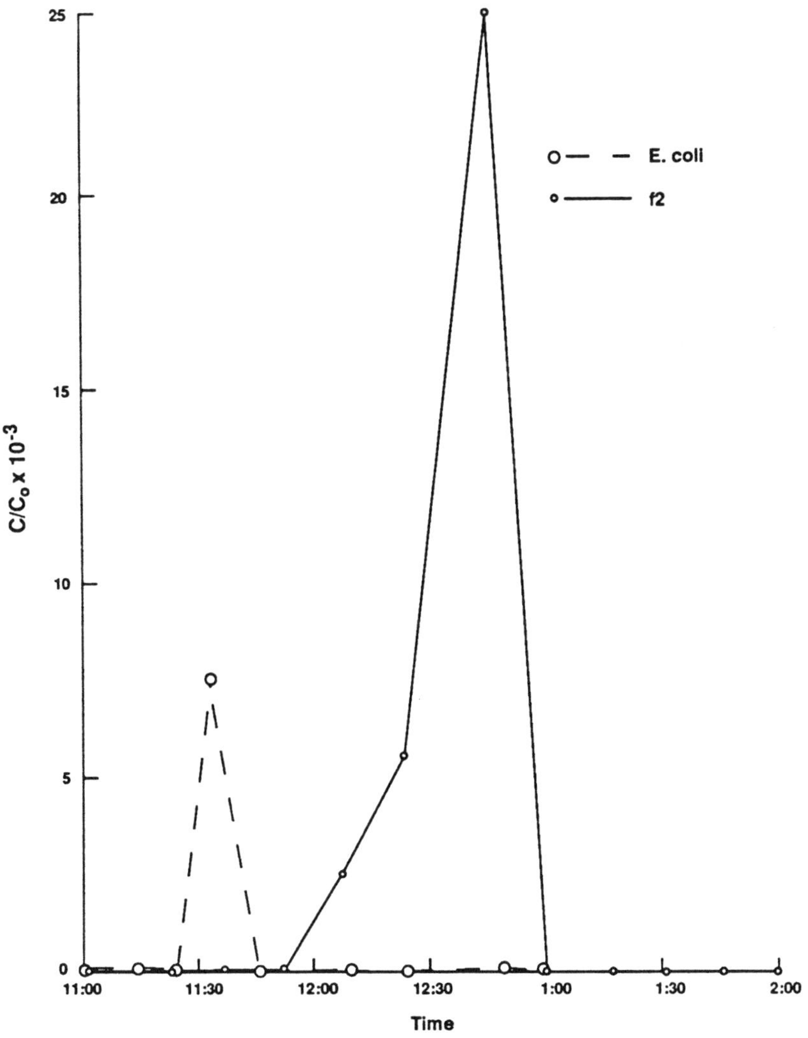

C_o = Conentration of tracer in injected water
C = Concentration measured in observation well

Figure 11–7. Breakthrough curves for *Escherichia Coli* and coliphage f2. *Source:* Bitton *et al.*, 1974. (Copyright Kluwer Academic Publishers, 1974. Reprinted with permission.)

11.3.2 Guidance on Inputs for, and Calculation of, Maximum Value

1. Hydrocarbon soil concentration; C_{HC}
   A maximum value of 10% (0.1 g/g dry soil) by weight of hydrocarbons in soil is assumed.
2. Partition coefficient, K_B
   The maximum K_B value found for hydrocarbons is 3.5×10^5 for benz[a] pyrene (Table 11-3). K_B values for lighter hydrocarbons are unavailable. This is a very high concentration factor that likely is applicable only in very dilute solutions. Somewhat arbitrarily, it is reasonable to limit cell uptake of hydrocarbons to fifty percent of the cell mass. In a soil with 10 percent mass fraction of hydrocarbons, this corresponds to K_B equal to five.
3. Bacterial soil population count, B_{soil}
   10^8 hydrocarbon degrading bacteria per gram of dry soil is the maximum value stated in Section 11.2.5. It is assumed that non-hydrocarbon degraders do not uptake hydrocarbons significantly.

11.3.3 Guidance on Inputs for, and Calculation of Average Value

1. Hydrocarbon soil concentration; C_{HC}
   An average value of 100 ppm (1.0×10^{-4} grams per gram of dry soil) is assumed.
2. Partition coefficient, K_B
   The partition coefficient for phenanthrene, 6.3×10^3, is arbitrarily chosen as an "average" value (Table 11-3).
3. Bacterial soil population count, B_{soil}
   Assuming that the likelihood of a given cell population is a Gaussian function of $\log_{10}(B_{soil})$ the "average" value of bacterial population count is 5×10^6 cells per gram of dry soil.

11.4 EXAMPLE CALCULATIONS

11.4.1 Storage Capacity Calculations

11.4.1.1 Maximum Value

Using equation 11.8 and the maximum values from Section 11.3.2, a maximum microbial hydrocarbon storage capacity, C_{HC}, may be calculated:

$$C_{HC}, \text{ microbes} = (0.1) \cdot (5) \cdot 10^8 \cdot (1.5 \times 10^{-12})$$
$$= 7.5 \times 10^{-5} \text{ grams hydrocarbon in or on microbes per gram of dry soil}$$

11.4.1.2 Average Value

Using equation 11.8 and "average" values cited in Section 11.3.3, an "average" microbial hydrocarbon storage capacity, C_{HC}, may be calculated.

$$C_{HC}, \text{ microbes} = (1.0 \times 10^{-4}) \cdot (6.3 \times 10^3) \cdot (5 \times 10^6) \cdot (1.5 \times 10^{-12})$$
$$= 4.7 \times 10^{-6} \text{ grams hydrocarbon in or on microbes per gram of dry soil}$$

11.4.2 Transformation Rate Calculations

As developed in Section 11.2.3, the form of the equation describing the rate of biodegradation of any organic contaminant depends upon both the cell population and the substrate concentration. Figure 11–5 shows the approximate ranges of initial substrate and cell number for each kinetic model.

1. Sample calculation: First order kinetics

If the substrate concentration is known to be 0.03 μg/mL and the initial cell population is approximately 10^7 microbes per mL, first order kinetics apply:

$$\frac{-dC}{dt} = k_3 C$$

For benzoate, k_3 = .0293 minutes^{-1} according to Table 11–5. Therefore, a volumetric rate of degradation can be calculated:

$$\frac{-dC}{dt} = (.0293)\,(.030) = 8.8 \times 10^{-4} \ \mu g/\text{min-mL}$$

2. Sample calculation: Monod kinetics with growth

If the substrate concentration is known to be 1.5 μg/mL, and X = 2.5 μg/mL, then Monod kinetics with growth apply. Rearranging equation 11.5 yields.

$$\frac{-dC}{dt} = \frac{\mu_{max}\, C\, X}{K_s + C}$$

For benzoate, $\mu_{max} = 7.2 \times 10^{-3}$ minutes^{-1} and $K_s = 0.45\ \mu g/mL$, according to Table 11-5. Therefore, a volumetric biodegradation rate can be calculated.

$$\frac{-dC}{dt} = \frac{(7.2 \times 10^{-3})\,(1.5)\,(2.5)}{(.45 + 1.5)}$$

$$= 0.013\ \mu g/\text{min-mL}$$

11.5 SUMMARY OF RELATIVE IMPORTANCE OF LOCUS

As stated in the introduction, the importance of locus no. 11 lies in its functions as the site of biodegradation. Biodegradation is the principal means by which subsurface organic pollutants can be mitigated naturally. The rate of natural biodegradation in the subsurface is usually severely limited by oxygen availability. The biodegradation rate can be increased by a large factor if bioremedial actions are taken.

Enhancement of biodegradation principally attempts to increase the availability of oxygen to soil and groundwater microbes. Bioremediation is a viable and advantageous option for abatement of leaked organic contaminants.

Attempts to predict rates of biodegradation by the use of mathematical models typically involve assumptions that may not be true in the underground environment. A better way to predict natural and enhanced biodegradation is actual laboratory or field microcosm studies that mimic actual environmental conditions. Decisions that affect the long term environmental condition of a leak site should consider enhanced biodegradation as a remedial alternative.

11.6 LITERATURE CITED

Alexander, M. 1977. Introduction to Soil Microbiology. 2nd Edition, John Wiley and Sons, New York.

Alexander, M. 1985. Biodegradation of Organic Chemicals. Environ. Sci. Technol., 19(2):106–111.

Alexander, M. 1986. Introduction to Soil Microbiology, John Wiley & Sons Inc., New York.

Atlas, R.M. 1981. Microbial Degradation of Petroleum Hydrocarbons: An Environmental Perspective. Microbiol. Rev., 45(1):180–209.

Bartha, R. and R.M. Atlas. 1977. The Microbiology of Aquatic Oil Spills. Adv. Appl. Microbiol., 22:225–226.

Baughman, G.L. and D.F. Paris. 1981. "Microbial Bioconcentration of Organic Pollutants from Aquatic Systems—A Critical Review." *In*: CRC Critical Reviews in Microbiology, January 1981. pp. 205–228.

Bell, J.P. and M. Tsezos. 1987. Removal of Hazardous Organic Pollutants by Biomass Adsorption. JWPCF, 59(4):191–198.

Bitton, G., S.R. Farrah, R.H. Ruskin, J. Butner and Y.J. Chou. 1983. Survival of Pathogenic and Indicator Organisms in Ground Water. Ground Water, 21(4):405–410.

Bitton, G., N. Lahav, and Y. Henis. 1974. Movement and Retention of Klebsiella Aerogenes in Soil Columns. Plant Soil, Vol 40, no. 373.

Bossert, I. and R. Bartha. 1984. The Fate of Petroleum in Soil Ecosystems. *In*: Petroleum Microbiology, R.M. Atlas (Ed.), McMillan Publishing Co., New York.

Corapcioglu, M.Y. 1984. Study and Prediction of Bacterial and Viral Contamination of Ground Water form Landfills and Septic Tank Systems. Project Completion Report, Grant No. 14-08-0001—G833/02, Dept. of Civil Engineering, University of Delaware, Newark, Delaware.

Harvey, R.W., L.H. George, R.L. Smith, and D.R. LeBlanc. 1989. Transport of Microspheres and Indigenous Bacteria through a Sandy Aquifer. Results of Natural- and Forced-Gradient Tracer Experiments. Environ. Sci. and Technol., 23:51–56.

Karickhoff, S.W., D.S. Brown and T.A. Scott. 1979. Sorption of Hydrophobic Pollutants on Natural Sediments. Water Research, Vol. 13, no. 241.

Krone, R.B., G.T. Orlob, and C. Hodgkinson. 1958. Sew. Indust. Wastes, Vol. 30, no. 1.

Leo, A.J. 1975. Calculation of Partition Coefficients Useful in the Evaluation of the Relative Hazards of Various Chemicals in the Environment. *In*: Structure—Activity Correlation in Studies of Toxicity and Bioconcentration with Aquatic Organisms, Veith, G.D. and Konasewich, D.G. (Eds.) International Joint Commission.

Savage, G.M., L.F. Dias and C.G. Golueke. 1985. Biocycle, 26(1):31–34.

Scott, C.C.L. and W.R. Finnerty. 1976. J. Bacteriology, 127(481–489).

Simkins, S. and M. Alexander. 1984. Models for Mineralization Kinetics with the Variables of Substrate Concentration and Population Density. Appl. Environ. Microbiol., 47:1299–1306.

Singer, M.E. and W.R. Finnerty. 1984. Microbial Metabolism of Straight-Chain and Branched Alkanes. *In*: Petroleum Microbiology, R. Atlas (Ed.) p. 1–60.

Smith, J.H., W.R. Mabey, N. Bohonos, B.R. Holt, S.S. Lee, T.W.

Chou, D.C. Bomberger, and T. Mill. 1978. Environmental Pathways of Selected Chemicals in Freshwater Systems. Part II, Laboratory Studies. U.S. EPA, Athens, GA.

Smith, M.S., G.W. Thomas, and R.E. White. 1983. Movement of Bacteria through Macropores to Ground Water. Kentucky Water Resource Institute, Lexington, KY. NTIS PB83-246546.

SECTION 12

Contaminants Dissolved in the Mobile Pore Water of the Unsaturated Zone

CONTENTS

List of Tables 310
List of Figures 310
12.1 Locus Description 313
12.1.1 Short Definition 313
12.1.2 Expanded Definition and Comments 313
12.2 Evaluation of Criteria for Remediation 313
12.2.1 Introduction 313
12.2.2 Mobilization/Remobilization 314
12.2.2.1 Partitioning into/onto Mobile Phase 314
12.2.2.2 Transport of/with Mobile Phase 314
Introduction 314
Definition of Mobile Water 315
Permanent Wilting Point 316
Porosity and Field Capacity 317
Hysteresis 319
Darcy's Law 319
Soil Moisture Profiles 322
Macropore Flow 324
Enhancement of Mobility 325
12.2.3 Fixation 326
12.2.3.1 Partitioning onto Immobile Phase 326
Impermeable Cover 326
Evaporation 326
Vacuum Extraction/Wind 326
Transpiration 327
12.2.3.2 Other Fixation Approaches 327
Freezing 327
Sorption 327
12.2.4 Transformation 327
12.2.4.1 Biodegradation 327

12.2.4.2 Chemical Oxidation 328
12.3 Storage Capacity in Locus 328
12.3.1 Introduction and Basic Equations 328
12.3.2 Guidance on Inputs for, and Calculation of, Maximum Value 328
12.3.2.1 Total Porosity 328
12.3.2.2 Water-Filled Porosity 329
12.3.2.3 Solubility 329
12.3.3 Guidance on Inputs for, and Calculation of, Average Value 329
12.3.3.1 Total Porosity 329
12.3.3.2 Water-Filled Porosity 329
12.3.3.3 Solubility 329
12.4 Example Calculations 329
12.4.1 Storage Capacity Calculations 329
12.4.1.1 Maximum Quantity 329
12.4.1.2 Average Quantity 330
12.4.2 Transport Rate Calculation 331
Depth of Infiltration Water Front 331
12.5 Summary of Relative Importance of Locus 331
12.5.1 Remediation 331
12.5.2 Loci Interactions 332
12.5.3 Information Gaps 333
12.6 Literature Cited 333
12.7 Other References 334

TABLES

12-1 Representative Porosity Values 319
12-2 Representative Values for Hydraulic Parameters 322

FIGURES

12-1 Schematic Cross-Sectional Diagram of Locus No. 12—Contaminants Dissolved in the Mobile Pore Water of the Unsaturated Zone 315
12-2 Schematic Representative of Important Transformation and Transport Processes Affecting Other Loci 316
12-3 Water Film Surrounding a Wet Solid Particle 317
12-4 Water-Holding Properties of Various Soils on the Basis of Their Texture 318
12-5 Moisture Content vs. Capillary Pressure Head 321

12–6 Changes of Soil Moisture with Depth During Infiltration into Sand with an Initial Moisture Content of 0.5 Percent 323
12–7 Soil Water Profiles During Drainage of a Silt Loam for Various Durations 325
12–8 Depth of Infiltration vs. Moisture Content at One Day 332

SECTION 12

Contaminants Dissolved in the Mobile Pore Water of the Unsaturated Zone

12.1 LOCUS DESCRIPTION

12.1.1 Short Definition

Contaminants dissolved in the mobile pore water of the unsaturated zone.

12.1.2 Expanded Definition and Comments

The definition of this locus includes water occupying a large fraction of the total porosity of the soil at certain times or places (e.g., after a heavy rainfall or above the capillary fringe). The pore water moves because of gravity and capillary tension, carrying dissolved contaminants with it. Individual contaminants, however, will be retarded by sorption to soil surfaces.

This locus excludes thin films of tightly-bound water of very low mobility; this situation is discussed in locus no. 3. Figures 12–1 and 12–2 present a schematic cross-sectional diagram of locus no. 12 and a schematic representation of the transformation and transport processes affecting other loci, respectively.

12.2 EVALUATION OF CRITERIA FOR REMEDIATION

12.2.1 Introduction

The evaluation of criteria for remediation of dissolved liquid contamination in locus no. 12 is based on the physical and chemical processes which control the movement of the contaminated mobile water through the unsaturated zone. These physical and chemical factors (e.g., hydraulic conductivity, capillary tension, and solubility) are discussed in Sections 6.2.2 and 6.2.3. The effects of these physical and chemical properties on the mobilization and fixation of contaminated water in locus no. 12 are discussed below.

12.2.2 Mobilization/Remobilization

12.2.2.1 Partitioning into/onto Mobile Phase

Contaminants in the water of locus no. 12 are, by definition, mobile. Contaminants may enter the mobile water of locus no. 12 by the dissolution of pure liquid contaminant in the unsaturated zone (locus no. 6). In the case of gasoline, those compounds which dissolve more readily into water are oxygenated additives (e.g., ethanol, methanol, MTBE), phenols, and simple aromatic hydrocarbons (e.g., benzene, toluene, and xylenes) (W. Lyman, pers. comm., 1987). Once dissolved into the mobile water, these contaminants move with the mobile water as a single phase. The process of dissolution of liquid contaminant into water is discussed in locus no. 7.

Mobile water of locus no. 12 may become contaminated through contact with polluted water in locus no. 3 as contaminated and pure waters mix. Contact with soil air (locus no. 1) containing volatilized liquid contaminant may also contaminate the mobile water of locus no. 12. The compounds which most readily partition from the air phase to the mobile water phase are those compounds with lower Henry's law constants (e.g., benzene, toluene, xylenes); those compounds with high Henry's law constants (e.g., 2-methylhexane, n-pentane, 2,2,5,5-tetramethylhexane) present in the mobile water of locus no. 12 partition more readily into the soil air of locus no. 1. The process of air/water partitioning is discussed in detail in locus no. 3.

Partitioning to the mobile water of locus no. 12 can result from contact with soil particles onto which the liquid contaminant adheres (locus no. 2) or into which the liquid contaminant has diffused (locus no. 10). Contact with soil particles serves also as a mechanism for removal of contaminants from the mobile water. Finally, the mobile water of locus no. 12 can become polluted by contact with contaminated water adhering to the soil particles (locus no. 3), with water containing contaminated microbiota (locus no. 11), or with water containing contaminated colloids (locus no. 9).

12.2.2.2 Transport of/with Mobile Phase

Introduction. Locus no. 12 concerns the movement of contaminants dissolved in the mobile water of the unsaturated zone. This discussion

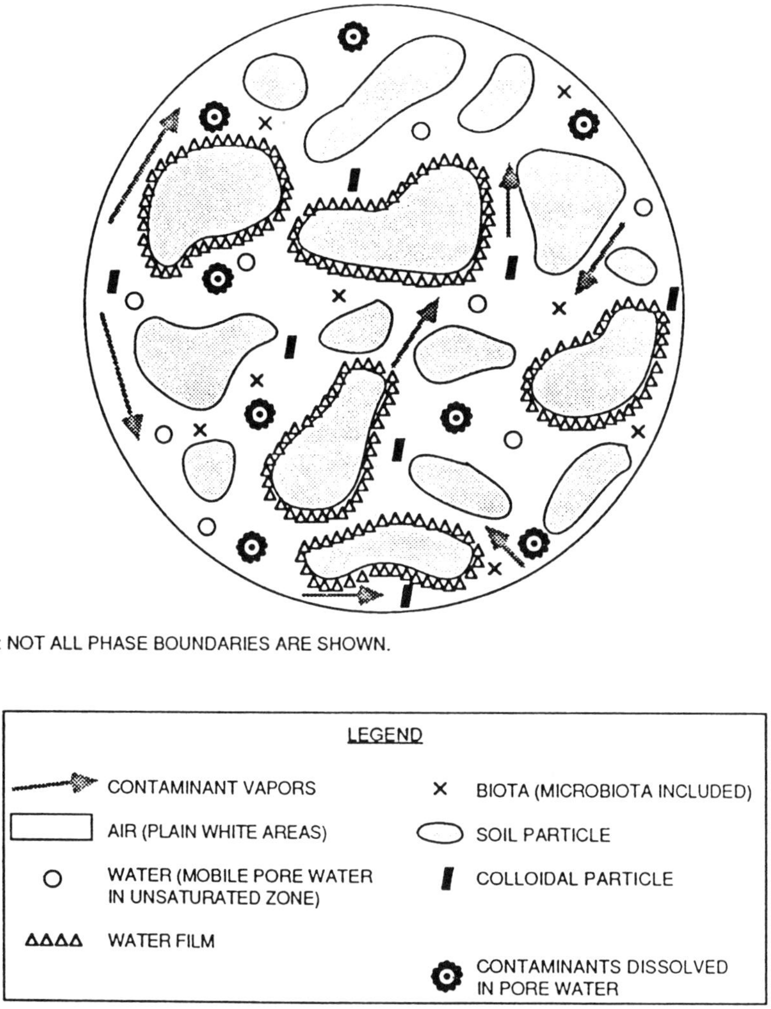

Figure 12–1. Schematic cross-sectional diagram of locus no. 12–contaminants dissolved in the mobile pore water of the unsaturated zone.

assumes that there are two immiscible fluid phases (i.e., air and water) present in the unsaturated zone. The definition of mobile water in the unsaturated zone and the influences on its movement in the unsaturated zone will be discussed below.

Definition of Mobile Water. In the unsaturated zone, a continuum in capillary tension exists between water held at zero capillary tension (i.e.,

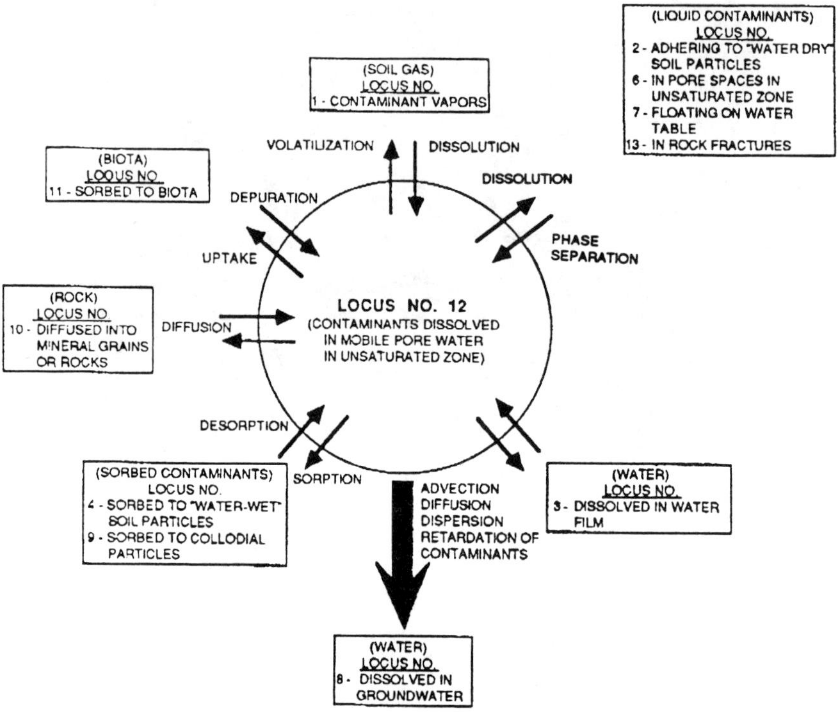

Figure 12–2. Schematic representation of important transformation and transport processes affecting other loci.

gravity drainage) and water held by very high capillary tensions to soil grain surfaces (Figure 12–3).

Water, held at gravity moves freely downward through the unsaturated zone; water held by high capillary tension is relatively immobile. Because of the range in capillary tension, a continuum in mobility also exists between water draining by gravity and immobile water held by capillary tension.

As a consequence of this continuum in water mobility, it is difficult to assign a natural boundary between mobile and immobile water. Nevertheless, an arbitrary lower limit to the volume of mobile water can be defined, below which the volume of water can be considered relatively immobile in the unsaturated zone.

Permanent Wilting Point. A convenient point for this arbitary boundary between mobile and immobile water is the permanent wilting point. The permanent wilting point is the soil moisture condition at which

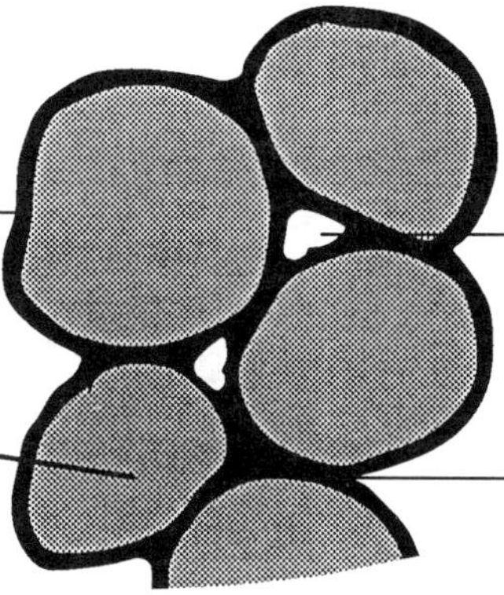

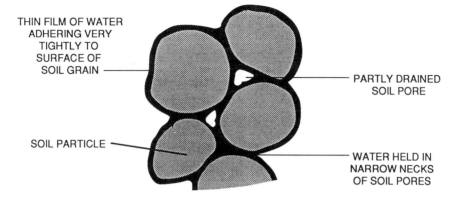

Figure 12–3. Water film surrounding a wet solid particle. *Source:* Adapted from Dunne and Leopold, 1978.

plants wilt. This occurs when plants can no longer remove water from the soil; the remaining water is too tightly held by the soil particles. Since most plants can exert up to 15 atm of capillary tension (155.1 m H_2O) (Dunne and Leopold, 1978), the volume of water remaining in the soil at capillary tensions equal to or greater than 15 atm will be considered immobile (locus no. 3). The immobile water occurs as a thin film adhering to the surface of soil grains (Figure 12–3).

Because of natural variation in size and distribution of grains of undisturbed soils, the volume of water held at the wilting point varies with soil type (Figure 12–4). Figure 12–4 indicates that sandy soils hold about five percent water by volume at the wilting point; at the fine end of the grain-size spectrum, clayey soils hold as a much as twenty five percent water by volume. The increase in water content with decreasing grain-size occurs because the smaller pores in finer soils hold water more firmly. The dependence of capillary tension on saturation is discussed in Section 6.2.2.1 of locus no. 6.

Porosity and Field Capacity. Figure 12–4 also shows the variation in porosity and field capacity with soil type; additional porosity data are given in Table 12–1. Rainwater infiltrates into the soil and fills empty pore spaces. The maximum amount of water that the soil can hold equals the non-solid volume of the soil, i.e., the total porosity; at this point the soil is saturated. After the rainstorm, the moisture content declines from saturation.

Drainage of the soil moisture under gravity continues until capillary tension balances gravity. At this point, drainage ceases and the soil is at

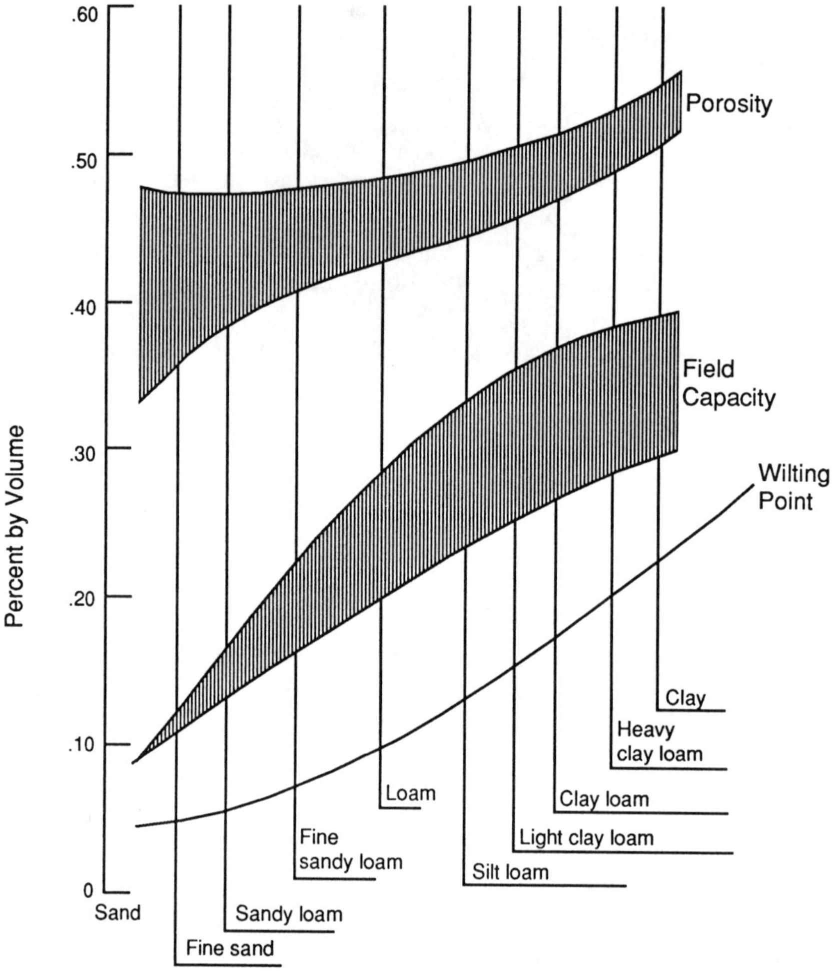

Figure 12–4. Water-holding properties of various soils on the basis of their texture. *Source:* Adapted from Dunne and Leopold, 1978.

field capacity. Porosity and field capacity define upper limits to the volume of mobile water in locus no. 12.

The lower limit on the volume of mobile water was arbitrarily set at 15 atm; the maximum volume of mobile water is defined by the porosity of the soil. These limits on the volume of mobile water range widely from 5 to 48.7 percent for sandy soils; the range of 20 to 55 percent is narrower for clayey soils (Figure 12–4).

The condition of maximum mobile water content occurs while rainwater is infiltrating into the soil and is transient. After the rainstorm has

Table 12–1. Representative Porosity Values

Material	Porosity, Percent	Material	Porosity Percent
Gravel, coarse	28[a]	Loses	49
Gravel, medium	32[a]	Peat	92
Gravel, fine	34[a]	Schist	38
Sand, coarse	39	Siltstone	35
Sand, medium	39	Claystone	43
Sand, fine	43	Shale	6
Silt	46	Till, predominantly silt	34
Clay	42	Till, predominantly sand	31
Sandstone, fine-grained	33	Tuff	41
Sandstone, medium-grained	37	Basalt	17
Limestone	30	Gabbro, weathered	43
Dolomite	26	Granite, weathered	45
Dune sand	45	Granite, weathered	45

a. These values are for repacked samples; all others are undisturbed.
Source: Pettyjohn and Haunslow (1982).

ceased, the soil water drains under gravity, until field capacity is reached. The water remaining in the soil at field capacity may still move under gravity and capillary tension. The range in mobile water volume between field capacity and at 15 atm limit for sandy soils is narrow, 5 to 9 percent; for clayey soils, the range is from 25 to 37 percent (Figure 12–4).

Hysteresis. The relationship between saturation of a soil and the capillary tension at which the water is held is not unique, but depends on the wetting history of the soil. More water is held in the soil during drainage than in wetting at the same capillary tension (Figure 12–5). This phenomenon is known as hysteresis.

Hysteresis is most pronounced in coarse-grained soils at low capillary tensions and has been attributed to several possible causes (Hillel, 1980). Although the discussion of the causes of hysteresis is beyond the scope of this report, the phenomenon is important because it affects unsaturated hydraulic conductivity as well as the moisture content of the soil.

Darcy's Law. The flow of water and transport of solutes in the unsaturated zone is governed by Darcy's law. This empirical law relates the discharge of water through a porous medium to the product of the hydraulic conductivity of the soil and the hydraulic gradient acting on the water. Hydraulic conductivity measures the resistance of a porous medium to water flow and depends on physical properties of the porous medium and of water; hydraulic gradient depends on gravity and pressure. A detailed discussion of Darcy's law and its variables is presented in Section 6.2.2.2 of locus no. 6.

Numerical and analytical models of water flow through the porous medium of soils are available which predict the transport of an infiltrating front or draining of water front through a soil. The three-dimensional equation for unsaturated water flow through a porous medium is (Hillel, 1980):

$$\frac{\partial \Theta}{\partial t} = -\frac{\partial}{\partial x}\left(K\frac{\partial \psi}{\partial x}\right) - \frac{\partial}{\partial y}\left(K\frac{\partial \psi}{\partial y}\right) - \frac{\partial}{\partial z}\left(K\frac{\partial \psi}{\partial z}\right) + \frac{\partial K}{\partial z} \qquad (12.1)$$

where Θ = moisture content (cm^3/cm^3)
t = time (sec)
x,y,z = space coordinates (cm)
ψ = capillary tension (cm H_2O)
K = hydraulic conductivity (cm/sec)

This is a difficult equation to solve in its three-dimensional form because it is non-linear and computationally cumbersome. Typically, a one-dimensional vertical form is solved (Gilliam et al., 1979; Giesel et al., 1972). The concern, however, is to understand how water moves vertically downward, since this water may recharge aquifers at depth. A relatively simple, analytical model can be derived to predict the vertical movement of water with time. This model is based on the one-dimensional continuity equation (Hillel, 1980):

$$\frac{\partial \Theta}{\partial t} = -\frac{\partial q}{\partial z} \qquad (12.2)$$

where Θ = moisture content (cm^3/cm^3)
q = discharge per unit area (cm/sec)
z = depth (cm)
t = time (sec)

Using the definition of a partial differential equation (Swokowski, 1975) and re-arranging equation 12.2:

$$\frac{dz}{dt} = \frac{q}{\Theta} \qquad (12.3)$$

The discharge of water through a porous medium is described by Darcy's law:

$$q = (k \cdot k_r/\mu) \cdot (dP/dz + \rho \cdot g \cdot dh/dz) \qquad (12.4)$$

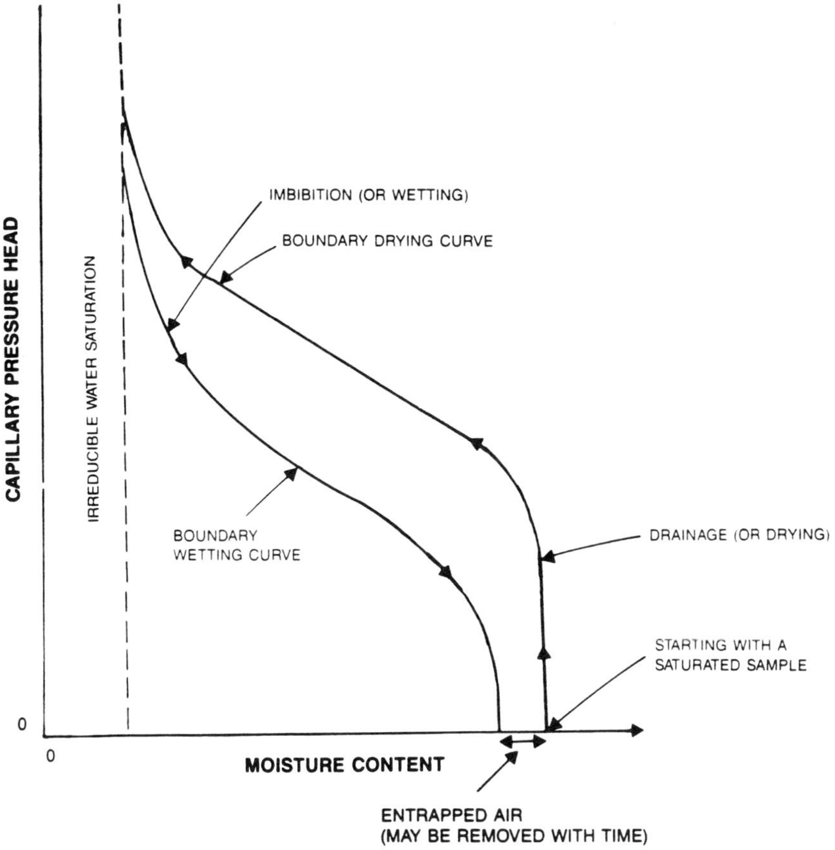

Figure 12–5. Moisture content vs. capillary pressure head. *Source:* Bear, 1979. (Copyright McGraw-Hill Inc., 1979. Reprinted with permission.)

where k = intrinsic permeability (cm^2)
k_r = relative permeability (dimensionless)
ρ = density of water, $\rho = 1.0\ g/cm^3$
g = gravity, $g = 980.7\ cm/sec^2$
μ = dynamic viscosity (g/cm.sec)
dP/dz = vertical pressure gradient ($g.cm.sec^{-2}/cm^3$)
dh/dz = vertical head gradient (cm/cm)

These parameters are described in detail in locus no. 6. If no pressure gradient exists, and a unit head gradient is assumed, equation 12.4 becomes:

Table 12–2. Representative Values for Hydraulic Parameters

Soil Texture	b	Ψ_s (cm)	Θ_s (cm^3/cm^3)	K_s (cm/min)
Sand	4.05	12.1	0.395	1.056
Loamy sand	4.38	9.0	0.410	0.938
Sandy loam	4.90	21.9	0.435	0.208
Silt loam	5.30	78.6	0.485	0.0432
Loam	5.39	47.8	0.451	0.0417
Sandy clay loam	7.12	29.9	0.420	0.0378
Silty clay loam	7.75	35.6	0.477	0.0102
Clay loam	8.52	63.0	0.476	0.0147
Sandy clay	10.4	15.3	0.426	0.0130
Silty clay	10.4	49.0	0.492	0.0062
Clay	11.4	40.5	0.482	0.0077

Source: Adapted from Clapp and Hornberger (1978).

$$q = \rho \cdot g \cdot k \cdot k_r/\mu = K_s \cdot k_r \qquad (12.5)$$

where K_s = saturated hydraulic conductivity (cm/sec)

An expression for the relative permeability is reported by Clapp and Hornberger (1978):

$$k_r = (\Theta/\Theta_s)^{2b+3} \qquad (12.6)$$

where Θ_s = moisture content at saturation (cm^3/cm^3)
b = grain-size distribution parameter (dimensionless)

Selected values of Θ_s,b,ψ_s, and K_s are given in Table 12–2. Substituting equation 12.5 and 12.6 into equation 12.3 gives:

$$dz/dt = (K_s/\Theta) \cdot (\Theta/\Theta_s)^{2b+3} \qquad (12.7)$$

which may be integrated to give the depth, z(t), of the infiltrating water front at time t:

$$z(t) = (K_s \cdot t/\Theta) \cdot (\Theta/\Theta_s)^{2b+3} \qquad (12.8)$$

This simple, analytical expression assumes that capillary forces are negligible compared to gravity transport in the downward vertical movement of water. This assumption will be most reasonable for a moderately well-saturated, sandy soil.

Soil Moisture Profiles. As described in detail in section 6.2.2.2, the hydraulic conductivity of the soil increases with moisture content to a maximum at saturation. As rainwater infiltrates into a soil, the moisture

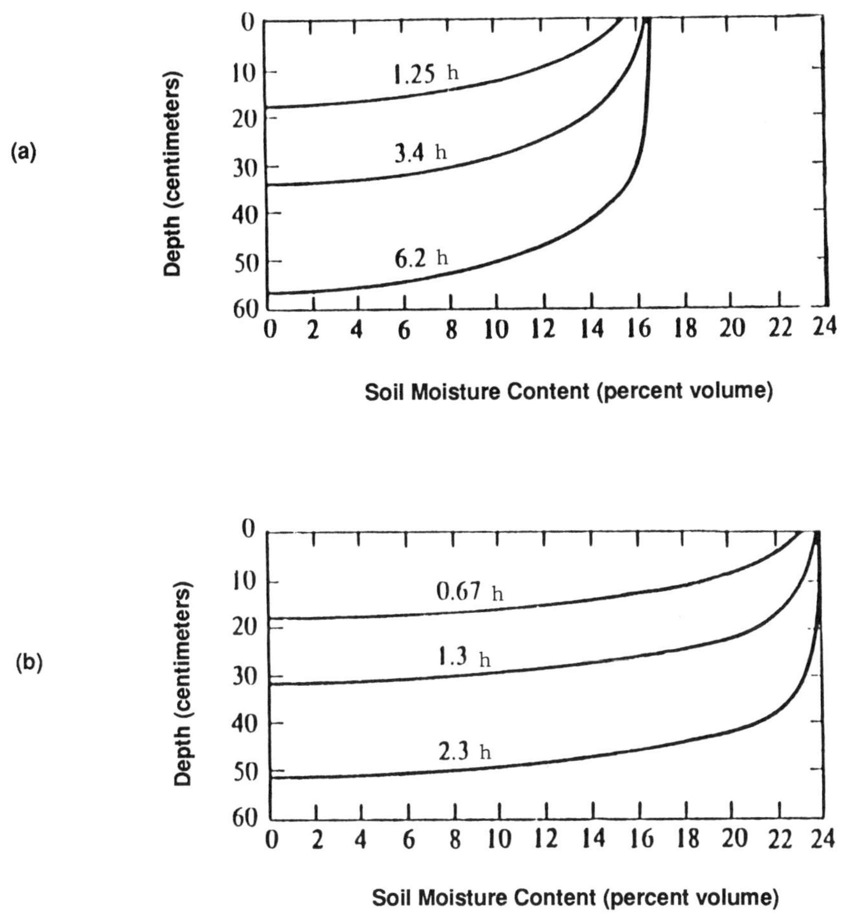

NOTE: Rainfall intensity in (b) is 3.8 times greater than in (a). As a result, the soil moisture content had to be greater (about 24 percent in (b) versus 16 percent in (a)) in order to transmit the water at the applied rate (saturated moisture content = 38 percent).

Figure 12–6. Changes of soil moisture with depth during infiltration into sand with an initial moisture content of 0.5 percent. *Source:* Rubin, 1966.

content increases as pores fill with water. At complete saturation, the infiltration rate is limited by the saturated hydraulic conductivity.

Figure 12–6 shows the change in soil moisture with depth during infiltration of water into a sand (Dunne and Leopold, 1978). The isochrones (i.e., lines of equal time) indicate that the time for water to reach a given depth decreases with increasing moisture content; a wetter soil allows faster infiltration.

The effect of rainfall intensity is seen also in Figure 12–6. Figure 12–6b was produced at a higher rainfall intensity than Figure 12–6a. The result is that the soil in Figure 12–6b holds more water and percolation is faster. If the rainfall intensity exceeds the saturated hydraulic conductivity, however, ponding of water on the soil surface and runoff occurs.

After infiltration ceases, percolation slows as the moisture is redistributed throughout the soil depth (Figure 12–5). Gravity drainage declines with time, until field capacity is reached (Figure 12–7).

Macropore Flow. It has been noted that natural soils commonly contain preferential flowpaths, known as macropores, which are formed by a variety of processes including activity of soil fauna, plant roots, natural soil piping, and cracks and fissures (Beven and Germann, 1982). Beven and Germann (1981) suggested that a macropore would be any soil pore where the capillary tension was larger than -0.1 kPa (0.1 cm H_2O); the diameter of this arbitrary macropore would be larger than 0.3 cm. Other definitions of macropores are reported by Beven and Germann (1982).

The presence of macropores in soil allows the transmission of water very quickly, by-passing the matrix of soil grains through which water moves more slowly by percolation. An example of the rapid flow of macropores was observed during a ponded infiltration experiment on a forest soil, in which the flow of water from an empty tree root located 2 meters from the measurement site occurred within 3 minutes after initiation of ponding (Levy, 1987). A common field method of determining the macropore and matrix flow of a natural soil is the measurement of infiltration of water into the soil under conditions of ponded water and under capillary tension (Watson and Luxmoore, 1986).

Macropores can transmit large volumes of water through the soil, even though they may comprise only a small fraction of the total soil volume. For example, measurements of macropore flow in a forest soil by Watson and Luxmoore (1986) indicate that 73 percent of the water ponded on the soil surface was carried by macropores. Calculations suggest that these macropores have a diameter of >0.1 cm and comprise less than 0.1 percent of the soil volume (Watson and Luxmoore, 1986).

The presence of macropores can reduce infiltration of water through the porous medium of the soil from the surface; a remedial program of artificial recharge may fail to clean the contaminated water from the soil as a result. The phenomenon of macropore flow is likely to occur during

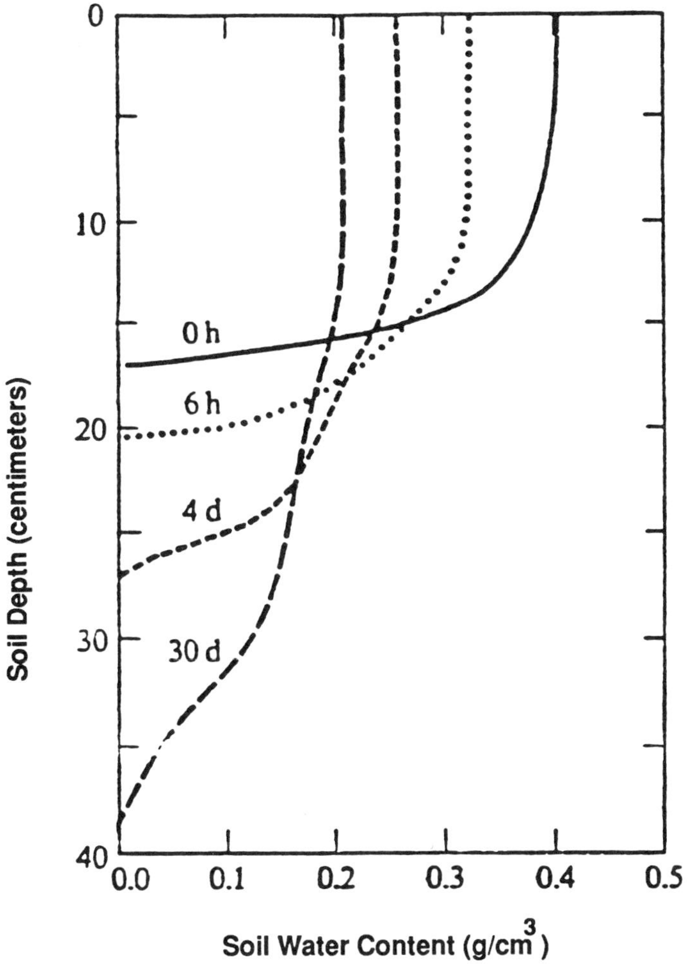

Figure 12–7. Soil water profiles during drainage of a silt loam for various durations, hours (h) or days (d). *Source:* Biswas et al., 1966.

high intensity rainstorms, increasing in importance to flow as the rainfall intensity increases.

Enhancement of Mobility. The mobility of water in locus no. 12 cannot be enhanced easily by changing the physical properties of the porous medium or of water. The reasons for this limitation is discussed in Section 6.2.2.3. The mobility of water in locus no. 12 can be enhanced, however, by increasing the moisture content of the unsaturated zone.

An increase in saturation in locus no. 12 by thc addition of clean water increases the hydraulic conductivity. The increase in hydraulic conductivity allows contaminated water to percolate more readily. Enough water must be added as clean infiltrate to enhance the mobility of the water present in locus no. 12. Too much clean water added to the soil in a short time period, however, will initiate macropore flow; much water may by-pass the contaminated soil matrix with no resulting remediation.

12.2.3 Fixation

12.2.3.1 Partitioning onto Immobile Phase

Several methods are possible alternatives by which the contaminants dissolved in the mobile water of locus no. 12 may become immobilized. These alternatives are described below in order of decreasing feasibility.

Impermeable Cover. Of the various fixation alternatives, emplacement of an impermeable cover over the land surface above a contaminated site is the most easily effected. The presence of the impermeable cover would prevent recharge of infiltrating rainwater to the unsaturated zone. Interdiction of recharge to locus no. 12 would prevent an increase in hydraulic conductivity with increased moisture content, preventing any further downward migration of contaminated water.

Evaporation (Details in Section 3). The reduction of hydraulic conductivity, and subsequent immobilization of contaminated water, can also be accomplished by the removal of water by evaporation. The removal of water from locus no. 12 by evaporation increases the capillary tension exerted by the porous medium on the remaining water. The remaining water in locus no. 12 will be held more firmly and will be consequently more immobile.

Evaporation of water from locus no. 12 naturally occurs if the soil surface is open to the atmosphere. Evaporation is most pronounced in sandy soils, and is largest during summer months when monthly average temperatures are the highest. Because of its dependence on temperature, evaporation could be enhanced by increasing the water temperature of locus no. 12; this enhancement may be accomplished by the pumping of dry, heated air into the ground through wells.

Vacuum Extraction/Wind. A more practical method of enhancing evaporative water loss from locus no. 12 would be to apply vacuum extraction to the soil, removing water-saturated air. More water evaporates into the new, dry soil air which replaces the evacuated, moist air. Evaporation increases, and the remaining water in locus no. 12 is immobilized.

Forced evaporation may result in the deposition of calcium carbonate and other naturally occurring salts in the upper sub-surface of the soil. Precipitation of these salts from evaporating water may reduce the hydraulic conductivity of the soil sub-surface with time. This phenomenon occurs naturally, producing a very hard, impermeable crust in salty, desert soils known as a caliche. The presence of caliche in a soil can prevent infiltration of water downward, acting in the same manner as an impermeable cover.

Transpiration. Moisture is removed from locus no. 12 by the plant cover on the soil surface, causing an increase in capillary tension exerted on the remaining water. The hydraulic conductivity of locus no. 12 could be reduced by covering the site with plant species that transpire large quantities of water (e.g., alfalfa, soybeans) (Rosenberg et al., 1983).

12.2.3.2 Other Fixation Approaches

The following fixation approaches are suggested for immobilization of contaminated water in locus no. 12, despite their apparent impracticality.

Freezing. Very simply, if the mobile water of locus no. 12 could be frozen, it would be effectively immobilized. Large-scale freezing of the unsaturated zone in areas other than in tundra must be highly unfeasible.

Sorption. It is unlikely that an enhancement of the sorptive properties of the porous medium of locus no. 12 could be achieved.

12.2.4 Transformation

12.2.4.1 Biodegradation (Details in Section 11)

Of all loci, locus no. 12 probably holds the greatest potential for biodegradation. Due to its location in the vadose zone and contact with infiltrated rainwater, the oxygen content is higher than in any other locus. Because the phase is aqueous and mobile, transport of all substrates (oxygen, carbon, minerals) is maximized. Nevertheless, oxygen or nitrogen concentration may still be rate-limiting. Since concentrations are subject to rainfall, wide variations in oxygen or other substrates may be encountered. Rates of natural biodegradation may be enhanced by the application of oxygenated waters containing hydrogen peroxide and nutrients through injection wells.

12.2.4.2 Chemical Oxidation (Details in Section 3.2.4.2)

Soil water provides the optimum mechanism for chemical oxidation of contaminants since this media is subject to microbial activities, is oxygenated by precipitation entering the system and by diffusion, and comes in contact with catalysts in the soil.

12.3 STORAGE CAPACITY IN LOCUS

12.3.1 Introduction and Basic Equations

The mass of liquid contaminant dissolved into the vadose zone water is calculated from:

$$m_w = S \cdot \Theta_w / 10^3 \tag{12.9}$$

where m_w = mass per unit volume of liquid contaminant dissolved in the vadose zone water (kg/m^3)
S = solubility of liquid contaminant in water (mg/L)
Θ_w = water-filled porosity (cm^3/cm^3)

The water-filled porosity in locus no. 12 is the difference between the total and air-filled porosity:

$$\Theta_w = \Theta_t - \Theta_a \tag{12.10}$$

where Θ_t = total soil porosity (cm^3/cm^3)
Θ_a = air-filled porosity (cm^3/cm^3)

12.3.2 Guidance on Inputs for, and Calculations of, Maximum Value

12.3.2.1 Total Porosity, Θ_t.

- Theoretical maximum porosity for sandy soils is 47.6%.
- User should refer to Table 12–1 or 12–2 for representative total porosity values for a variety of soils.

12.3.2.2 Water-Filled Porosity, Θ_w.

- The maximum quantity of dissolved liquid contaminant held in locus no. 12 occurs when the water-filled porosity is at a maximum; the air-filled porosity is at a minimum. It is assumed that the water content in the soil is the total porosity of the soil.
- Water-filled porosity ranges from 0 to Θ_t.

12.3.2.3 Solubility, S.

- User should supply the solubility of liquid contaminant.
- For gasolines with no oxygenated additives, a value of 200 mg/L for solubility is suggested. Refer to locus no. 3 for detailed information on solubility.

12.3.3 Guidance on Inputs for, and Calculations of, Average Value

12.3.3.1 Total Porosity, Θ_t.

- User should refer to Table 12–1 or 12–2 for representative total porosity values for a variety of soils.

12.3.3.2 Water-Filled Porosity, Θ_w.

- The average quantity of dissolved liquid contaminant held in locus no. 12 occurs when the water-filled porosity is at field capacity. Suggest using 0.09 for water-filled porosity for a sandy soil at field capacity.
- Water-filled porosity ranges from 0 to Θ_t.

12.3.3.3 Solubility, S.

- User should supply the solubility of liquid contaminant.
- For gasolines with no oxygenated additives, a value of 100 mg/L for solubility is suggested as half the value reported in Section 12.3.2. Refer to locus 3 for detailed information on solubility.

12.4 EXAMPLE CALCULATIONS

12.4.1 Storage Capacity Calculations

12.4.1.1 Maximum Quantity

For this sample calculation, a sandy soil is assumed. Since we wish to maximize the amount of dissolved contaminant liquid, we maximize the water content in the soil at the total porosity of the soil. For sandy soils, the water content at the total porosity is about 39.5% (Clapp and Hornberger, 1978). The air-filled porosity is zero.

$$\Theta_W = \Theta_t - \Theta_a$$
$$\Theta_t = 0.395$$
$$\Theta_a = 0.00$$
$$\Theta_w = 0.395$$

In computing the maximum quantity of dissolved liquid in locus no. 12, it is assumed that the solubility is at a maximum. A value of 200 mg/L for S as reported in Section 12.3.2 for gasoline will be used to calculate the maximum mass per unit volume of dissolved contaminant liquid held in locus no. 12 from equation 12.9.

$$m_w = S \cdot \Theta_w/10^3$$
$$m_w = (200 \text{ mg/L}).(0.395)/10^3$$
$$m_w = 0.08 \text{ kg/m}^3$$

12.4.1.2 Average Quantity

A sandy soil is also assumed for the calculation of the average quantity of dissolved contaminant liquid held in locus no. 12. It is assumed that the water-filled porosity is at field capacity. For a sandy soil, the water-filled porosity at field capacity is about 9%. As shown in Table 12–2, the total porosity is 39.5%.

$$\Theta_w = \Theta_t - \Theta_a$$
$$\Theta_t = 0.395$$
$$\Theta_a = 0.09$$
$$\Theta_w = 0.305$$

In computing the average quantity of dissolved liquid in locus no. 12, it is assumed that the solubility is half that reported in Section 12.3.2. A value of 100 mg/L for S will be used to calculate the average mass per unit volume of dissolved contaminant liquid held in locus no. 12:

$$m_w = S.\Theta_w/10^3$$
$$m_w = (100 \text{ mg/L}).(0.305)/10^3$$

$$m_w = 0.03 \text{ kg/m}^3$$

12.4.2 Transport Rate Calculation

Depth of Infiltrating Water Front. The depth of an infiltrating water front can be estimated using equation 12.8.

$$z(t) = \left(\frac{K_s t}{\Theta}\right) \left(\frac{\Theta}{\Theta_s}\right)^{2b+3} \tag{12.11}$$

This equation was derived under the following assumptions: no pressure gradient; unit hydraulic head; negligible capillary forces; and, uniform vertical moisture distribution. If time is one day, and appropriate values from Table 12–2 are used, values for the depth of an infiltrating water front for different types of soil can be found. Figure 12–8 shows graphically the effect of moisture content on the depth of infiltration for three soil types. As can be expected, sand allows for deeper penetration than the other soil types. Also illustrated is the important role the moisture content plays in the depth of infiltration.

12.5 SUMMARY OF RELATIVE IMPORTANCE OF LOCUS

12.5.1 Remediation

Contaminated mobile water in the unsaturated zone can contaminate groundwater (locus no. 8) by mixing. Partitioning between the water and air phases may contaminate the soil air (locus no. 1). The extent of contamination of the mobile water depends on the solubility of the liquid contaminant in water.

The relative importance of locus no. 12 depends on the solubility of the liquid contaminant and the moisture content of the soil. If locus no. 12 holds a maximum mass per unit volume, it probably holds more mass per unit volume than loci nos. 4,9,10 and 11. Locus no. 12 should hold the same mass per unit volume of water as loci no. 3 and 8. Because locus no. 12 holds more water than locus no. 3, it will also hold more mass; similarly, locus no. 12 holds less mass than locus no. 8. Under the right conditions, (e.g., low volatility, low Henry's law constant, locus no. 12 may be as important as locus no. 1. When the contaminant liquid is sparingly soluble in water, locus no. 12 will hold less mass per unit volume than loci nos. 2,5,6,7 and 13.

If the contaminated mobile water in locus no. 12 occupies a large fraction of the unsaturated zone porosity, and if the soil permeability is

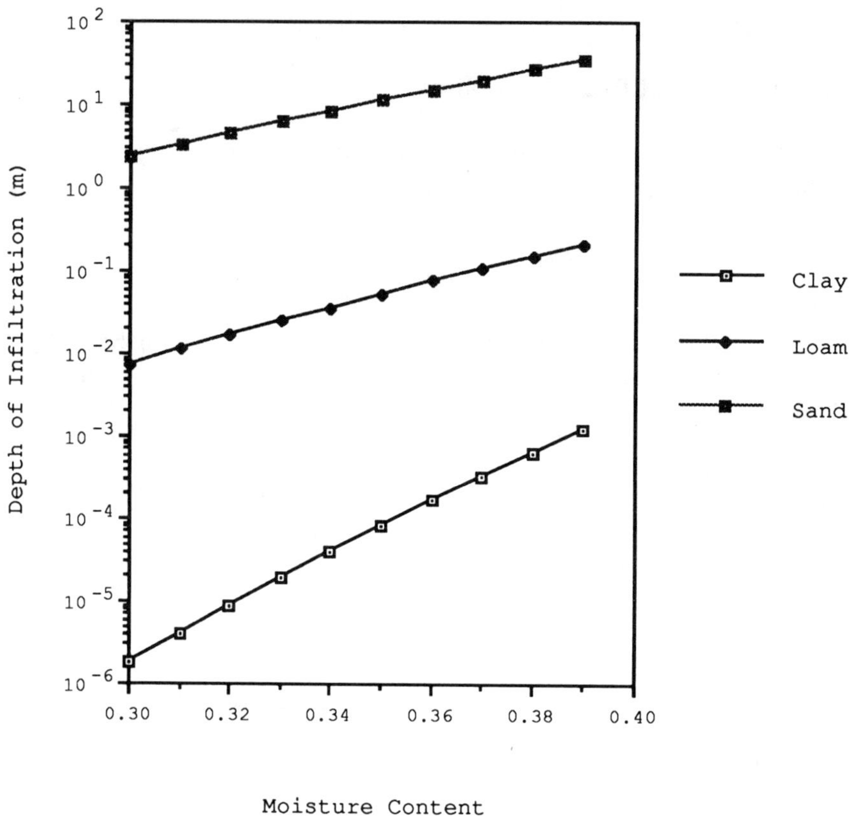

Figure 12–8. Depth of infiltration vs. moisture content at one day.

sufficiently high, then the contaminated water may be fairly mobile. Pumping may remove much of the contaminated water in this case.

12.5.2 Loci Interactions

Dissolution from liquid contaminant in locus no. 6 represents the primary mechanism for partitioning into this locus. Contaminants may partition into the air phase (locus no. 1) if the Henry's law constants are high enough. Soil particles may attenuate contamination by retardation and sorption (loci nos. 2 and 4). If the contaminated mobile water is in sufficient quantity, it may migrate vertically, contaminating the groundwater (locus no. 8). It may be aided in this regard by mobile colloids (locus no. 9).

12.5.3 Information Gaps

Research into the following area is important to better define the fate of contaminated unsaturated soil water:

1. Attenuation of hydrocarbons from water to soil particles.
2. Exchange rates between air and water phases.

12.6 LITERATURE CITED

Bear, J. 1979. Hydraulics of Groundwater. McGraw-Hill, Inc., New York.

Beven, K. and P.F. Germann. 1981. Water Flow in Soil Macropores, 2, A Combined Flow Model. Journal of Soil Science, 32:15–29.

Beven, K. and P.F. Germann. 1982. Macropores and Water Flow in Soils. Water Resources Research, 18(5):1311–1325.

Biswas, T.D., D.R. Nielsen and J.W. Bigger. 1966. Redistribution of Soil Water after Infiltration. Water Resources Research, 2:513–524.

Clapp, R.B. and G.M. Hornberger. 1978. Empirical Equations for Some Soil Hydraulic Properties. Water Resources Research, 14(4):601-604.

Dunne, T. and L.B. Leopold. 1978. Water in Environmental Planning. W.H. Freman & Co.

Giesel, W., M. Renger, and O. Strebel. 1972. Numerical Treatment of the Unsaturated Water Flow Equation: Comparison of Experimental and Computed Results. Water Resources Research, 9(1):174–177.

Gilliam, R.W., A. Klute, and D.F. Heermann. 1979. Measurement and Numerical Simulation of Hysteretic Flow in a Heterogeneous Porous Medium. Soil Science Society of America Journal, 43(6):1061–1067.

Hillel, D. 1980. Fundamentals of Soil Physics. Academic Press.

Levy, B.S. 1987. Kinematic Wave Approximation to Solute Transport in a Forest Soil. M.S. Thesis, University of Virginia.

Pettyjohn, W.A. and A.W. Haunslow. 1982. Organic Compounds and Groundwater Pollution. In: Proceedings of 2nd National Symposium on Aquifer Restoration and Ground Water Monitoring. D.M. Nielsen (editor). National Water Well Association.

Rosenberg, N.J., B.L. Verma, and S.B. Verma, 1983. Microclimates: the Biological Environment. 2nd ed. Wiley—Interscience.

Rubin, J. 1966. Theory of Rainfall Uptake by Soils Initially Drier than their Field Capacity and its Applications. Water Resources Research, 2:739–750.

Swokowski, E.W. 1975. Calculus with Analytic Geometry. Prindle, Weber and Schmidt.

Watson, K.W. and R.J. Luxmoore. 1986. Estimating Macroporosity in a Forest Watershed by Use of a Tension Infiltrometer. Soil Science Society of America Journal, 50:578–582.

12.7 OTHER REFERENCES

Baehr, A.L., G.E. Hoag, and M.C. Marley. 1989. Removing Volatile Contaminants from the Unsaturated Zone by Inducing Air-phase Transport. Journal of Contaminant Hydrology, 4, 1–26.

Baek, N.H., L.S. Clesceri, and N.L. Clesceri. 1989. Modeling of Enhanced Biodegradation in Unsaturated Soil Zone. Journal of Environmental Engineering, 115(1):150–172.

Bouyoucos, G.L. 1915. Effect of Temperature on the Movement of Water Vapor and Capillary Moisture in Soils. Journal of Agricultural Research, 5:141–172.

Hutzler, N.J., J.S. Gierke, and L.C. Krauss. 1989. Movement of Volatile Organic Chemicals in Soils. In: Reactions and Movement of Organic Chemicals in Soils, Soil Science Society of America Special Publication No. 22. pp. 373–403.

Jury, W.A., W.F. Spencer, and W.J. Farmer. 1983. Behavior Assessment Model for Trace Organics in Soil, I. Model Description. Journal of Environmental Quality, 12(4).

Kreamer, D.K., G.M. Thompson, E.S. Simpson, K. Stetzenbach, and A. Long, 1985. Seasonal, Diurnal, and Storm-related Changes in Concentrations of Trichloroethylene (TCE) and Other Volatile Organic Compounds in the Unsaturated Zone. Report for USDI, USGS—Arizona Water Resources Research Center.

Marrin, D.L. 1984. Investigation of Volatile Contaminants in the Unsaturated Zone above TCE-polluted Groundwater. U.S. EPA, #CR811018-01-0.

Wagenet, R.L., J.L. Hutson, and J.W. Biggar. 1989. Simulating the Fate of a Volatile Pesticide in Unsaturated Soil: A Case Study with DBCP. Journal of Environmental Quality, 18(1):78–84.

Wierenga, P.J. and M.T. Van Genuchten, 1989. Solute Transport through Small and Large Unsaturated Soil Columns. Ground Water, 27(1):35–42.

Wilson, J.T., C.G. Enfield, W.J. Dunlap, R.L. Cosby, D.A. Foster, and L.B. Baskin, 1981. Transport and Fate of Selected Organic Pollutants in a Sandy Soil. Journal of Environmental Quality, 10(4):501–506.

SECTION 13

Liquid Contaminants in Fractured Rock or Karstic Limestone in Either the Unsaturated or Saturated Zone

CONTENTS

List of Figures 335
13.1 Locus Description 337
    13.1.1 Short Definition 337
    13.1.2 Expanded Definition and Comments 337
13.2 Evaluation of Criteria for Remediation 337
    13.2.1 Introduction 337
        13.2.1.1 Fractures and Dissolution Zones 339
            Distinction Between Primary and Secondary Porosity 339
            Sources and Types of Fractures 340
            Importance of Locus 340
            Fracture Characteristics 340
            Fracture Density 341
            Fracture Width 341
            Fracture Orientation and Connectivity 342
            Fracture Porosity 342
            Level of Present Understanding 343
    13.2.2 Mobilization/Remobilization 344
        13.2.2.1 Partitioning into Mobile Phase 344
        13.2.2.2 Transport of/with Mobile Phase 344
            Saturated Flow 344
            Unsaturated Flow 346
    13.2.3 Fixation 346
        13.2.3.1 Partitioning onto Immobile (Stationary) Phase 346
        13.2.3.2 Other Fixation Approaches 348
    13.2.4 Transformation 348
        13.2.4.1 Biodegradation 348
        13.2.4.2 Chemical Oxidation 348

13.3 Storage Capacity in Locus 349
13.3.1 Introduction 349
13.3.2 Guidance on Inputs for, and Calculation of, Maximum Value 349
13.3.3 Guidance on Inputs for, and Calculation of, Average Values 350
Unsaturated Zone 350
13.4 Example Calculations 351
13.4.1 Storage Capacity Calculations 351
13.4.1.1 Maximum Quantity 351
13.4.1.2 Average Quantity at Unsaturated Conditions 351
13.4.1.3 Average Quantity at Saturated Conditions 351
13.4.2 Transport Rate Calculations 352
Parallel-Plate Approximation 352
13.5 Summary of Relative Importance of Locus 353
13.5.1 Remediation 353
13.5.2 Loci Interactions 353
13.5.3 Information Gaps 354
13.6 Literature Cited 354
13.7 Additional Reading 355

FIGURES

13-1 Schematic Cross-Sectional Diagram of Locus No. 13—Liquid Contaminants in Rock Fractures in Either the Unsaturated or Saturated Zone 338
13-2 Schematic Representation of Important Transformation and Transport Processes Affecting Other Loci 339
13-3 Map of Fracture Spacings and Orientations from Borehole Records 341
13-4 Depiction of Strike, Dip Direction, and Dip 343
13-5 Saturation Versus Pressure Head for Fractures and Rock Matrix 347
13-6 Effective Flow Areas in Fractures as a Function of Pressure 347

SECTION 13

Liquid Contaminants in Fractured Rock or Karstic Limestone in Either the Unsaturated or Saturated Zone

13.1 LOCUS DESCRIPTION

13.1.1 Short Definition

Liquid contaminants in fractured rock or karstic limestone in either the unsaturated or saturated zone.

13.1.2 Expanded Definition and Comments

The presence of liquid contaminant in the fractures of rock or the dissolution cavities of karstic limestone distinguishes this locus from loci nos. 2, 5, 6 and 7, which discuss liquid contaminants present in granular, consolidated and unconsolidated porous media. Essentially, this locus discusses the presence of liquid contaminants in the secondary porosity of consolidated rocks; the presence of liquid contaminants in the primary porosity of consolidated rocks is discussed in locus no. 10.

Figure 13–1 presents a schematic cross-section of the locus; Figure 13–2 is a schematic representation of the transformation and transport processes affecting other loci.

13.2 EVALUATION OF CRITERIA FOR REMEDIATION

13.2.1 Introduction

The presence of fractured rock or karstic limestone at a leaking UST site may mean greater difficulty in locating, modeling, and remediating the contamination than if no fractions or dissolution zones were present. Rocks with fracture zones or dissolution zones are typically very permeable, and leaked product can travel very quickly from the spill site.

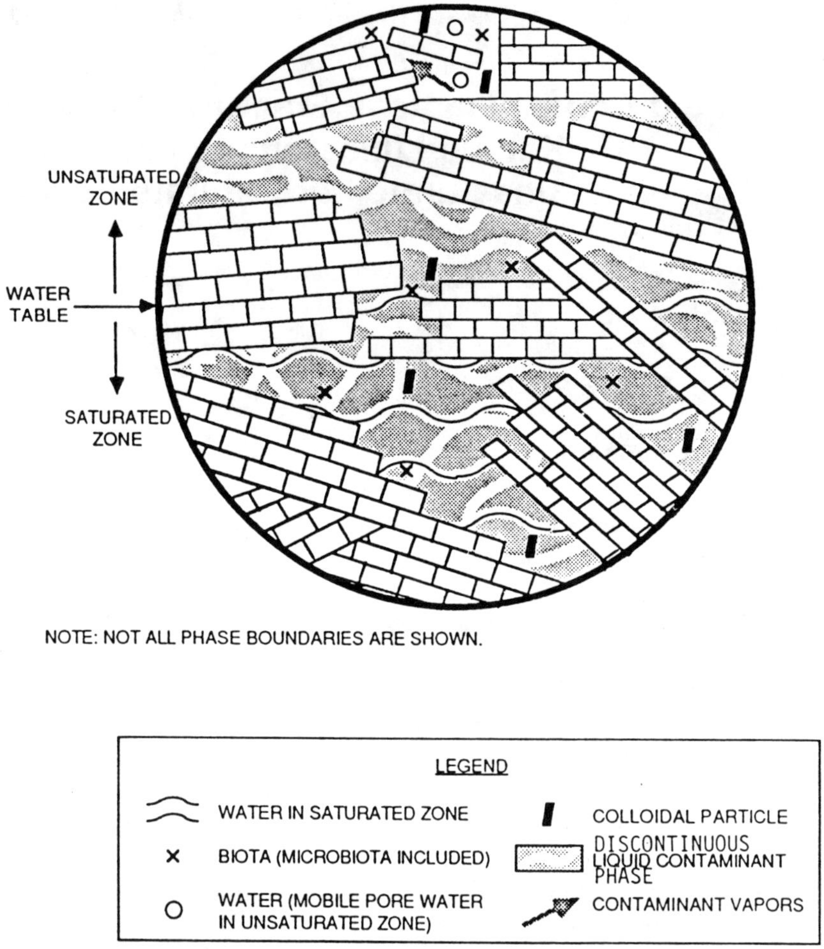

Figure 13–1. Schematic cross-sectional diagram of locus no. 13–liquid contaminants in rock fractures in either the unsaturated or saturated zone.

Many fracture or dissolution zones are highly anisotropic, and thus cannot be simulated using models which assume an isotropic medium.

Remediation is also made more difficult by the presence of fractures and dissolution zones. Recovery of free product is generally very low. Garrett (1987) reports recoveries from six sites that range up to only 70 gallons; up to 6000 gallons were recovered from eleven sand and gravel sites. Among available remedial solutions, most often used are pumping

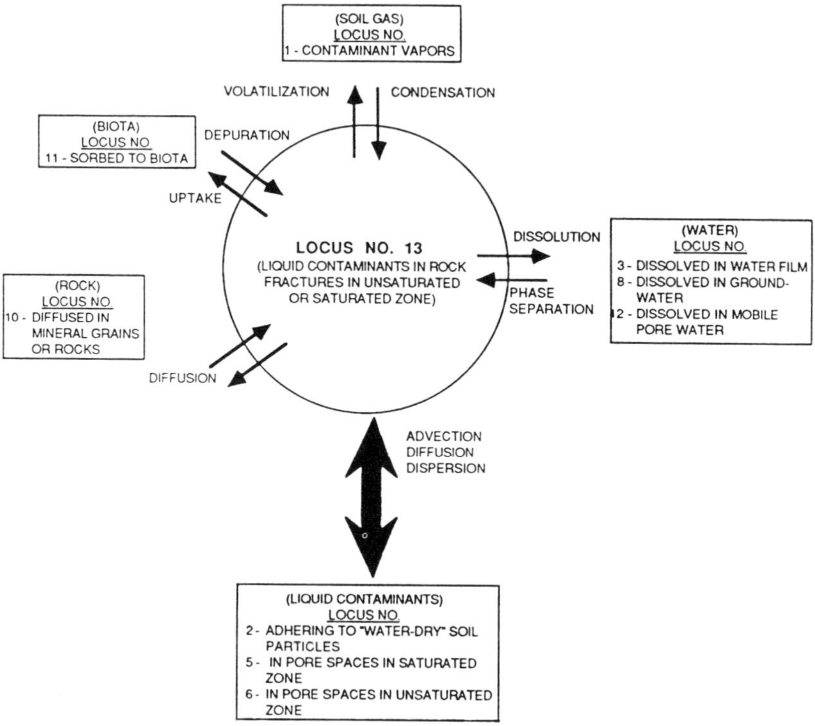

Figure 13–2. Schematic representation of important transformation and transport processes affecting other loci.

methods, which induce a gradient to remove product or water-product mixtures; biodegradation is also used.

13.2.1.1 Fractures and Dissolution Zones

Distinction Between Primary and Secondary Porosity. Total porosity is the volume of void space divided by the total unit volume of a rock. Total porosity may be divided into primary porosity and secondary porosity. Primary porosity is porosity due to void space in the rock matrix. Secondary porosity is void space due to fractures or dissolution zones.

Locus no. 13 describes rock fractures and dissolution zones, focusing on secondary porosity. Locus no. 10, which discusses the rock matrix, concerns primary porosity. Crystalline rocks, such as granite, may have primary porosities which range from 0–10 percent. Typical secondary porosities are 0.001–1 percent (Freeze and Cherry, 1979). These very small values for secondary porosities have important implications for transport within fractures. This is discussed in Section 13.2.2.2.

Sources and Types of Fractures. Secondary porosity is formed after deposition and induration of the rock (Duguid and Lee, 1977); primary porosity is formed during deposition of a sedimentary rock or the solidification of an igneous rock. Secondary porosity generally results from dissolution and geochemical weathering, or fracturing by tectonic action. The range in size of fractures and dissolution zones is very large; microcracks may be as small as a few microns, whereas limestone caverns in karst topography may be tens of meters deep and wide. This variation in size contributes to the anisotropic nature of transport in fractured rocks and karstic limestone.

Importance of Locus. The importance of fracture and dissolution zones in the fields of earthquake engineering, geothermal energy, and deep underground burial of radioactive waste has been long acknowledged. Fractures and dissolution zones may be equally important with regard to groundwater contamination. Where a rock contains both primary and secondary porosity, the fractures and dissolution zones typically provide the greater permeability (sometimes by several orders of magnitude) due to larger openings (Tang et al., 1981). Results published by Garrett (1987) show that many contaminated well incidents occur in fractured bedrock, after only a few gallons or tens of gallons of gasoline have been spilled. Garrett concludes that fracture zones are more vulnerable than sand and gravel aquifers to groundwater pollution and should be given more weight in hazardous waste site ranking systems, such as DRASTIC (U.S. EPA, 1985).

Fracture Characteristics. Although the following discussion focuses on the characterization of fracture systems in rock, it may be extended to describe the dissolution zones of karstic limestones. Fracture systems in rocks may be described by several properties, including fracture density and spacing, fracture width, and fracture orientation and connectivity. The properties of the fractures control, to a large extent, the flow and transport of water and contaminants through the fractured rock. Understanding mass transport through rock fractures and modeling such systems must be based on a good understanding of these properties. The difficulty in gaining such an understanding, however, has thus far impeded a complete understanding of flow of liquid contaminants in locus no. 13.

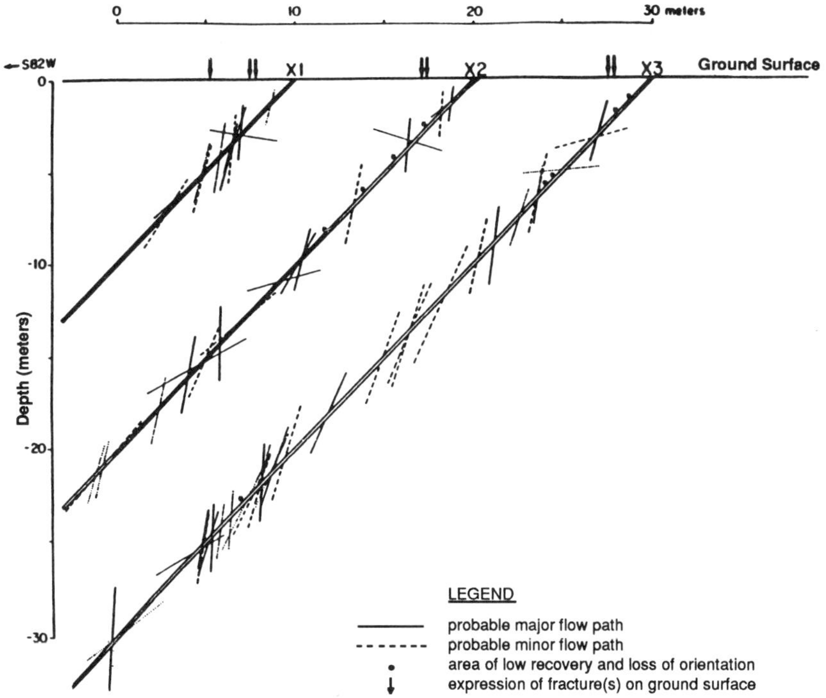

NOTE: Each fracture intersected by a borehole is shown at the angle of intersection.

Bearing and plunge of boreholes ≅ 162°/45 SW

Figure 13–3. Map of fracture spacings and orientations from borehole records. *Source:* Rasmussen and Evans, 1987.

Fracture Density. Fracture density is the number of fractures per unit rock volume and has a direct relationship to the permeability of a rock zone. Highly fractured rocks have high fracture densities and are more permeable than less fractured rocks, all other factors being equal. Fracture spacing is typically determined by measuring the number of fractures intersecting a rock core of a given length. An example of a diagram showing fracture intersections in a geologic cross-section is shown in Figure 13-3.

Fracture Width. Fracture width refers to the distance between fracture walls measured perpendicular to the walls. The width varies along a fracture from zero (where the fracture walls are in contact) to some maximum value. This maximum value may be very large, as in the case of karst limestone dissolution cavities or large joints; typical reported values are usually on the order of tens of microns to one millimeter (10^{-5} to 10^{-3}m). Fractures typically contain asperities (irregularities) that make the effective width smaller than the maximum opening at some places.

Fracture Orientation and Connectivity. The spatial orientation of a fracture or fracture plane is described by its strike and dip (see Figure 13–4). The strike of the fracture refers to the surficial expression of the fracture upon a horizontal, flat land surface. The orientation of the strike of a fracture is measured as an angle from north. For example, a fracture with a strike of N30°E has a surficial expression whose orientation is 30 degrees east of north.

The dip of the fracture refers to the angle from horizontal of the fracture plane and is measured perpendicular to the direction of strike. Continuing from the example, since the strike of the fracture is northeast-southwest. The dip is either to the northwest or the southeast. For example, if the dip is 60°SE, it dips 60° from horizontal to the southeast.

Characterizing fracture orientation is important because it affects the flow direction and fracture connectivity; it may be delineated by surficial mapping of fractures in bedrock outcrops or by interpretation of borehole geophysical logs.

Connectivity is extremely important in determining the hydraulic conductivity of a fracture network, because fractures that do not connect to other fractures or to a permeable rock matrix do not transmit fluid. Connectivity is a property which must be described indirectly. As such, the information collected on connectivity tests (e.g., pump tests, natural and artificial tracers) or by mapping is inferential and probably incomplete.

Fracture Porosity. The fracture porosity is the fraction of a unit volume that is occupied by fracture openings. A more important parameter is the effective porosity, which is less than or equal to the total porosity, and which includes only connected fractures. Thus, void volume formed by non-connected fractures is not included.

Freeze and Cherry (1979) report effective porosities on the order of 0.001–1 percent. Data from several other published reports (Garrett, 1988; Duguid and Lee, 1977; Snow, 1968; and Hendry et al., 1986) all fall within this range, with most porosities on the order of 0.01–0.1 percent. Torbarov (1976) reports porosities in karst regions that range from one to two percent.

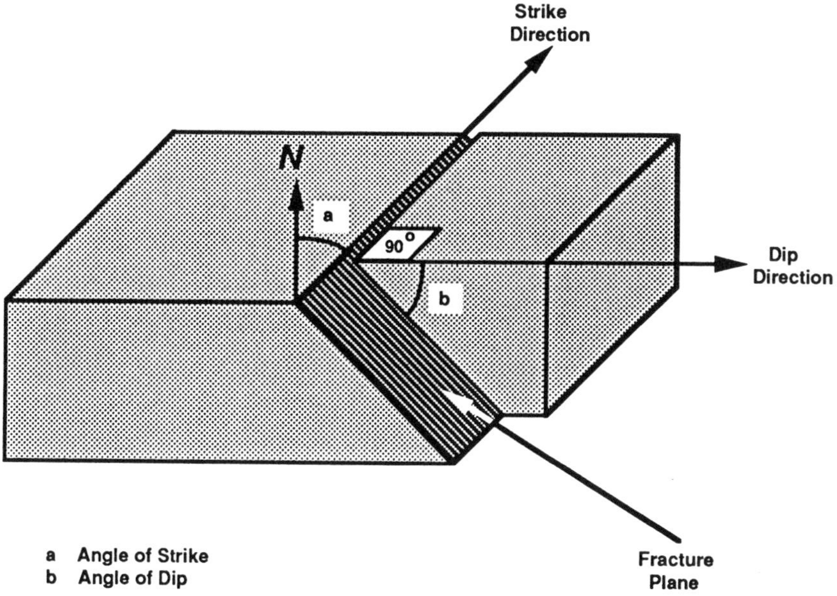

Figure 13–4. Depiction of strike, dip direction, and dip.

Level of Present Understanding. Despite the potential importance of locus no. 13 in groundwater contamination scenarios, relatively little is known about the flow of liquid contaminants in fractures. The fluid mechanics of multi-phase flow, as opposed to single-phase flow, is especially in need of field experiments to allow researchers to develop and validate flow models applicable to leaking UST sites. Most of the research that has been conducted on flow of contaminants in fractured rocks in recent years on this topic has been due to the need for siting of high-level radioactive waste repositories and understanding the transport of radionuclides through fractures. This research may be possibly applied to the problem of liquid hydrocarbon transport.

Flow models developed for sand or gravel aquifers generally assume homogeneity and isotropy in a porous medium. These assumptions do not hold for fractured rock; the models are not applicable for fracture-flow conditions unless the rock is so extensively fractured so as to represent effectively a porous medium.

13.2.2 Mobilization/Remobilization

13.2.2.1 Partitioning into Mobile Phase

In this discussion, contaminants are assumed to be mobile in the fractures (i.e., contaminants are already in the mobile phase in locus no. 13). While some researchers (Wang and Narasimhan, 1985) have reported that fractures can serve as barriers to flow under conditions of high capillary tension (high negative pressure), fractures can increase mobility due to characteristically high permeabilities.

Liquid contaminant from locus no. 13 may also partition into water or air, by dissolution or volatilization (see Figure 13–2). Only liquid contaminant in the unsaturated zone may volatilize, whereas liquid contaminant in either the saturated or unsaturated zones may dissolve into water. Volatilization is discussed in Section 1; dissolution is discussed in Sections 8 and 12.

13.2.2.2 Transport of/with Mobile Phase

According to Wang and Narasimhan (1985), transport of contaminant through the fractures of rocks with high primary porosity depends highly on the pressure within the fractures. At high capillary pressure characteristic of unsaturated conditions, fractures are barriers to flow. At saturation, however, the fractures carry most of the contaminant flow through the rock. Where fractures comprise all the porosity of a geologic structure, all flow occurs in those fractures. Transport of volatilized liquid contaminant or of dissolved liquid contaminant (locus no. 1 and loci nos. 8 and 12, respectively) also occurs through fractures. Transport is treated separately for the saturated and unsaturated zones.

Saturated Flow. A lack of understanding of the actual flow mechanics of fracture flow impedes a better knowledge of transport phenomenon in locus no. 13. Most research on fracture flow relates to radionuclide-containing water; little has been published regarding systems containing water and petroleum or other contaminants, or multi-phase flow in general.

A potential source of major error is the application of Darcy's law to multi-phase flow in fractures. While conclusive evidence is not yet available, this assumption is probably not valid (personal communication from J.L. Wilson, New Mexico Technology, 1988). This uncertainty notwithstanding, several researchers have developed theories regarding flow through fractures. The main results are summarized below.

On a regional scale, researchers have assumed that Darcy's law can be used to describe flow through fractures:

$$\bar{V} = \frac{K_{sat,\,w}}{\Theta_r} \cdot \frac{dh}{dl} \qquad (13.1)$$

where $\bar{V}$ = groundwater velocity (m/sec), $K_{sat,w}$ = hydraulic conductivity (m/sec), dh/dl = hydraulic gradient (meter/meter) and Θ_r = transport porosity of rock (dimensionless).

Because the effective fracture porosity is typically very low (10^{-3} to 10^{-5}), groundwater velocity in fractures is much larger than in soils and other porous media; velocities maybe on the order of meters to tens of meters per day or more, making liquid contaminant very mobile in this locus. Section 13.4 gives a sample calculation for this equation.

Many researchers (Snow, 1968; Wang and Narasimhan, 1985) have assumed that individual fractures can be modeled as parallel-plate flow. Under this assumption, the flow is described by:

$$Q = \frac{g \cdot w \cdot \Sigma b^3}{12 \cdot \nu} \cdot \frac{dh}{dl} \qquad (13.2)$$

where: Q = volumetric flow rate through fractures (m^3/sec)
b = fracture width (meters)
g = gravitational constant (9.8 m/sec^2)
w = lateral width of fracture (meters)
ν = kinematic viscosity (m^2/sec)
dh/dl = hydraulic gradient (meter/meter)

The summation includes all fractures, which are not of equal width. The use of this equation makes several assumptions which may be erroneous: all apertures remain constant in width throughout their length; fluid does not influence the medium through which it flows; all the fluid channels remain parallel to each other and to the fracture; and, flow is non-turbulent, single-phase, non-compressible, and Newtonian. Wang and Narasimhan (1985) report that this cubic law overestimates flow in cases where there is a high percentage of contact points, or points where stress is transferred, within the fractures. Thus, the application of parallel plate theory to liquid contaminant transport seems to be limited.

Some work has attempted to relate flow in fractures to viscous, gravitational, and capillary forces in order to derive flow equations (Wilson and Conrad, 1984; Hunt et al., 1986). Thus far, a complete theory based on this approach has not been published.

Unsaturated Flow. Wang and Narasimhan (1985) discuss the interaction of the fracture network and a porous rock matrix under conditions of capillary tension (i.e., in the unsaturated zone). The authors argue that, while the fractures transport most of the flow under saturated conditions, the rock matrix transports most of the flow when conditions of capillary tension exist. Under these conditions, Wang and Narasimhan (1985) postulate that fractures act as barriers to flow.

Rasmussen and Evans (1987) report that the unsaturated hydraulic conductivity of a fractured rock is a function of saturation. As capillary tension increases, the degree of saturation decreases (see Figure 13–5). At higher capillary tension, water or liquid contaminant is held in smaller fractures which have a higher surface area to volume ratio. As a result, the smaller fractures have an increased resistance to flow and a lower hydraulic or fluid conductivity. Therefore, there is an inverse relationship between capillary tension, and saturation and hydraulic conductivity.

The increase in capillary tension also decreases the effective flow area because of a lowered fluid content in the fractures (see Figure 13–6). The decrease in cross-sectional area of flow reduces the ability of the fractured rock to transmit fluids.

For rock with low or non-existent primary porosity in the unsaturated zone, fractures provide the only means of transport for fluid flow. Flow under this condition is analogous to rain drops on a windshield; flow is a function of interfacial surface tensions, capillary forces, wettability patterns, and gravity (personal communication, J.L. Wilson, New Mexico Technology, 1988).

13.2.3 Fixation

13.2.3.1 Partitioning onto Immobile (Stationary) Phase

Liquid contaminants in locus no. 13, as a mobile phase, may become stationary by either diffusing into the rock matrix (locus no. 10) or by means of a high capillary tension (high negative pressure). Liquid contaminant held in locus no. 10, however, (as explained in Section 10.3) is not entirely stationary. Liquid contaminants my also adhere to the walls of the fractures. Unlike typical soils, however, the ratio of surface area to volume is very small and relatively little of the liquid contaminant will be trapped. Also, inducing a suction in the fractures may result in the removal of liquid contaminant from the fractures into the matrix (as explained earlier) but will not immobilize liquid contaminant. Thus,

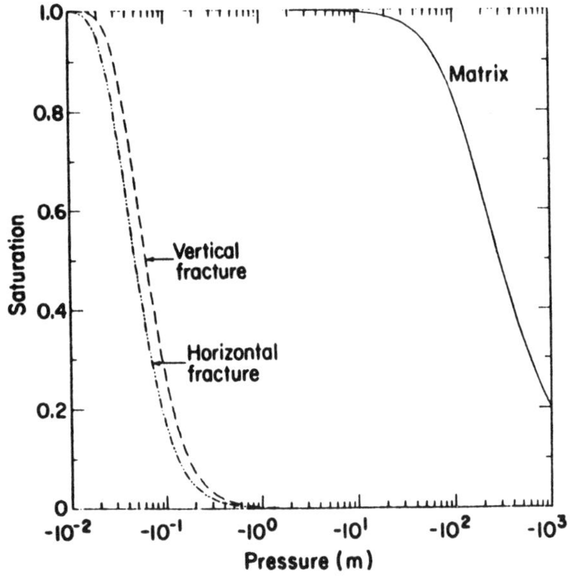

Figure 13–5. Saturation versus pressure head for fractures and rock matrix. *Source:* Wang and Narasimhan, 1985.

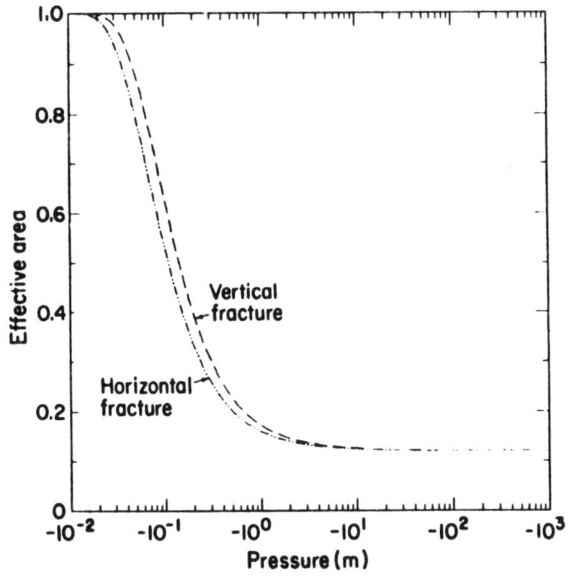

Figure 13–6. Effective flow areas in fractures as a function of pressure. *Source:* Wang and Narasimhan, 1985.

liquid contaminants in this locus generally do not partition to a stationary phase.

13.2.3.2 Other Fixation Approaches

The immobilization of liquid contaminants in rock fractures is not generally attempted. The very high mobility of liquids in rock fractures, as discussed in Section 13.2.2, makes it infeasible to trap and indefinitely hold liquid contaminants in this locus.

Several techniques to fix contaminants in a particular locus are known, including stabilization, slurry walls, vitrification, and foam stabilization. Stabilization techniques generally use a siliceous material to bind material into a solid mass. For fractures, it is difficult to transmit the cementing agent throughout the zone to be stabilized, and therefore this is not usually done. Slurry walls are often used to control the flow of groundwater. In a fractured rock zone, it is difficult to seal off all possible routes of contaminant migration to ensure long-term stabilization. Vitrification uses electric fields to turn a soil or rock mass into a glass-like material. This technique may prove technically feasible, but at the present the technique is costly and leaves the site unusable for months to years due to high temperatures. Foam stabilization is currently in the experimental stage (personal communication, S. Benson, Lawrence Berkeley Laboratory 1988). This technique involves injecting foam into the soil pores and allowing solidification of the foam. The permeability is greatly decreased. This technology is well-advanced, but may never be appropriate for aquifer situations due to the constituents of the foam.

13.2.4 Transformation

13.2.4.1 Biodegradation (Details in Section 11)

As with other bulk liquid loci, biodegradation in locus no. 13 is probably insignificant. Biodegradation is largely mediated by aerobic bacteria which require oxygen as well as mineral nutrients. For this reason, liquid contaminants from locus no. 13 are most likely to be biodegraded only after they have been dispersed and dissolved into one of the aqueous phase loci.

13.2.4.2 Chemical Oxidation (Details in Section 3.2.4.2)

Rocks may have secondary clay mineral coatings or contain primary minerals which could catalyze chemical oxidation.

13.3 STORAGE CAPACITY IN LOCUS

13.3.1 Introduction

In general, the storage capacity in rock fractures is very small, and is limited by the pore space constituted by the fractures. For this locus we are assuming that the primary porosity is negligible; all the available void volume is constituted by the fractures. For calculation of the amount stored in the fractures, the total porosity is appropriate; for transport equations, the effective hydraulic porosity was used. This quantity, which equals the total porosity minus the porosity due to non-connected fractures, describes the porosity available for liquid transport through a rock zone. Non-connected fractures ("dead zones"), however, function as storage; the total porosity is thus used in this section.

13.3.2 Guidance on Inputs for, and Calculation of, Maximum Value

The porosity of the fractures determines the volume of liquid contaminant stored in locus no. 13. The fracture porosity forms an upper-bound for the amount of liquid contaminant potentially held in this locus. As stated previously, the fracture porosity is typically less than one percent. In regions of karst limestone, where solutions channels and caverns exist, porosities of one to three percent are possible (Duguid and Lee, 1977) on a regional scale. Locally, however, the fracture porosity may be considerably higher.

The maximum amount of liquid contaminant held in this locus is determined by multiplying the fracture porosity by the density of the liquid contaminant:

$$m_r = \Theta_r \cdot \rho_l \tag{13.3}$$

where m_r = mass of liquid contaminant per unit volume of rock (kg/m^3)
Θ_r = total porosity of rock fractures
ρ_l = liquid contaminant density (kg/m^3)

For the calculation of the maximum quantity of liquid contaminant held in locus no. 13, it is assumed that there are no losses of liquid contaminant from volatilization, dissolution, or microbial degradation, and that the fracture porosity is occupied by only liquid contaminant.

13.3.3 Guidance on Inputs for, and Calculation of, Average Values

While the fracture porosity places an upper limit on the volume of liquid contaminant which may be stored in locus no. 13, the actual volume stored in the fractures may be less than this due to the presence of air and water in the fractures.

Unsaturated Zone. In this case both air and water phases are present in the fracture and are presumed to be in direct contact with the liquid contaminant. The air allows liquid contaminant to volatilize; dissolution of liquid contaminant into the water phase occurs.

This case is characteristic of the condition where a slug of liquid contaminant passes a fracture zone, leaving liquid contaminant held in residium. No guidance was found for typical values of residual saturation in fractures, but it would depend on fracture width, interfacial tension of the liquid contaminant, and losses due to microbial degradation, dissolution, and volatilization, etc.

The average mass of liquid contaminant in the fractured rock is:

$$m_r = m_{res} - m_w - m_v - m_b - m_l \tag{13.4}$$

where m_r = contaminant mass per unit volume of rock (kg/m^3)
m_{res} = residual mass of liquid contaminant per unit volume of rock (kg/m^3)
m_w = contaminant mass in water per unit volume of rock (kg/m^3)
m_v = contaminant mass volatilized per unit volume of rock (kg/m^3)
m_b = contaminant mass lost to microbial degradation per unit volume of rock (kg/m^3)
m_l = contaminant mass oxidized by chemical reaction per unit volume of rock (kg/m^3)

The mass of liquid contaminant lost to dissolution in water is:

$$m_w = S \cdot \Theta_w / 10^3 \tag{13.5}$$

where S = solubility of liquid contaminant in water (mg/L)
Θ_w = water-filled porosity of rock (cm^3/cm^3)

The mass of liquid contaminant lost to volatilization in air is:

$$m_v = \rho_{va} \cdot \Theta_a \tag{13.6}$$

where ρ_{va} = vapor density of volatilized liquid contaminant—air mixture (kg/m^3)

Θ_a = air-filled porosity (cm^3/cm^3)

13.4 EXAMPLE CALCULATIONS

13.4.1 Storage Capacity Calculations

13.4.1.1 Maximum Quantity

The maximum quantity of liquid contaminant held in locus no. 13 is calculated assuming a fracture porosity of 0.05% and assumes that the liquid contaminant is the synthetic gasoline described in locus no. 7. Assuming a liquid density of 0.713 g/cm^3, the maximum amount of liquid contaminant held in locus no. 13 is:

$$m_r = (0.0005) \cdot (713 \text{ kg/m}^3)$$
$$= 0.36 \text{ kg of liquid contaminant per m}^3 \text{ of fractured rock}$$

13.4.1.2 Average Quantity at Unsaturated Conditions

In this example, the values for fracture porosity and liquid density are retained for comparison. Additionally, it is assumed that the losses of liquid contaminant to microbial degradation and chemical reaction are negligible. The residual, water-filled, and air-filled porosities are 0.0003, 0.0001, and 0.0001, respectively. Values of solubility and vapor density are assumed to be 130 mg/L and 1600 g/m^3, respectively. The mass of liquid contaminant per unit volume of fractured rock is:

$$m_r = (713 \text{ kg/m}^3) \cdot (0.0003) - (130 \text{ mg/L}) \cdot (10^{-4}) \cdot (10^{-3}) - (10^{-4}) \cdot (1.6 \text{ kg/m}^3)$$
$$m_r = 0.21 \text{ kg of liquid contaminant per m}^3 \text{ of fractured rock}$$

13.4.1.3 Average Quantity at Saturated Conditions

A similar calculation may be made for saturated conditions, under the same assumptions. Here, the fraction of fracture porosity occupied by air is zero. The mass of liquid contaminant per unit volume of fractured rock is:

$$m_r = (0.0003) \cdot (713 \text{ kg/m}^3) - (130 \text{ mg/L}) \cdot (10^{-4}) \cdot (10^{-3})$$
$$m_r = 0.21 \text{ kg of liquid contaminant per m}^3 \text{ of fractured rock}$$

13.4.2 Transport Rate Calculations

As stated earlier, the flow velocity of fluids in fractures can be very large relative to that in sand and gravel aquifers. In the following example, it is assumed that:

$K_{sat,l}$	=	fluid conductivity	= 10^{-7} m/sec
dh/dl	=	gradient	= 0.1 m/m
Θ_r	=	fracture porosity	= 0.0005
$\bar{V}_l$	=	velocity of immiscible fluid, m/sec	

Then, using equation 13.1,

$$\bar{V} = \frac{K_{sat,l}}{\Theta_r} \cdot \frac{dh}{dl}$$
$$= (10^{-7}\ m/sec)\cdot(0.1)/0.0005$$
$$= 2 \times 10^{-5}\ m/sec$$
$$= 1.73\ m/day$$

This is a high velocity compared to typical groundwater flow, using conservative values; in the case where the fracture porosity is only 0.0001, which is the low end of the range quoted in Freeze and Cherry (1979), the velocity would be 8.64 m/day. The high seepage velocities in fracture zones are cited by Garrett (1987) as the main reason that aquifers in fractured rock zones are so sensitive to contamination, even after only a few gallons or tens of gallons leaked. These very high flow rates may also be responsible for the non-Darcian flow patterns (personal communication, J.L. Wilson, New Mexico Technology, 1988).

Parallel-Plate Approximation. Equation 13.2 may not be entirely appropriate, but it may estimate the quantity of flow occurring through a single fracture. Assume that the following values characterize a fracture:

dh/dl	=	hydraulic gradient	= 1 cm/cm
b_a	=	aperture opening	= 0.01 cm
w	=	lateral width	= 0.1 cm
ν_l	=	kinematic viscosity of liquid contaminant	= 6.36×10^{-3} cm²/sec
g	=	gravitational constant	= 980.7 cm/sec²

$$Q = \frac{(980.7\ cm/sec^2) \cdot (0.01cm)^3 \cdot (0.1cm)}{(12) \cdot (6.36 \times 10^{-3}\ cm^2/sec)}$$

$$Q = 1.28 \times 10^{-9}\ m^3/sec$$
$$= 0.11\ liters/day$$
$$= 10.7\ gallon/year$$

This represents the flow of the synthetic gasoline blend through a single idealized fracture with aperture of 100 microns, a typical value for fractured igneous rock (Snow, 1969) with a unit gradient. Large joints and conduits will transmit a far greater volume of fluid.

13.5 SUMMARY OF RELATIVE IMPORTANCE OF LOCUS

13.5.1 Remediation

Contamination in fractures is highly mobile and therefore relatively difficult to remediate. Where the primary porosity is negligible, the fractures provide the principal pathway for flow. The relatively large size of fracture openings permit fluid to move quickly away from a leak site, contaminating large areas in a short time.

The first step to remove liquid contaminant from this locus is usually to induce a local hydraulic gradient. In this manner, free product may be removed as well as pure liquid product blobs within the groundwater. Because of the low porosity and high permeability of the fracture network, relatively little free product is typically recovered. Groundwater pumping with above-ground treatment by air stripping or activated carbon is another method that is often used for the saturated zone.

Immobilization measures are generally not undertaken for this locus. The high mobility of contaminants in the fractures and the lack of knowledge regarding fracture connectivity lower the feasibility of these actions.

13.5.2 Loci Interactions

Figure 13–2 shows the major interactions among locus no. 13 and the other loci. Liquid contaminant is highly mobile and can interact with almost all of the other loci. Bulk transport to fractures from soil, or vice versa, is important; contaminant may diffuse into loci nos. 2, 5, or 6, or gather as a free product lens (locus no. 7); contaminant may adsorb onto the rock face (locus no. 4); or contaminant may be adsorbed to micells traveling through the fractures (locus no. 9). Also, biodegradation and inorganic transformations may occur in this locus. Volatilization of contaminant (locus no. 1) may occur; or contaminants may dissolve into water (locus nos. 3, 8, or 12). Important interactions among locus no. 13 and others are discussed in the appropriate sections of this report.

13.5.3 Information Gaps

In general, the behavior and mechanics of multi-phase flow through fractures is poorly understood. This topic has yet to attract widespread attention among researchers. In particular need of research are the following areas:

(1) field and laboratory experiments with actual fracture networks to develop a comprehensive data base for fracture porosities, permeabilities, aperture sizes, and other basic data;
(2) the formulation of appropriate flow models to use to represent flow through fractures;
(3) the influence of wettability patterns, interfacial tensions, capillary pressure, gravitational and viscous forces on flow through fracture networks;
(4) the interactions between fractures and the rock matrix in both the saturated and unsaturated zones; and,
(5) data and a better understanding of fractures, including methods to determine connectivity of fractures.

13.6 LITERATURE CITED

Duguid, J.O. and P.C.Y. Lee. 1977. Flow in Fractured Porous Media. Water Resource Research, 13(3):558–566.

Freeze, R.A. and J.A. Cherry. 1979. Groundwater. Prentice-Hall, Inglewood Cliffs, N.J.

Garrett, Peter. 1987. Comparative Vulnerability to Oil Spills of Bedrock versus Sand and Gravel Aquifers. *In*: Proceedings of Focus on Eastern Regional Ground Water Issues. NWWA.

Hendry, M.J., J.A. Cherry, and E.I. Wallick. 1986. Origin and Distribution of Sulfate in a Fractured Till in Southern Alberta, Canada. Water Resources Research, 22(1): 45–61.

Hunt, J.R., N. Sitar, and K.S. Udell. 1986. Organic Solvents and Petroleum Products in the Subsurface: Transport and Cleanup. Univ. Calif. (Berkeley), UCB-SEERHL Report No. 86–11. Sanitary Engineering and Environmental Health Research Laboratory.

Rasmussen, T.C. and D.D. Evans. 1987. Unsaturated Flow and Transport Through Fractured Rock Related to High-Level Waste Repositories, Final Report. Prepared for Nuclear Regulatory Commission, Washington, D.C.

Snow, D.T. 1968. Rock Fracture Spacings, Openings, and Porosities. Journal of the Soil Mechanics and Foundations Division, Proc. ASCE, 94:73–91.
Snow, D.T. 1969. Anisotropic Permeability of Fractured Media. Water Resources Research, 5(6):1273–1289.
Tang, D.H., E.O. Frind, and E.A. Sudicky. 1981. Contaminant Transport in Fractured Porous Media: Analytical Solution for a Single Fracture. Water Resources Research, 17(3): 555–564.
Torbarov, K. 1976. Estimation of Permeability and Effective Porosity in Karst on the Basis of Recession Curve Analysis. Proc. of the U.S.-Yugoslavian Symposium. Karst Hydrology and Water Resources, Volume I. Water Resources Publications, Fort Collins, CO.
U.S. EPA. 1985. DRASTIC: A Standardized System for Evaluating Ground Water Pollution Potential Using Hydrogeologic Settings, EPA/600/2-85/018.
Wang, J.S.Y. and T.N. Narasimhan. 1985. Hydrologic Mechanisms Governing Fluid Flow in a Partially Saturated, Fractured, Porous Medium. Water Resources Research, 21(12):1861–1874.
Wilson, J.L. and S.H. Conrad. 1984. Is Physical Displacement of Residual Hydrocarbons a Realistic Possibility in Aquifer Restoration? *In*: Proceedings of Petroleum Hydrocarbons and Organic Chemicals in Groundwater. NWWA/API.

13.7 ADDITIONAL READING

Flow Hydraulics in Fractured and Karstic Media

Long, J.C.S., J.S. Remer, C.R. Wilson, and P.A. Witherspoon. 1982. Porous Media Equivalents for Networks of Discontinuous Fractures. Water Resources Research, 18(3):645–658.
Moore, G.K. 1973. Hydraulics of Sheetlike Solution Cavities. Ground Water, 11(4):4–11.
Tsang, Y.W. and C.F. Tsang. 1987. Channel Model of Flow Through Fractured Media. Water Resources Research, 23(3):467–479.

Characterization of Fractured and Karstic Media

Silliman, E. and R. Robinson. 1989. Identifying Fracture Interconnections between Boreholes Using Natural Temperature Profiling: The Conceptual Basis. Ground Water, 27(3):393–402.
Smart, C.C. 1988. Artificial Tracer Techniques for the Determination of the Structure of Conduit Aquifers. Ground Water, 26(4):445–453.

Taylor, R.W. and A.H. Fleming. 1988. Characterizing Jointed Systems by Azimuthal Resistivity Surveys. Ground Water, 26(4):464–474.

Contaminant Transport in Fractured and Karstic Media

Andersson, J. and R. Thunvik. 1986. Predicting Mass Transport in Discrete Fractures, Water Resources Research, 22(13):1941–1950.

Grisak, G.E. and J.F. Pickens. 1980. Solute Transport Through Fractured Media, 1. The Effect of Matrix Diffusion. Water Resources Research, 16(4): 719–730.

Rasmuson, A. 1985. Analysis of Hydrodynamic Dispersion in Discrete Fracture Networks using the Method of Moments. Water Resources Research, 21(11):1677–1683.

Ross, B. 1986. Dispersion in Fractal Fracture Networks. Water Resources Research, 22(5): 823–827.

Simmleit, N. and R. Herrmann. 1987. The Behavior of Hydrophobic, Organic Micropollutants in Different Karst Water Systems. Water, Air, and Soil Pollution, 34:97–109.

Sudicky, E.A. and E.O. Frind. 1982. Contaminant Transport Porous Media: Analytical Solutions for a System of Parallel Fractures. Water Resources Research, 18(6):1634–1642.

Models of Flow and Transport in Fractured and Karstic Media

Andersson, J., A.M. Shapiro, and J. Bear. 1984. A Stochastic Model of a Fractured Rock Conditioned by Measured Information. Water Resources Research, 201(1):79–88.

Endo, H.K., J.C.S. Long, C.R. Wilson, and P.A. Witherspoon. 1984. A Model for Investigation Mechanical Transport in Fractured Networks. Water Resources Research, 20(10):1390–1400.

Oda, M. 1986. An Equivalent Continuum Model for Coupled Stress and Fluid Flow Analysis in Jointed Rock Masses. Water Resources Research, 22(13):1845–1856.

GLOSSARY

Anaerobe: An organism that does not require air or free oxygen to maintain its life process.

Abiotic: Referring to the absence of living organisms.

Advection: The process of transfer of fluids (vapors or liquids) through a geologic formation in response to a pressure gradient which may be caused by changes in barometric pressure, water table levels, wind fluctuations, or rainfall percolation. Advection can result from a thermal gradient caused by a heat source. See Darcy's law.

Aliphatic: Of or pertaining to a broad category of carbon compounds distinguished only by a straight, or branched, open chain arrangement of the constituent carbon atoms. The carbon-carbon bonds may be saturated or unsaturated.

Anisotropy: The dependence of property upon direction of measurement (e.g., hydraulic conductivity, porosity, compressibility, dispersion, etc.).

Aromatic: Of or pertaining to organic compounds that resemble benzene in chemical behavior.

Bentonite: A colloidal clay, largely made up of the mineral sodium montmorillonite, a hydrated aluminum silicate.

Brownian Movement: Random movement of molecules or colloids suspended in a fluid.

Biodegradation: A process by which microbial organisms transform or alter through enzymatic action the structure of chemicals introduced into the environment.

Biomass: The amount of living matter in a given area or volume.

Biota: A term that encompasses the spectrum of living things within a given area.

Biotic: Of or pertaining to life and living organisms. Induced by the actions of living organisms.

Caliche: A layer of calcium and magnesium carbonate deposited in a near-surface soil horizon by evaporating soil water or groundwater. It may occur as a soft thin soil horizon, as a hard thick bed just below the land surface, or as a surface layer exposed by erosion.

Capillarity: The pressure difference across the interface of two immiscible liquids or between a liquid and a solid. In a porous medium, capillarity is zero in the saturated zone and less than zero in the vadose zone. See surface tension, saturated zone, vadose zone.

Capillary Fringe: The zone of a porous medium above the water table within which the porous medium is saturated but is at less than atmospheric pressure. The capillary fringe is considered to be part of the vadose zone, but not of the unsaturated zone.

Catalyst: A substance that alters the rate of a chemical reaction and may be recovered, essentially unaltered, in form and amount at the end of the reaction.

Chemisorption: A process in which weak chemical bonds are formed between gas or liquid molecules and a solid surface.

Clay: A mineralogical term describing a family of aluminosilicate minerals formed by the decomposition of primary minerals (e.g., micas and feldspars) and re-composed into clay minerals. A textural term referring to particles <2 μm in diameter.

Coalescence: The bonding of welded materials into one body, or the uniting by growth in one body as particles, gas or liquid.

Colloid: Particles having dimensions of 10–10,000 angstroms (1–1000 nanometers) and which are dispersed in a different phase, such as a fluid or liquid.

Complex: A combination of two or more atoms into a molecular species, usually charged, and existing in water or some other fluid.

Compressibility: The change in volume of a porous medium in response to an applied stress which is counterbalanced by the incompressibility of the saturating fluid and the granular skeleton of the porous medium which it saturates. Compressibility has units of inverse force (Pa–1).

Constituents: An essential part or component of a system or group: examples are an ingredient of a chemical system, or a component of an alloy.

Containment: The prevention of the spreading of oil or other hazardous materials by the placing of booms or physical barriers and the use of absorbents, gelling, or herding agents or other materials to restrain, entrap, and collect a spill.

Corrective Action: The removal of chemicals and/or contaminated soils, objects and groundwater, from a site, or other clean-up activities designed to restore the local environment to acceptable conditions. Corrective actions may include, for example, vacuum extraction of the vadose zone, soil washing, and the extraction and treatment of contaminated groundwater.

Cytoplasm: The protoplasm of an animal or plant cell external to the nucleus.

Darcy's Law: An empirical relationship between hydraulic gradient and the viscous flow of water in the saturated zone of a porous medium under conditions of laminar flow. The flux of vapors through the voids of the vadose zone can be related to pressure gradient through the air permeability by Darcy's law. See hydraulic conductivity, air permeability, hydraulic gradient, pressure gradient, laminar flow, vadose zone, saturated zone.

Dehydrogenation: Removal of hydrogen from a compound.

Density: The amount of mass of a substance per unit volume of the substance (g/cm^3).

Desorption: The process of removing a sorbed substance by the reverse of adsorption or absorption.

Diffusion: The process whereby the molecules of a compound in a single phase equilibrate to a zero concentration gradient by random molecular motion. The flux of molecules is from regions of high concentration to low concentration and is governed by Fick's Second Law. See Fick's Second Law, effective diffusion coefficient.

Dispersion: The process by which a substance or chemical spreads and dilutes in flowing groundwater or soil gas. On a microscale, dispersion is due to mixing within individual pores, mixing between pore channels, and mixing due to molecular diffusion. At larger scales, geologic heterogeneity and anisotropy cause dispersion. Dispersion has units of squared length per time (cm^2/sec).

Dissociation: Separation of a molecule into two or more fragments (atoms, ions, radicals) by collision with a second body or by the absorption of electromagnetic radiation.

Dissolution: Dissolving of a material in a liquid solvent (e.g., water).

Dynamic Viscosity: The measure of internal friction of a fluid that resists shear within the fluid; the constant of proportionality between a shear stress applied to liquid and the rate of angular deformation within the liquid, having units of mass per length per time (gm/cm sec).

Effective Diffusion Coefficient: The constant of proportionality in Fick's Second Law which is dependent on tortuosity, porosity, and moisture content and properties of the diffusing compound, having units of squared length per time (cm^2/sec). See tortuosity, porosity, moisture content.

Emulsification: The process of dispersing one liquid in a second immiscible liquid.

Entrainment: A process in which solid particles or liquid droplets are, by force of friction with a passing fluid (e.g., air or water), lifted from a resting place and carried along with the flowing fluid.

Fick's Second Law: An equation relating the change of concentration with time due to diffusion to the change in concentration gradient with distance from the source of concentration. See diffusion, effective diffusion coefficient.

Field Capacity: The percentage of water remaining in the soil 2 or 3 days after gravity drainage has ceased from saturated conditions.

Fluid Conductivity: The constant of proportionality in Darcy's law relating the rate of flow of a fluid through a cross-section of porous medium in response to a hydraulic gradient. Fluid conductivity is a function of the intrinsic permeability of a porous medium and the kinematic viscosity of the fluid which flows through it. Fluid conductivity has units of length per time (cm/sec).

Flux: The rate of movement of mass through a unit cross-sectional area per unit time in response to a concentration gradient or some advective force, having units of mass per area per time (g/cm^2*sec).

Freundlich Isotherm: An expression relating the equilibrium between the sorbed chemical concentration and its aqueous phase concentration through a power law.

Fugacity: A function used as an analog of the partial pressure in applying thermodynamics to real systems; at a constant temperature, it is proportional to the exponential of the ratio of the chemical potential of constituent of a system divided by the product of the gas constant and the temperature, and it approaches the partial pressure as the total pressure of the gas approaches zero.

Fulvic Acid: A term of varied usage but usually referring to the mixture of organic substances remaining in solution upon acidification of a dilute alkali extract from the soil. Thus, fulvic acids are soluble under all pH conditions.

Funicular Zone: A narrow band above the groundwater table bounded above by the capillary rise in the slimmest continuous pores, and bounded below by the capillary rise in the widest pores. (See Figure 7–4).

Half-life: The time required for half of a substance to decay or alter by a process, (e.g., radioactive decay, biodegradation, volatilization, photolysis, hydrolysis, oxidation, etc.).

Henry's Law: The relationship between the partial pressure of a compound and the equilibrium concentration in the liquid through a constant of proportionality known as Henry's law constant. See partial pressure.

Heterogeneity: The dependence of property upon location of measurement (e.g., hydraulic conductivity, porosity, compressibility, dispersion, etc.). Heterogeneity may be due to grain size trends, stratigraphic contacts, faults, and vertical bedding.

Homogeneity: The independence of property with location of measurement (e.g., hydraulic conductivity, porosity, compressibility, dispersion, etc.).

Humic acid: That fraction of humic substances that is not soluble in water under acid conditions (below pH 2), but becomes soluble at greater pH.

Humic substances: A general category of naturally occurring, biogenic heterogeneous organic substances that can generally be characterized as being yellow to black in color, of high molecular weight, and refractory. Humic substances include humin, humic acids and fulvic acids.

Humin: That fraction of humic substances that is not soluble in water at any pH value.

Hydraulic Conductivity: The constant of proportionality in Darcy's law relating the rate of flow of water through a cross-section of porous medium in response to a hydraulic gradient. Also known as the coefficient of permeability, hydraulic conductivity is a function of the intrinsic permeability of a porous medium and the kinematic viscosity of the water which flows through it. Hydraulic conductivity has units of length per time (cm/sec).

Hydraulic Gradient: The change in piezometric head between two points divided by the horizontal distance between the two points, having dimensions of length per length (cm/cm). See piezometric head.

Hydraulic Head: A measure of mechanical energy per unit weight density of water as the sum of elevation head and pressure head, having units of length (cm).

Hydrocarbon: One of a very large group of chemical compounds composed only of carbon and hydrogen; the largest source of hydrocarbons is from petroleum crude oil.

Hydrolysis: Decomposition or alteration of a chemical substance by reaction with water.

Hydrophobic: Lacking an affinity for, repelling, or failure to absorb or dissolve in, water.

Hysteresis: The dependence of the state of a system on direction of the process leading to it; a non-unique response of a system to stress, responding differently when the stress is released. Compressibility, moisture content, soil adsorption and unsaturated hydraulic conductivity exhibit hysteretic behavior.

Illite: A group of 2:1, non-expanding, hydrous potassium-magnesium aluminosilicate clay minerals of intermediate properties between kaolinite and montmorillonite.

Immersion: Placement into or within a fluid, usually water.

Infiltration: The downward movement of water through a soil from rainfall or from the application of artificial recharge in response to gravity and capillarity.

Intrinsic Permeability: A measure of the ease with which a porous medium transmits a fluid which is independent of the properties of the fluid and which has units of squared length (cm^2).

Isotropy: The independence of a property with direction of measurement (e.g., hydraulic conductivity, porosity, compressibility, dispersion, etc.).

Kaolinite: A 1:1, non-expanding, hydrous aluminosilicate clay mineral formed by the alteration of micas and feldspars.

Kinematic Viscosity: The ratio of the dynamic viscosity of a fluid to its density, having units of squared length per time (cm^2/sec).

Locus (pl., Loci): In this report, the term locus is used to refer to one of 13 generic contaminated environments in the subsoil region. Each locus is defined by considering: (1) its position relative to the groundwater table (above, on, below); (2) the phase of the contaminant (liquid, vapor, aqueous solution, sorbed to soil); and (3) the nature of the local natural matter (unconsolidated sediments, fractured rock, flowing water).

Macropore: A large pore in a porous medium which may be formed by physical phenonema or biological activity, and through which water, or other fluids, flows solely under the influence of gravity, unaffected by capillarity.

Meniscus: The curved surface of a liquid between solid boundaries (e.g., capillary tube, mineral grains) which is caused by the surface tension of the liquid. The geometry of the meniscus is affected by the surface tension of the liquid, the difference in density between the liquid and overlying air, the distance between the solid boundaries, and the hydrophobic nature of the liquid.

Mica: A family of platy, igneous and metamorphic, aluminosilicate minerals which weather and form clays. Micas separate readily into thin sheets or flakes.

Microbe: A microorganism, especially a bacterium.

Microorganisms: Microscopic organisms including bacteria, protozoans, yeast, fungi, viruses and algae.

Mobilization: The process or processes by which a liquid contaminant in a locus is made more mobile in the locus or more transferable between loci.

Moisture Content: The amount of water lost from the soil upon drying to a constant weight, expressed as the weight per unit weight of dry soil or as the volume of water per unit bulk volume of the soil. For a fully saturated medium, moisture content equals the porosity; in the vadose zone, moisture content ranges between zero and the porosity value for the medium. See porosity, vadose zone, saturated zone.

Mole Fraction: The ratio of the number of moles of a substance in a mixture or solution to the total number of moles of all the components in the mixture or solution.

Monovalent: A radical or atom whose valency is 1.

Montmorillonite: A 2:1, expanding, hydrous magnesium aluminosilicate clay mineral exhibiting pronounced swelling-shrinkage behavior and high plasticity and cohesion.

Oxidation: A chemical reaction that increases the oxygen content of a compound, or raises the oxidation state of an element.

Oxidation Potential: The difference in potential between an atom or ion and the state in which an electron has been removed to an infinite distance from this atom or ion.

Partial Pressure: The portion of total vapor pressure in a system due to one or more constituents in the vapor mixture.

Percolation, Soil Water: The downward movement of water through soil. Especially, the downward flow of water in saturated or nearly saturated soil at hydraulic gradients of the order of 1.0 or less.

Phreatic Level: The groundwater level that would be seen in an observation well projecting down into an aquifer. (See Figure 7–4.)

Polymerization: The bonding of two or more monomers to produce a polymer.

Porosity: The volume fraction of a rock or unconsolidated sediment not occupied by solid material but usually occupied by water and/or air. Porosity is a dimensionless quantity.

Redox: A chemical reaction in which an atom or molecule losses/gains electrons to/from another atom or molecule. Also called oxidation-reduction. Oxidation is the loss of electrons; reduction is the gain in electrons.

Reduction: A chemical reaction in which an atom of molecule gains electrons. Sometimes results by reaction of the substance with hydrogen.

Refractory: A nonspecific characteristic of some chemicals implying resistance to biodegradation or other degradation or treatment processes.

Residence Time: The average time that water remains in a porous medium or a particle remains in a reservoir. Residence time is calculated as the ratio of reservoir volume to total water inflow rate having units of time (sec).

Residual Saturation: The amount of water or oil remaining in the voids of a porous medium and held in an immobile state by capillarity and dead-end pores.

Salinity: The quantity of anions and cations in water, usually between 33 and 37 parts per thousand in sea water.

Sinter: A chemical sedimentary rock deposited by precipitation from mineral waters, especially siliceous sinter and calcareous sinter.

Slurry: A thick mixture of liquid, especially water, and any of several finely divided substances, such as cement or clay particles.

Solubility: The maximum amount of mass of a compound that will dissolve into a unit volume of solvent, usually water, having units of mass per volume (gm/cm^3).

Sorption: A general term used to encompass the process of absorption, adsorption, ion exchange, and chemisorption.

Specific Gravity: The ratio of the weight of a given volume of the material at 4°C (or some stated temperature) to the weight of an equal volume of distilled water. Materials with specific gravity of less than 1 will float on water; materials with specific gravity over 1 will sink in water.

Surface Tension: A measure of the interfacial tension due to molecular attraction between two fluids in contact or between a liquid in contact with a solid, having units of mass per squared time (dyne/cm).

Surfactant: Also Surface Active Agent. A chemical material which provides a "linking action" between two materials, such as oil and water, which normally resist mixing or readily joining in solution. Surfactants provide the emulsification forces which allow oil and water to mix and remain in either oil-in-water or water-in-oil solutions.

Tortuosity: The ratio of path length through a porous medium to the straight-line flow path which describes the geometry of the porous medium. Tortuosity is a dimensionless parameter which ranges in value from 1 to 2.

Transpiration: The release of water withdrawn from the soil by plants during photosynthesis and other life processes.

Unsaturated Zone: The portion of a porous medium, usually above the water table in an unconfined aquifer, within which the moisture content is less than saturation and the capillary pressure is less than atmospheric pressure. The unsaturated zone does not include the capillary fringe.

Vadose Zone: The portion of a porous medium above the water table within which the capillary pressure is less than atmospheric and the moisture content is usually less than saturation. The vadose zone includes the capillary fringe.

Vapor Pressure: The equilibrium pressure exerted on the atmosphere by a liquid or solid at a given temperature. Also a measure of a substance's propensity to evaporate or give off flammable vapors. The higher the vapor pressure, the more volatile the substance.

Vermiculite: A 2:1, hydrous ferro-magnesium aluminosilicate clay mineral similar in structure and property to montmorillonite.

Vitrification: Formation of a glassy or noncrystalline material.

Volatilization: The process of transfer of a chemical from the water or liquid phase to the air phase. Solubility, molecular weight, and vapor pressure of the liquid and the nature of the air-liquid/water interface affect the rate of volatilization. See solubility, vapor pressure.

Water Table: The water surface in an unconfined aquifer at which the fluid pressure in the voids is at atmospheric pressure.

Wilting Point: The point at which a plant wilts, no longer able to withdraw water from a soil or sediment to support transpiration processes and retain turgor pressure.

Index

Abandonment, 195–196
Abiotic factors, 88, 89, 90, 95, 357. *See also* specific types
Abiotic transformation, 88–98, 121
Absorption, 34, 286, 287. *See also* Sorption
Acetic acid, 291
Acetone, 119, 291
Adhesion, 58
Adsorption, 115. *See also* Sorption
  on dry soils, 34
  exchange, 244
  of gases, 34–37
  of organic acids, 246
  of soil microbiota sorbed contaminants, 286–287
  of vapors, 37–38
  of water table floating liquid contaminants, 196
Advection
  defined, 29, 357
  of dissolved contaminants, 80–81, 103
  equation for, 30
  of groundwater dissolved contaminants, 213, 216, 218–219
  of mineral grain diffused contaminants, 260, 261, 262, 264, 272–273
  pressure-driven, 30
  of residual liquid contaminants, 149
  of vapor contaminants, 15, 29–30, 32, 44
Advection-diffusion equation, 215
Advection-dispersion equation, 213, 216
Advective velocity, 264, 268, 271
Aerobic biodegradation, 283
Aerobic degradation, 144, 283
Air diffusion coefficient, 28, 33
Air-filled porosity, 40, 42, 69, 171, 174, 175
Air space porosity, 66
Air-water transport, 82, 102
Alcohols, 89, 94. *See also* specific types
Aldehydes, 291. *See also* specific types
Aliphatics, 88, 89, 94, 357. *See also* specific types
Alkaline agents, 56
Alkanes, 17, 91, 94, 168
n-Alkanes, 91
Alumina, 60
Amino acids, 94. *See also* specific types
Amphiboles, 95
Anaerobe, defined, 357
Anaerobic biodegradation, 211
Anionic surfactants, 65
Anisotropy, 357
Aromatics, 83, 87, 88, 156, 168, 314. *See also* specific types
  defined, 357

solubility of, 138
Ascorbate, 97
Average value calculations
for colloid attached contaminants, 251, 252, 275, 276
for dissolved contaminants, 100
for fractured rock liquid contaminants, 350–351
for groundwater dissolved contaminants, 228–229
for mobile pore water dissolved contaminants, 329, 330–331
for non-aqueous phase liquids, 67
for residual liquid contaminants, 145–147, 170–173, 174–176
for soil microbiota sorbed contaminants, 304
for vapor contaminants, 41–42
for water table floating liquid contaminants, 199–200
for water-wet soil particle contaminants, 122–123, 124

Bentonite, 357
Benzene, 92, 101, 117, 118, 119, 224
bulk transport of, 125
desorption of, 124
kinematic viscosity of, 189
solubility of, 221
sorption of, 34
storage capacity of, 123
transport of, 125
vapor sorption of, 43
Benzene, toluene, ethylbenzene and xylenes (BTEX), 87, 102
partitioning of, 156
solubility of, 138, 222, 227
vapor pressure of, 17
volatilization and sorption of, 83
Benzopyrene, 92
BET. *See* Brunauer-Emmett-Teller
Biodegradation
aerobic, 283
anaerobic, 211
of colloid attached contaminants, 250
defined, 357
of dissolved contaminants, 87–88
factors affecting, 297–300
of fractured rock liquid contaminants, 348, 353
of groundwater dissolved contaminants, 210, 211, 224–226
of hydrocarbons, 95
kinetics of, 291–297
mechanisms of, 291
microbial, 39
of mineral grain diffused contaminants, 273–274, 278
of mobile pore water dissolved contaminants, 327–328
mobility and, 300–301
of non-aqueous phase liquids, 66
pathway of, 291
pH and, 300

of residual liquid
contaminants, 137, 144,
149, 169
of soil microbiota sorbed
contaminants, 284,
291-300
temperature and, 298
of vapor contaminants,
38-39
of water table floating liquid
contaminants, 196-197
of water-wet soil particle
contaminants, 111, 121
Biological removal, 284
Biomass, defined, 358
Bioremediation, 285. *See also*
Remediation
Bioremoval, 284
Biota, defined, 358
Biotic properties, 88, 358. *See
also* specific types
Biotite, 260
Blob mobility, 140
Boiling points, 23, 36, 37
Boundary conditions, 25
Brownian diffusion, 249
Brownian motion, 263, 357
Brunauer-Emmett-Teller (BET)
equation, 37
BTEX. *See* Benzene, toluene,
ethylbenzene and xylenes
Bulk density, 100, 244, 268
Bulk transport, 114, 125-126,
149, 194, 284
Buoyancy forces, 139
n-Butane, 17, 36, 168
Butyl alcohol, 221
By-passing, 134

Caliche, 358
Capillaries, 57, 70
Capillarity, 62, 358
Capillary forces, 62, 134, 135,
138, 139, 171
fractured rock liquid
contaminants and, 345
water table floating liquid
contaminants and, 191
Capillary fringe, 57, 191, 358
Capillary number, 139, 148
Capillary pressure, 63
Capillary pressure gradients, 65,
165-166
Capillary retention, 184
Capillary suction, 80
Capillary tension, 62, 63, 165,
172, 175, 259
fractured rock liquid
contaminants and, 346
mobile pore water dissolved
contaminants and, 313,
315, 316
saturation and, 165, 173
Capillary trapping, 137
Carbohydrates, 95
Carbon, 60, 116, 118
dissolved organic, 83, 221,
223, 242, 245
distribution of, 245
as limiting substrate, 296
in mobile pore water dissolved
contaminants, 327-328
Carbonate rocks, 140
Carbon-based sorption, 117,
118-119
Carbon-hydrogen bonds, 92
Carbon magnetic resonance
(CMR), 94
Catalysts, defined, 358
Catalytic transformation, 169
Catechol, 95
Cation surfactants, 65
Cell growth equation, 294
Cellulose, 94

Chemical addition, 60
Chemical enhancers, 66
Chemical interaction forces, 249
Chemical oxidation
  of colloid attached contaminants, 250
  of fractured rock liquid contaminants, 348
  of groundwater dissolved contaminants, 226
  of mineral grain diffused contaminants, 274
  of mobile pore water dissolved contaminants, 328
  of non-aqueous phase liquids, 66
  of residual liquid contaminants, 144, 169
  of vapor contaminants, 40
  of water table floating liquid contaminants, 197
  of water-wet soil particle contaminants, 121
Chemical retardation, 44
Chemisorption, 244, 358
Chlorinated hydrocarbons, 89
Chlorinated phenols, 120
Clay, defined, 358
CMR. *See* Carbon magnetic resonance
Coalescence, 358
Colloidal humic substances, 246
Colloidal particles, 119, 221
  bound, 241
  contaminants attached to. *See* Colloid attached contaminants (locus 9)
  defined, 240, 358
  free, 239, 241, 246, 247
  hydrophilic, 242
  inorganic, 243
  organic, 243
  types of, 242, 243
Colloid attached contaminants (locus 9), 239–254
  average value calculations for, 251, 252
  biodegradation of, 250
  chemical oxidation of, 250
  defined, 239
  fixation of, 249–250
  immobilization of, 243
  importance of, 253–254
  information gaps on, 253–254
  interactions of, 253
  maximum value calculations for, 252
  mobilization of, 243–249
  remediation of, 240–250, 253
    fixation and, 249–250
    mobilization and, 243–249
    remobilization and, 243–249
    transformation and, 250
  remobilization of, 243–249
  resorption of, 249
  retardation of, 244
  sorption of, 247
  storage capacity of, 251, 252
  transformation of, 239, 243, 250
  transport of, 239
Colloid:water partition coefficient, 245
Compaction, 59, 63, 70
Competitive sorption, 120
Complexation, 88, 358
Compressibility, 359
Concentration gradients, 24, 81, 82, 264, 266–268, 272
Condensation, 51, 55, 56

Conductivity
  fluid, 162, 163, 187, 200, 361
  hydraulic. *See* Hydraulic conductivity
  increase in, 218–219
Constituents, 359
Contact angle, 64
Containment, 359
Contaminant-filled porosity, 171
Continuity equation, 320
Corrective action, 262–262, 359. *See also* Remediation
Cytoplasm, 359

Darcian advective velocity, 271
Darcy's law
  defined, 163–164, 359
  fractured rock liquid contaminants and, 344, 345
  mobile pore water dissolved contaminants and, 319–322
  non-aqueous phase liquids and, 57
  residual liquid contaminants and, 163–167
  vapor contaminants and, 29–30
  water table floating liquid contaminants and, 187, 189
DDT, 246
Decomposition of ozone, 92
Degradation, 283. *See also* Biodegradation
Dehydrogenase, 90
Dehydrogenation, 90, 359
Density, 66, 67, 168, 189, 216
  bulk, 100, 244, 268
  defined, 359
  differences in, 133
  fracture, 341
  of gasoline, 68, 69
  liquid, 158, 162, 163
  of soil, 115
  vapor, 41–42, 173
Dentrifying bacteria, 91
Desorbing agents, 268. *See also* specific types
Desorption
  defined, 359
  of groundwater dissolved contaminants, 221
  of mineral grain diffused contaminants, 262
  potential for, 119
  time required for, 124–125
  of water-wet soil particle contaminants, 112, 113, 115, 124–125, 1111
Dichlorobenzene, 34
Diesel fuel, 90
Diffusion
  Brownian, 249
  defined, 24, 359–360
  of dissolved contaminants, 81–82, 102
  equation for, 30
  Fick's laws of. *See* Fick's diffusion laws
  fracture, 268
  of groundwater dissolved contaminants, 217–218
  hydraulic conductivity and, 268–270
  intraparticle, 113
  of mineral grain diffused contaminants. *See* Mineral grain diffused contaminants
  molecular, 31, 32, 213, 263

between primary and secondary porosity, 265–268
pure, 259
of residual liquid contaminants, 149, 156
of soil microbiota sorbed contaminants, 284
transverse, 273
in unsaturated zone, 270–271
of vapor contaminants, 15, 24–26, 28, 30
volatilization and, 156
in water, 81–82
of water table floating liquid contaminants, 196
of water-wet soil particle contaminants, 113
Diffusion coefficients, 28, 81
air, 28, 33
effective, 29, 266, 360
generic, 272
Diffusion flux, 101–102
Diffusivity, 113, 272. *See also* Diffusion
Dislodgement, 138–140
Dispersion
defined, 360
of groundwater dissolved contaminants, 213, 216, 217
hydrodynamic, 31, 213
mechanical, 31, 32, 213
of mineral grain diffused contaminants, 269
transverse, 269
of vapor contaminants, 31–32, 44
Displacement, 55–56, 65–66, 138–140
Dissociation, 97, 360
Dissolution, 7, 97
defined, 360
equilibrium, 138
of fractured rock liquid contaminants, 337, 338, 339–343
of groundwater dissolved contaminants, 209
mass lost to, 175–176
of non-aqueous phase liquids, 55, 61, 65–66
pH and, 186
pressure and, 186
of residual liquid contaminants, 136, 147, 148, 157, 175–176
salinity and, 186
temperature and, 186
of vapor contaminants, 34
of water table floating liquid contaminants, 184, 185–197
of water-wet soil particle contaminants, 109
Dissolved contaminants (locus 3), 77–103
advection of, 80–81, 103
average value calculations for, 100
biodegradation of, 87–88
complexation of, 88
concentrations of, 99
defined, 77
dehydrogenation of, 90
diffusion of, 81–82, 102
elimination of, 89
fixation of, 87
hydrolysis of, 89
importance of, 102–103
information gaps on, 103
interactions of, 103
maximum value calculations for, 99–101

mixing of, 80–81
in mobile pore water. *See* Mobile pore water dissolved contaminants
mobilization of, 80–87
partitioning of, 82–83, 87
photo-oxidation of, 89
polymerization of, 93–98
porosity of, 82
redox reactions of, 90–91
remediation of, 78–98, 102–103
fixation and, 87
mobilization and, 80–87
remobilization and, 80–87
transformation and, 87–98
remobilization of, 80–87
sorption of, 83, 87, 103
storage capacity of, 98–101
transformation of, 87–98, 88–98
transport of
with mobile phase, 87
to mobile pore water, 80–82
rate of, 101–102
to soil gas, 82–85
volatilization of, 82–85, 103
Dissolved organic carbon (DOC), 83, 221, 223, 242, 245, 247, 248, 251
Dissolved oxygen, 211
DOC. *See* Dissolved organic carbon
Dodecane, 125, 126
Dynamic viscosity, 60, 162, 168, 189, 360

Effective diffusion coefficients, 29, 266, 360
Effective permeability, 56
Electron acceptors, 91
Electron-deficient substrates, 90
Electronic interactions, 80
Electrostatic forces, 249
Elimination, 89
Emulsification, 61, 360
Emulsifiers, 61. *See also* specific types
Enhanced oil recovery (EOR), 143
Entrainment, 133, 360
Enzymes, 90. *See also* specific types
EOR. *See* Enhanced oil recovery
Equilibrium, 87, 110, 114
Equilibrium dissolution, 138
Equilibrium partitioning, 82–83, 288
Equilibrium sorption, 111
Equilibrium vapor concentration, 41, 42
Equilibrium vapor pressure, 44
Ethanol, 99, 138, 186, 227, 314
Evaporation, 326
Exchange adsorption, 244
Extracellular enzymes, 90
Extraction
soil vapor, 44
vacuum, 33, 44, 82, 85, 102, 326-327

Fatty acids, 291
Fick's diffusion laws, 83–85, 217
first, 25, 265
second, 25, 26, 27, 265, 266, 272, 360
Field capacity, 79, 171, 172, 175, 317-319, 361
Fixation
of colloid attached contaminants, 249-250

of dissolved contaminants, 87
of fractured rock liquid contaminants, 346–348
of groundwater dissolved contaminants, 223–224
of mineral grain diffused contaminants, 273
of mobile pore water dissolved contaminants, 326–327
of non-aqueous phase liquids, 55, 65–66
of residual liquid contaminants, 144, 167–168
of vapor contaminants, 34–38
of water table floating liquid contaminants, 194–196
of water-wet soil particle contaminants, 121
Fluid conductivity, 162, 163, 187, 200, 361
Fluid pressure drop, 57
Fluorene, 120
Flux, 24, 25, 101–102, 268, 361
Foam stabilization, 348
Fracture diffusion, 268
Fractured rock liquid contaminants (locus 13), 337–354
average value calculations for, 350–351
biodegradation of, 348, 353
chemical oxidation of, 348
defined, 337
dissolution of, 337, 338, 339–343
fixation of, 346–348
immobilization of, 353
importance of, 340, 343, 353–354
information gaps on, 354
interactions of, 353–354
mobilization of, 344–346
partitioning of, 344, 346–348
remediation of, 327–348, 353
fixation and, 346–348
mobilization and, 344–346
remobilization and, 344–346
transformation and, 348
remobilization of, 344–346
saturated flow of, 344–345
storage capacity of, 349–351
transformation of, 348
transport of, 344–346
unsaturated flow of, 346
in unsaturated zone, 350–351
volitilization of, 353–354
Fracture mass flux, 268
Fracture porosity, 342, 345
Fracturing, 260, 265, 268, 270. *See also* Fractured rock liquid contaminants (locus 13)
characteristics of, 340
connectivity of, 342
density of, 341
hydraulic, 143, 218–219
orientation of, 342
sources of, 340
types of, 340
width of, 341–342
Free product recovery, 338
Free radicals, 91
Freezing, 327
Freundlich equation, 112, 288, 289, 361
Fugacity, 361
Fulvic acids, 94, 95, 361
Funicular zone, 361

Gas adsorption, 34–37

Gaussian distribution, 231
Generic diffusion coefficient, 272
Geochemical weathering, 260
Glucose oxidation, 90
Grain diffused contaminants. *See* Mineral grain diffused contaminants (locus 10)
Grain-size distribution, 57, 58, 172, 175, 218, 219
Gravitational sedimentation, 249
Gravity, 29, 30, 31, 56, 61, 65, 165
  colloid attached contaminants and, 240
  fractured rock liquid contaminants and, 345
  mobile pore water dissolved contaminants and, 319
  transport driven by, 30–31, 44
Gravity gradient, 165
Groundwater dissolved contaminants (locus 8), 209–234
  advection of, 213, 216, 218–219
  average value calculations for, 228–229
  biodegradation of, 210, 211, 224–226
  chemical oxidation of, 226
  defined, 209
  desorption of, 221
  diffusion of, 217–218
  dispersion of, 213, 216, 217
  dissolution of, 209
  fixation of, 223–224
  immobilization of, 210–211
  importance of, 232–234
  information gaps on, 233–234
  interactions of, 232–233
  maximum value calculations for, 226–228
  mobilization of, 210, 212–223
  partitioning of, 209–210, 223–224
  remediation of, 209–226, 232
    fixation and, 223–224
    mobilization and, 212–223
    remobilization and, 212–223
    transformation and, 224–226
  remobilization of, 212–223
  retardation of, 219–223
  solubility of, 229
  sorption of, 219, 223
  storage capacity of, 226–229
  transformation of, 209, 210, 224–226
  transport of, 209
  volatilization of, 209
Groundwater velocity, 213, 223
Grout curtains, 196

Half-life, 361
Heat activation of soil, 58, 59
Hemicellulose, 94
Henry's law, 44, 82–83, 314, 332, 361
Heterocyclic compounds, 95
Heterogeneity, 141, 362
Hexadecane, 284
n-Hexane, 289
Homogeneity, 362
Humic acids, 95, 362
Humic substances, 88, 93, 94, 95, 243, 246, 362. *See also* specific types
Humin, 362

Humus, 91
Hydraulic conductivity, 59
  defined, 362
  diffusion and, 268–270
  of groundwater dissolved contaminants, 218
  of mobile pore water dissolved contaminants, 313, 319, 322, 323, 326
  of residual liquid contaminants, 158–159, 160–161, 163, 166
  saturated, 322, 323
  of water table floating liquid contaminants, 187, 189
Hydraulic fracturing, 143, 218–219
Hydraulic gradients, 218
  changes in, 163
  defined, 362
  naturally-occurring, 219
  for residual liquid contaminants
    in saturated zone, 134, 136, 137, 138, 140, 142, 143, 148–149
    in unsaturated zone, 163
  for water table floating liquid contaminants, 200
Hydraulic head, 362
Hydraulic removal, 140–142
Hydrocarbons, defined, 362
Hydrodynamic dispersion, 31, 213
Hydrodynamic forces, 249. *See also* specific types
Hydrogen bonding, 80, 92
Hydrogen peroxide, 92–93
Hydrolysis, 89, 363
Hydrophilic colloidal particles, 242
Hydrophilic compounds, 186. *See also* specific types
Hydrophobic collidal particles, 242
Hydrophobic compounds, 112, 113, 115, 117, 246, 363. *See also* specific types
Hydrophobicity, 289
Hydroquinone, 95, 97
Hydrostatic pressure, 61, 63
Hydroxides, 243
Hysteresis, 115, 319, 363

Illite, 260, 363
Immersion, 363
Immobilization
  of colloid attached contaminants, 243
  of fractured rock liquid contaminants, 353
  of groundwater dissolved contaminants, 210–211
  of mineral grain diffused contaminants, 259
  of non-aqueous phase liquids, 55–65, 70
  of residual liquid contaminants, 137
  of vapor contaminants, 34
Impedance, 102
Impermeable covers, 326
Importance
  of colloid attached contaminants, 253–254
  of dissolved contaminants, 102–103
  of fractured rock liquid contaminants, 340, 343, 353–354
  of groundwater dissolved contaminants, 232–234

of mineral grain diffused contaminants, 277–278
of mobile pore water dissolved contaminants, 331–333
of non-aqueous phase liquids, 70–71
of residual liquid contaminants, 148–150, 176–177
of soil microbiota sorbed contaminants, 306
of vapor contaminants, 43–45
of water table floating liquid contaminants, 200–202
of water-wet soil particle contaminants, 126–127
Impulse input, 26
Infiltration, 363
Information gaps
on colloid attached contaminants, 253–254
on dissolved contaminants, 103
on fractured rock liquid contaminants, 354
on groundwater dissolved contaminants, 233–234
on mineral grain diffused contaminants, 278
on mobile pore water dissolved contaminants, 333
on non-aqueous phase liquids, 70–71
on residual liquid contaminants, 150, 177
on vapor contaminants, 45
on water table floating liquid contaminants, 202
on water-wet soil particle contaminants, 127
Inorganic salts, 120
Interactions
of colloid attached contaminants, 253
of dissolved contaminants, 103
of fractured rock liquid contaminants, 353–354
of groundwater dissolved contaminants, 232–233
of mineral grain diffused contaminants, 277–278
of mobile pore water dissolved contaminants, 332
of non-aqueous phase liquids, 70
of soil microbiota sorbed contaminants, 284
of vapor contaminants, 44–45
of water-wet soil particle contaminants, 126
Interception, 249
Interfacial tension, 56, 57, 64, 65, 80
residual liquid contaminants and, 140, 142, 143, 149
Intergranular space contaminants. *See* Mineral grain diffused contaminants (locus 10)
Intramineral space contaminants. *See* Mineral grain diffused contaminants (locus 10)
Intraparticle diffusion, 113
Intrinsic permeability, 140, 158–159, 162, 168, 218, 363

Ionization constant, 245
Iron, 97
Irreversibility, 115
Isobutane, 17, 36, 156–157, 168
Isopentane, 17, 168
Isotropy, 363
ITF. *See* Interfacial tension

Kaolinite, 260, 363
Karstic limestone contaminants. *See* Fractured rock liquid contaminants (locus 13)
Kerosene, 60, 90
Ketones, 89. *See also* specific types
Kinematic viscosity, 60, 159, 160, 161
  of benzene, 189
  defined, 363
  of water, 189
  of water table floating liquid contaminants, 187, 189, 191
Kinetics, 113, 284, 289–290
  of biodegradation, 291–297
  of sorption, 113

Lignin, 94
Limestone contaminants. *See* Fractured rock liquid contaminants (locus 13)
Liquid contaminants. *See also* specific types
  in fractured rock. *See* Fractured rock liquid contaminants (locus 13)
  in karstic limestone. *See* Fractured rock liquid contaminants (locus 13)
  residual. *See* Residual liquid contaminants
  on water-dry soil particles in unsaturated zone. *See* Non-aqueous phase liquids
  water table floating. *See* Water table floating liquid contaminants (locus 7)
Liquid pressure gradients, 56
Liquid surface tension, 55, 56
Loams, 160, 166
Loci, 3, 4, 6, 7. *See also* specific types
  1. *See* Vapor contaminants
  2. *See* Non-aqueous phase liquids
  3. *See* Dissolved contaminants
  4. *See* Water-wet soil particle contaminants
  5. *See* Residual liquid contaminants, in saturated zone
  6. *See* Residual liquid contaminants, in unsaturated zone
  7. *See* Water table floating liquid contaminants
  8. *See* Groundwater dissolved contaminants
  9. *See* Colloid attached contaminants
  10. *See* Mineral grain diffused contaminants
  11. *See* Soil microbiota sorbed contaminants
  12. *See* Mobile pore water dissolved contaminants
  13. *See* Fractured rock liquid contaminants
  defined, 363

Macropore flow, 324–325

Macropores, 364
Mass, 175
Mass balance equation, 294
Mass flux, 268
Mass transport, 24, 85, 149
Maximum value calculations
for colloid attached contaminants, 252, 275, 276
for dissolved contaminants, 99–101
for fractured rock liquid contaminants, 349
for groundwater dissolved contaminants, 226–228
for mineral grain diffused contaminants, 274–275, 276
for mobile pore water dissolved contaminants, 328–330
for non-aqueous phase liquids, 67, 68–69
for residual liquid contaminants, 145, 169–170
for soil microbiota sorbed contaminants, 304–305
for vapor contaminants, 40–41
for water table floating liquid contaminants, 198–199
for water-wet soil particle contaminants, 122
Mechanical dispersion, 31, 32, 213
Mechanical plowing, 63
Meniscus, 364
Metal hydroxides, 243
Metal oxides, 242
Methanol, 119, 138, 186, 227, 314
2-Methylhexane, 314
Methyl tertiary butyl ether (MTBE), 99, 119, 186, 222, 227, 228, 314
solubility of, 138, 221
Micas, 260, 364
Microbes, 364
Microbial uptake of hydrocarbons, 286–290
Microbiota sorbed contaminants. *See* Soil microbiota sorbed contaminants (locus 11)
Microorganisms, 364
Micro-scale partitioning, 7
Migration, 259, 260
Mineral grain diffused contaminants (locus 10), 259–278
advection of, 260, 261, 262, 264, 272–273
average value calculations for, 275, 276
biodegradation of, 273–274, 278
chemical oxidation of, 274
defined, 259
desorption of, 262
dispersion of, 269
fixation of, 273
immobilization of, 259
importance of, 277–278
information gaps on, 278
interactions of, 277–278
maximum value calculations for, 274–275, 275, 276
migration of, 259, 260
mobilization of, 262–273
pore spaces and, 260
remediation of, 259–274, 277
fixation and, 273
immobilization and, 259

mobilization and, 262–273
remobilization and, 262–273
transformation and, 273–274
remobilization of, 262–273
retardation of, 272
in secondary pore water, 275
sorption of, 260, 261, 262, 271–272
storage capacity of, 272, 274–276
transformation of, 273–274
in unsaturated zone, 270–271
Mixing, 80–81
Mobile pore water, defined, 315–316
Mobile pore water dissolved contaminants (locus 12), 313–333
average value calculations for, 329, 330–331
biodegradation of, 327–328
chemical oxidation of, 328
defined, 313
field capacity of, 317–319
fixation of, 326–327
importance of, 331–333
information gaps on, 333
interactions of, 332
macropore flow and, 324–325
maximum value calculations for, 328–330
mobility of, 325–326
mobilization of, 314–326
porosity of, 317–319
remediation of, 313–328, 331–332
fixation and, 326–327
mobilization and, 314–326
remobilization and, 314–326
transformation and, 327–328
remobilization of, 314–326
soil moisture profiles and, 322–324
solubility of, 313, 329
sorption of, 327
storage capacity of, 328–331
transformation of, 327–328
water-filled porosity of, 328, 329
Mobility. *See also* Mobilization
biodegradation and, 300–301
blob, 140
enhancement of, 167, 325–326
of mobile pore water dissolved contaminants, 325–326
of non-aqueous phase liquids, 65
reduction of, 60–61
Mobility ratio, 55–56
Mobilization. *See also* Mobility
of colloid attached contaminants, 243–249
defined, 364
of dissolved contaminants, 80–87
enhancement of, 32–33
of fractured rock liquid contaminants, 344–346
of groundwater dissolved contaminants, 210, 212–223
of mineral grain diffused contaminants, 262–273
of mobile pore water dissolved contaminants, 314–326

of non-aqueous phase liquids, 55–65
of residual liquid contaminants, 135
in saturated zone, 138–143
in unsaturated zone, 156–167
temperature and, 33
of vapor contaminants, 16–33
of water table floating liquid contaminants, 184–192
of water-wet soil particle contaminants, 111–121
Moisture in soil, 22–23, 100, 322–324, 364
Molecular diffusion, 31, 32, 213, 263
Molecular transformation, 291
Mole fraction, 364
Monovalent, 364
Montmorillonite, 260, 364
MTBE. *See* Methyl tertiary butyl ether

Naphthalene, 91, 120
2-Naphthol, 91
NAPL. *See* Non-aqueous phase liquids
Nitrogen, 121, 274, 328
Non-aqueous phase liquids (NAPL, locus 2), 51–71
average value calculations for, 67
biodegradation of, 66
chemical oxidation of, 66
condensation of, 55, 56
defined, 51–52
displacement of, 55–56, 65–66
dissolution of, 55, 61, 65–66
fixation of, 55, 65–66
immobilization of, 55–65, 70
importance of, 70–71
information gaps on, 70–71
interactions of, 70
maximum value calculations for, 67, 68–69
mobility of, 65
mobilization of, 55–65
partitioning of, 55–56
porosity of, 66, 67
remediation of, 52–66, 70
fixation and, 65–66
immobilization and, 55–65
mobilization and, 55–65
partitioning and, 55–56, 65–66
transformation and, 66
sorption of, 56
storage capacity of, 66–67, 68–69
transformation of, 52, 66
transport of, 52, 55
of mobile phase, 56–65
rate of, 69–70
viscosity of, 59–65, 70
volatilization of, 55, 56, 59, 70
Non-carbon based sorption, 118
Non-Darcian flow patterns, 352
Non-Fickian processes, 217

n-Octane, 140, 148, 289
Octanol-water partition coefficients, 115, 116, 247, 289
Oil-wet case, 51
Olivines, 95
Organic acids, 245, 246. *See also* specific types
Organic carbon, 83, 221, 223, 242, 245, 247, 248, 251

Organic carbon partition coefficient, 245
Organic oxidation, 90
Oxidants, 66, 90, 91, 169. *See also* specific types
Oxidation, 90
  catalysts for, 91
  chemical. *See* Chemical oxidation
  defined, 364
  of glucose, 90
  of hydroquinone, 97
  organic, 90
  photo-, 40, 89
Oxidation inhibitors, 90
Oxidation potential, 365
Oxygen, 91, 121
  dissolved, 211
  as limiting substrate, 296
  in mobile pore water dissolved contaminants, 327–328
  supply of, 144
  transport of, 274
Oxygenated additives, 99
Oxygenated hydrocarbons, 91
Ozonation, 91–92
Ozone, 66, 91–92, 169

Parallel-plate approximation, 352–353
Partial differential equations, 320
Partial vapor pressure, 22, 365
Particle size distribution, 58, 113, 114
Partition coefficients, 115, 119, 121, 288, 290
  for colloid attached contaminants, 244
  colloid:water, 245
  octanol-water, 115, 116, 247, 289
  organic carbon, 245
Partitioning, 7, 8. *See also* specific types
  into air, 156
  of colloid attached contaminants, 243–246, 249–250
  of dissolved contaminants, 82–83, 87
  equilibrium, 82–83, 288
  of fractured rock liquid contaminants, 344, 346–348
  of groundwater dissolved contaminants, 209–210, 212, 223–224
  of hydrophobic compounds, 246
  onto immobile phase, 34–38, 65–66, 87, 121, 144
    of colloid attached contaminants, 249–250
    of fractured rock liquid contaminants, 346–348
    of groundwater dissolved contaminants, 223–224
    of mineral grain diffused contaminants, 273
    of mobile pore water dissolved contaminants, 326–327
    of water table floating liquid contaminants, 195–196
  micro-scale, 7
  of mineral grain diffused contaminants, 262–273, 273
  into mobile phase, 55–56, 138, 156–157
    of colloid attached contaminants, 243–246

of fractured rock liquid contaminants, 344
of groundwater dissolved contaminants, 212
of mineral grain diffused contaminants, 262–273
of mobile pore water dissolved contaminants, 314
of vapor contaminants, 16–23
of water table floating liquid contaminants, 184–192
of water-wet soil particle contaminants, 111–120
of mobile pore water dissolved contaminants, 314, 326–327
of non-aqueous phase liquids, 55–56
of residual liquid contaminants, 138, 156–157
of soil microbiota sorbed contaminants, 288, 289
of vapor contaminants, 16–23, 44
onto immobile phase, 34–38
into water, 156–157
water:colloid, 243, 249
of water table floating liquid contaminants, 184–197
water-wet, 55
of water-wet soil particle contaminants, 109, 111–120, 121
Partitioning coefficients, 272
Peclet number, 31, 32
n-Pentane, 314
Percolating water, 78, 79
Percolation, 365
Permanent wilting point, 316–317
Permeability
effective, 56
intrinsic, 140, 158–159, 162, 168, 218, 363
low, 219
relative, 161–162, 163, 168, 322
soil, 44, 57, 57–59
Peroxides, 66, 169
Peroxy radicals, 91
Pesticides, 112. *See also* specific types
pH, 186, 300
Phase separation, 212
Phenolic compounds, 95. *See also* specific types
Phenols, 89, 120, 314. *See also* specific types
Phosphorous, 121
Photo-oxidation, 40, 89
Phreatic level, 365
Phyllosilicates, 260
Physiochemical-phase loci, 4, 6, 7. *See* Loci; specific types
Plowing, 63
Polychlorinated biphenyls, 133
Polycyclic aromatic hydrocarbons, 89
Polymerization, 93–98, 121, 144, 365
Polymers, 56, 60, 65. *See also* specific types
Pore diameter, 69
Porosity, 171, 175
air-filled, 40, 42, 69, 171, 174, 175
air space, 66
available, 170
contaminant-filled, 171

defined, 365
of dissolved contaminants, 82
fracture, 342, 345
increase in, 63
of mobile pore water dissolved contaminants, 317–319
of non-aqueous phase liquids, 66, 67
primary, 265–268, 274–275, 339–340, 344
of rocks, 260
secondary, 260, 265–268, 272, 339–340
soil, 58, 59, 66, 68, 80, 115, 244
storage, 266
storage capacity and, 66
total, 40, 41, 66, 170, 174, 328
water-filled, 40, 41, 100, 170, 171, 174
Preferential sorption, 55
Pressure, 57. *See also* specific types
capillary, 63
differences in, 62
dissolution and, 186
drop in, 57
drops in, 60, 61–65
hydrostatic, 61, 63
partial, 22, 365
static, 62
vapor. *See* Vapor pressure
Pressure differentials, 63
Pressure-driven advection, 30
Pressure gradients, 32, 33, 61, 62, 80
ambient, 164, 165
artificial, 165
capillary, 65, 165–166
external, 165
liquid, 56
mobile pore water dissolved contaminants and, 321
Pressure-temperature correlations, 22
Primary porosity, 265–268, 274–275, 339–340, 344
Proteins, 94. *See also* specific types
Pumping, 176, 192, 196, 338–339
Pump and treat method, 210
Pure chemical vapor pressure, 17–22, 44
Pure diffusion, 259
Pyroxenes, 95

Recovery, 143, 338
Redox reactions, 90–91, 365
Reduction, 97, 365
Reduction/oxidation process, 90
Refractory, 365
Relative permeability, 161–162, 163, 168, 322
Relative sorption coefficient, 119
Remediation, 7, 8. *See also* specific types
bio-, 285
of colloid attached contaminants. *See* under Colloid attached contaminants
of dissolved contaminants. *See* under Dissolved contaminants
of fractured rock liquid contaminants. *See* under Fractured rock liquid contaminants

of groundwater dissolved contaminants. *See* under Groundwater dissolved contaminants
of mineral grain diffused contaminants. *See* under Mineral grain diffused contaminants
of mobile pore water dissolved contaminants. *See* under Mobile pore water dissolved
of non-aqueous phase liquids. *See* under Non-aqueous phase liquids (NAPL)
of residual liquid contaminants. *See* under Residual liquid contaminants
of soil microbiota sorbed contaminants. *See* under Soil microbiota sorbed contaminants
of vapor contaminants. *See* under Vapor contaminants
of water table floating liquid contaminants. *See* under Water table floating liquid
of water-wet soil particle contaminants. *See* under Water-wet soil particle contaminants
Remobilization
of colloid attached contaminants, 243–249
of dissolved contaminants, 80–87
of fractured rock liquid contaminants, 344–346
of groundwater dissolved contaminants, 212–223
of mineral grain diffused contaminants, 262–273
of mobile pore water dissolved contaminants, 314–326
of residual liquid contaminants
in saturated zone, 138–143
in unsaturated zone, 156–167
of vapor contaminants, 16–33
of water table floating liquid contaminants, 184–192
of water-wet soil particle contaminants, 111–121
Removal, 140–142, 284
Residence time, 365
Residual liquid contaminants
in saturated zone (locus 5), 133–150
advection of, 149
average value calculations for, 145–147
biodegradation of, 137, 144, 149
chemical oxidation of, 144
defined, 133
diffusion of, 149
displacement of, 138–140
dissolution of, 136, 147, 148
fixation of, 144
hydraulic removal of, 140–142
immobilization of, 137
importance of, 148–150
information gaps on, 150
interactions of, 149

maximum value calculations for, 145
mobilization of, 135, 138–143
partitioning of, 144
physical dislodgement of, 138–140
physical removal of, 136
remediation of, 133–144, 148–149
fixation and, 144
mobilization and, 138–143
partitioning and, 138, 144
remobilization and, 138–143
transformation and, 144
remobilization of, 138–143
storage capacity of, 145–147
transformation of, 137, 144
transport of
bulk, 149
with mobile phase, 138–143
rate of, 147–148
in unsaturated zone (locus 6), 155–177
average value calculations for, 170–173, 174–176
biodegradation of, 169
chemical oxidation of, 169
defined, 155
diffusion of, 156
dissolution of, 157, 175–176
fixation of, 167–168
importance of, 176–177
interactions of, 177
maximum value calculations for, 169–170
mobilization of, 156–167
partitioning of, 156–157
remediation of, 155–169, 176–177
fixation and, 167–168
mobilization and, 156–167
partitioning and, 156–157
transformation and, 169
remobilization of, 156–167
storage capacity of, 169–173, 174–176
transformation of, 169
transport of
enhancement of, 162–163
of mobile phase, 157–167
rate of, 176
volatilization of, 172, 177
Residual mass, 175
Residual saturation, 139, 141, 365
Resorption, 249
Retardation, 215
of bulk transport, 125–126
chemical, 44
of colloid attached contaminants, 244
diffusivity, 272
of groundwater dissolved contaminants, 219–223
of mineral grain diffused contaminants, 272
Retention, 67, 69, 102
Reversible exchange mechanisms, 113
Rock bulk density, 268

Rock capacity factor, 265, 266, 268
Rock diffused contaminants. *See* Mineral grain diffused contaminants (locus 10)
Rock porosity, 260

Salinity, 83, 186, 366
Saperites, 270
Saturated flow, 344–345
Saturation, 139, 141, 162, 165, 173, 296, 365
Saturation vapor pressure, 156
Secondary porosity, 260, 265–268, 272, 339–340
Sedimentation, 249
Sediment sizes, 168
Shear, 60
Shearing stress, 60
Silica-alumina humus complexes, 88
Silicates, 64, 243
Sinter, 366
Slurry, 366
Slurry walls, 348
Snap-off, 134
Soil
  bulk density of, 100
  compaction of, 59, 63, 70
  density of, 115
  heat activation of, 58, 59
  iron in, 97
  moisture in, 22–23, 100, 322–324, 364
  permeability of, 44, 57–59, 65
  porosity of, 58, 59, 66, 68, 80, 115, 244
  sorption of, 37–38, 271
  surface area of, 58, 65
Soil gas vapor contaminants. *See* Vapor contaminants (locus 1)
Soil microbiota sorbed contaminants (locus 11), 283–306
  absorption of, 286, 287
  adsorption of, 286–287
  average value calculations for, 304
  biodegradation of, 284, 291–300
  bioremediation of, 285
  defined, 283–284
  diffusion of, 284
  importance of, 306
  interactions of, 284
  maximum value calculations for, 304–305
  microbial uptake of, 286–290
  partitioning of, 288, 289
  remediation of, 284–302
    biodegradation and, 291–300
    microbial uptake and, 286–290
    viability and, 301–302
  storage capacity of, 302–305
  transformation of, 291
  transport of, 284
  viability of, 301–302
Soil vapor extraction, 44
Soil water percolation, 365
Solubility, 99, 120, 173
  of additive-free gasoline, 185
  of benzene, 221
  of BTEX, 222, 227
  defined, 366
  enhancement of, 229
  of gasoline, 185, 221
  of gasoline constituents, 138

of groundwater dissolved contaminants, 229
of hydrocarbons, 227
of mobile pore water dissolved contaminants, 313, 329
of MTBE, 138, 221
2-naphthol, 91
of n-octane, 140
variability of, 83
water, 83, 91, 117
Solute-solid partitioning coefficient, 272
Sorption, 44, 51, 58. *See also* Absorption; Adsorption
capacity for, 34, 262
carbon-based, 117, 118–119
of colloid attached contaminants, 247
competitive, 120
defined, 111, 366
diffusivity and, 272
of dissolved contaminants, 83, 87, 103
equilibrium, 111
of groundwater dissolved contaminants, 219, 223
isotherms of, 37
kinetics of, 113
mechanisms of, 111
of mineral grain diffused contaminants, 260, 261, 262, 271–272
of mobile pore water dissolved contaminants, 327
of non-aqueous phase liquids, 56
non-carbon based, 118
of organic acids, 245
potential for, 120
preferential, 55
soil, 37–38, 271
soil moisture and, 23
total, 119
vapor, 43
volatilization combined with, 83
of water-wet soil particle contaminants, 110, 111, 113, 115, 118
Sorption coefficients, 112, 115, 117, 119, 120, 122, 266, 272
Sorption equilibrium, 87
Specific gravity, 366
Stabilization, 348
Static pressure, 62
Step input, 27–28
Storage capacity
of colloid attached contaminants, 251, 252, 275–276
of dissolved contaminants, 98–101
of fractured rock liquid contaminants, 349–351
of groundwater dissolved contaminants, 226–229
of mineral grain diffused contaminants, 272, 274–276
of mobile pore water dissolved contaminants, 328–331
of non-aqueous phase liquids, 66–67, 68–69
of residual liquid contaminants, 145–147, 169–173, 174–176
of soil microbiota sorbed contaminants, 302–305
of vapor contaminants, 40–43

of water table floating liquid contaminants, 197–200
of water-wet soil particle contaminants, 122–124
Storage porosity, 266
Substrates, 87, 90, 91, 274, 294. *See also* specific types
Sugars, 94
Surface area, 58, 65, 138
Surface tension, 61, 62, 63
defined, 366
of gasoline, 69
interfacial, 57, 64
liquid, 55, 56
temperature and, 63
Surfactants, 56, 61, 64, 65, 136. *See also* specific types
anionic, 65
cation, 65
defined, 366

TBA. *See* 1,2,4-Trichlorobenzene
TCA. *See* 1,1,1-Trichloroethane
Temperature, 82
biodegradation and, 298
changes in, 160
dissolution and, 186
mobilization and, 33
physical properties and, 160
sorption coefficients and, 120
surface tension and, 63
vapor pressure and, 17, 22
viscosity and, 60, 63
Temperature-pressure correlations, 22
Tertiary recovery, 60
Tetrachloroethylene, 133
2,2,5,5-Tetramethylhexane, 314
Thermodynamics, 111, 284
Toluene, 92
Tortuosity, 28–29, 81, 101, 217, 366
Total porosity, 40, 41, 66, 170, 174
Total sorption, 119
Total vapor pressure, 22
Transformation, 7, 8. *See also* specific types
abiotic, 88–98, 121
catalytic, 169
of colloid attached contaminants, 239, 243, 250
of dissolved contaminants, 87–98
of fractured rock liquid contaminants, 348
of groundwater dissolved contaminants, 209, 210, 224–226
of mineral grain diffused contaminants, 273–274
of mobile pore water dissolved contaminants, 327–328
molecular, 291
of non-aqueous phase liquids, 52, 66
rate of, 305–306
of residual liquid contaminants, 137, 144, 169
of soil microbiota sorbed contaminants, 291, 305-306
of vapor contaminants, 38–40
of water table floating liquid contaminants, 196–197
of water-wet soil particle contaminants, 111, 121
Transpiration, 327, 366

Transport, 8, 9
air-water, 82, 102
bulk, 114, 125–126, 149, 194, 284
of colloid attached contaminants, 239, 246–249
of dissolved contaminants
with mobile phase, 87
to mobile pore water, 80–82
rate of, 101–102
to soil gas, 82–85
enhancement of, 162–163
of fractured rock liquid contaminants, 344–346, 352–353
gravity, 30–31, 44
of groundwater dissolved contaminants, 209
with mobile phase, 212–223
rate of, 216–217, 229–232
of hydrocarbons, 8
mass, 24, 85, 149
of mineral grain diffused contaminants, 273, 276–277
with mobile phase, 23–33, 87, 121, 138–143
of colloid attached contaminants, 246–249
of dissolved contaminants, 87
of fractured rock liquid contaminants, 344–346
of groundwater dissolved contaminants, 212–223
of mineral grain diffused contaminants, 273
of mobile pore water dissolved contaminants, 314–325
of residual liquid contaminants, 138–143, 157–167
of water table floating liquid contaminants, 187–192
of water-wet soil particle contaminants, 121
to mobile pore water, 80–82
of mobile pore water dissolved contaminants, 314–325, 331
of nitrogen, 274
of non-aqueous phase liquids, 52, 55, 56–65, 69–70
of oxygen, 274
rate of, 69–70
for dissolved contaminants, 101–102
for fractured rock liquid contaminants, 352–353
for groundwater dissolved contaminants, 216–217, 229–232
for mineral grain diffused contaminants, 276–277
for mobile pore water dissolved contaminants, 331
for non-aqueous phase liquids, 69–70
for residual liquid contaminants, 147–148, 176
for water table floating liquid contaminants, 200
for water-wet soil particle contaminants, 124–126

of residual liquid contaminants
bulk, 149
enhancement of, 162–163
with mobile phase, 138–143, 157–167
rate of, 147–148, 176
to soil gas, 82–85
of soil microbiota sorbed contaminants, 284
of substrates, 274
of vapor contaminants, 23–33, 43, 45
of water table floating liquid contaminants, 187–192, 200
of water-wet soil particle contaminants, 121, 124–126, 125–126
Transverse diffusion, 273
Transverse dispersion, 269
1,2,4-Trichlorobenzene (TBA), 34, 227
1,1,1-Trichloroethane (TCA), 89

Uniformity coefficient, 219
Unsaturated zone, defined, 366–367

Vacuum, 33
Vacuum extraction, 33, 44, 82, 85, 102, 326–327
Vadose zone, defined, 367
Van der Waals forces, 249
Vapor contaminants (locus 1), 15–45
adsorption of, 37–38
advection of, 29–30, 32, 44
average value calculations for, 41–42
biodegradation of, 38–39
chemical oxidation of, 40
Darcy's law and, 29–30
defined, 15
diffusion of, 24–26, 28, 30
dispersion of, 31–32, 44
dissolution of, 34
fixation of, 34–38
immobilization of, 34
importance of, 43–45
information gaps on, 45
interactions of, 44–45
maximum value calculations for, 40–41
mobilization of, 16–33
partitioning of, 16–23, 44
onto immobile phase, 34–38
photo-oxidation of, 40
remediation of, 15–40, 43–44
fixation and, 34–38
mobilization and, 16–33
partitioning and, 16–23, 34–38
transformation and, 38–40
transport and, 23–33
remobilization of, 16–33
sorption of, 43
storage capcity of, 40–43
tortuosity of, 28–29
transformation of, 38–40
transport of, 23–33, 43, 45
viscosity of, 33
volatilization of, 16, 22–23, 25, 30, 44
Vapor density, 41–42, 173
Vapor pressure, 33, 44, 55
defined, 367
equilibrium, 44
for hydrocarbons, 17
partial, 22, 365
pure chemical, 17–22, 44

saturation, 156
temperature and, 17, 22
total, 22
variability of, 83
water solubility ratio to, 83
Velocity, 213, 223, 264, 268, 269, 271, 352
Vermiculites, 260, 367
Viability, 301–302
Viscosity, 57
altering of, 65
defined, 60
dynamic, 60, 162, 168, 189, 360
kinematic. *See* Kinematic viscosity
liquid, 65, 69, 158
of non-aqueous phase liquids, 59–65, 70
reduction in, 63
temperature and, 60, 63
of vapor contaminants, 33
Viscosity extenders, 60
Viscosity index improvers, 60
Viscous forces, 56, 139, 345. *See also* specific types
Vitrification, 348, 367
Volatility, 44
Volatilization
defined, 367
diffusion and, 156
of dissolved contaminants, 82–85, 103
of fractured rock liquid contaminants, 353–354
of groundwater dissolved contaminants, 209
of non-aqueous phase liquids, 55, 56, 59, 70
physicochemical property changes and, 23
rate of, 83–85
of residual liquid contaminants, 156, 172, 177
soil moisture and, 22–23
from solution, 82–85
sorption combined with, 83
of vapor contaminants, 16, 22–23, 25, 30, 44
of water table floating liquid contaminants, 184, 189
weathering and, 23
Volume, 141

Water:colloid partitioning, 243, 249
Water-filled porosity, 40, 41, 100, 170, 171, 174
of mobile pore water dissolved contaminants, 328, 329
Water film-dissolved contaminants. *See* Dissolved contaminants
Water solubility, 83, 91, 117
Water soluble fraction (WSF), 227, 229
Water table, defined, 367
Water table floating liquid contaminants (locus 7), 183–202
abandonment of, 195–196
adsorption of, 196
average value calculations for, 199–200
biodegradation of, 196–197
chemical oxidation of, 197
defined, 183
diffusion of, 196
dissolution of, 184, 185–197
fixation of, 194–196
importance of, 200–202
information gaps on, 202

interactions of, 201-202
maximum value calculations for, 198-199
partitioning of, 184-197
remediation of, 184-197, 200-201
fixation and, 194-196
mobilization and, 184-192
remobilization and, 184-192
transformation and, 196-197
storage capacity of, 197-200
transformation of, 196-197
transport of, 187-192, 200
volatilization of, 184, 189
Water-wet partitioning, 55
Water-wet soil particle contaminants (locus 4), 109-127
average value calculations for, 122-123, 124
biodegradation of, 111, 121
chemical oxidation of, 121
defined, 109
deorption of, 115
desorption of, 111, 112, 113, 124-125
diffusion of, 113
dissolution of, 109
fixation of, 121
importance of, 126-127
information gaps on, 127
interactions of, 126
maximum value calculations for, 122
mobilization of, 111-121
partitioning of, 109
onto immobile phase, 121
into mobile phase, 111-120
remediation of, 110-121, 126
fixation and, 121
mobilization and, 111-121
remobilization and, 111-121
remobilization of, 111-121
sorption of, 110, 111, 113, 115, 118
storage capacity of, 122-124
transformation of, 111, 121
transport of, 121, 124-126
Weathering, 23, 88, 260
Wet chemistry oxidative degradation studies, 94
Wilting point, 40, 42, 79, 174, 316-317, 367
Wind, 326-327
WSF. *See* Water soluble fraction

Xylene, 92